Springer Proceedings in Mathematics & Statistics

Volume 306

Springer Proceedings in Mathematics & Statistics

This book series features volumes composed of selected contributions from workshops and conferences in all areas of current research in mathematics and statistics, including operation research and optimization. In addition to an overall evaluation of the interest, scientific quality, and timeliness of each proposal at the hands of the publisher, individual contributions are all refereed to the high quality standards of leading journals in the field. Thus, this series provides the research community with well-edited, authoritative reports on developments in the most exciting areas of mathematical and statistical research today.

More information about this series at http://www.springer.com/series/10533

Naokant Deo · Vijay Gupta · Ana Maria Acu ·
P. N. Agrawal
Editors

Mathematical Analysis I: Approximation Theory

ICRAPAM 2018, New Delhi, India, October 23–25

Springer

Editors
Naokant Deo
Department of Applied Mathematics
Delhi Technological University
New Delhi, Delhi, India

Vijay Gupta
Department of Mathematics
Netaji Subhas University of Technology
New Delhi, Delhi, India

Ana Maria Acu
Department of Mathematics
Lucian Blaga University of Sibiu
Sibiu, Romania

P. N. Agrawal
Department of Mathematics
Indian Institute of Technology Roorkee
Roorkee, Uttarakhand, India

ISSN 2194-1009 ISSN 2194-1017 (electronic)
Springer Proceedings in Mathematics & Statistics
ISBN 978-981-15-1152-3 ISBN 978-981-15-1153-0 (eBook)
https://doi.org/10.1007/978-981-15-1153-0

Mathematics Subject Classification (2010): 41-XX, 26-XX, 30-XX, 46-XX, 47-XX

This Springer imprint is published by the registered company Springer Nature Singapore Pte Ltd.
The registered company address is: 152 Beach Road, #21-01/04 Gateway East, Singapore 189721, Singapore

Dedicated to Professor Niranjan Singh

"Not everything that can be counted counts, and not everything that counts can be counted"

—Albert Einstein

Mathematics is a game of logic and there was no one better at deciphering it than Prof. Niranjan Singh—a pioneer in his field with a heart of an innovator. Always eager to learn and teach, Prof. Singh taught at Kurukshetra University, Haryana, India, for 32 years, focusing on areas of analysis and related fields. He was instrumental in the development of innovative and creative culture in a large number of academic institutions in and around Kurukshetra, thereby bringing a significant change in the teaching methodologies of postgraduate courses.

Apart from being an exemplary mathematician, Prof. Singh was a dedicated social reformist. He was a firm advocate for the use of Hindi language, as a result of which he taught and wrote in Hindi, inspiring its use. Professor Singh authored many books on mathematics of which Beej Ganit, written in 1979, was awarded the gold medal by Sahitya Akademi—a national organization dedicated to the promotion of literature in the languages of India. Insisting on the fact that money shouldn't define the caliber of any student, he firmly promoted the optimal utilization of resources and cost reduction in higher education. His ideology became his strength and helped many educationalists understand the relevance of an unerring education system. His journey led him to be the head of Bhartiya Shikshan Mandal, through which he traveled around India and influenced society.

Professor Singh is remembered by his peers for his academic excellence, research, aptitude, dedication to work, human values, and behavior. His will always be an inspiration to budding professors and mathematicians as he was the embodiment of everything mathematics.

> "Carve your name on hearts, not tombstones. A legacy is etched into the minds of others and the stories they share about you."
>
> —Shannon Alder

Preface

The international conference on "Recent Advances in Pure and Applied Mathematics 2018 (ICRAPAM-2018)" was organized by the Department of Applied Mathematics, Delhi Technological University, Delhi, India, during October 23–25, 2018. This international conference was organized in the memory of our beloved late Prof. Niranjan Singh who worked at the Department of Mathematics, Kurukshetra University, Haryana, India. Professor Singh, a well-known mathematician, worked in the area of summability analysis and did commendable work during the time.

The purpose of the conference was to bring together mathematicians from all over the world working on recent developments in pure and applied mathematics to present their research, exchange new ideas, discuss challenging issues, foster future collaborations, and provide exposure young researchers. The proceedings consist of two volumes, and the first volume is devoted to the papers on approximation theory and related areas. It is an outcome of the invited lectures and research papers presented during the conference. It also includes some articles by the invited speakers, who could not attend the conference, like Prof. Ioan Rasa, Technical University of Cluj-Napoca, Romania; Prof. Ali Aral, Kirikkale University, Turkey; Prof. Harun Karsli, Abant Izzet Baysal University, Bolu, Turkey; Prof. V. Ravichandran, NIT, Tiruchirappalli, India; and Prof. Tarun Das, University of Delhi, Delhi, India.

A total of 180 research papers were presented by young researchers in diversified areas. To maintain the quality of the work, each of these papers was reviewed by two carefully chosen global subject experts. Based on their recommendations, 22 papers were selected for inclusion in Volume I of the proceedings.

Papers in the first volume of the proceedings include areas of approximation theory which cover the estimation of convergence behavior of generalized Durrmeyer-type operators, Lupas–Kantorovich operators, certain exponential type operators due to Ismail–May, α-Bernstein–Kantorovich operator, linear operators based on PED and IPED, and Bernstein–Chlodowsky operators and other generalizations of known operators. Some papers are devoted to the study on fixed point theory, holomorphic functions, summability theory, and analytic functions, which are extended to the topics on iterative approximation, fuzzy setting, uniqueness, starlikeness, statistical convergence, advances in distributional chaos theory, etc. The authors of these papers have carefully described the problem and discussed appropriate methods to obtain the solution.

These proceedings will be a valuable source for young as well as experienced researchers in mathematical sciences. The keynote speaker was Prof. Margareta Heilmann, University of Wuppertal, Germany. The plenary speakers were Prof. Antonio-Jesús López-Moreno, Universidad de Jaen, Spain; Prof. Wutiphol Sintunavarat, Thammasat University, Rangsit Center, Thailand; and Prof. Voichita Radu, Babeş-Bolyai University, Cluj-Napoca, Romania. We are thankful to all the speakers who very kindly accepted our invitation talks in the conference.

We are thankful to all the funding agencies: Science and Engineering Research Board (SERB), Government of India, New Delhi; Third phase of Technical Education Quality Improvement Programme (TEQIP-III), Government of India and Government of NCT, New Delhi, for the partial financial support to make the conference successful.

We wish to thank Prof. Yogesh Singh, Vice-Chancellor, DTU, Delhi, India, for his constant encouragement, motivation, guidance, and support. Thanks are also due to Prof. Anu Lather and Prof. S. K. Garg, Pro-Vice-Chancellors of DTU, Delhi for their moral support.

We are grateful to the members of the screening committee, registration committee, publication committee, academic program committee, finance committee, and the advisory committee who put in a lot of hard work to make this event a huge success. Special thanks are due to Dr. Nilam and Dr. Sivaprasad Kumar Shanmugam as co-convenor.

Subject experts from all over the world contributed to the peer-review process. We express our heartfelt gratitude to them for spending their precious time in reviewing the papers.

We will have achieved our goal if the readers find this volume useful and informative for their research. We are thankful to Springer for publishing the proceedings of the conference.

Thanks to all the research students of the department of applied mathematics, DTU, especially to Neha, Ram Pratap, Lipi, Nav Shakti Mishra, and Sandeep Kumar for their hard work during the conference.

New Delhi, India — Naokant Deo
New Delhi, India — Vijay Gupta
Sibiu, Romania — Ana Maria Acu
Roorkee, India — P. N. Agrawal

Contents

About the Editors

Naokant Deo is a professor at the Department of Applied Mathematics, Delhi Technological University, India. He completed his Ph.D. in Mathematics from Guru Ghasidas University, Bilaspur, India. His areas of research include approximation theory and real analysis. Professor Rao is a recipient of the CAS-TWAS Fellowship awarded by the Chinese Academy of Sciences, Beijing, China, and International Centre for Theoretical Physics, Trieste, Italy. He is an active member of academic bodies such as Indian Mathematical Society, India, Research Group in Mathematical Inequalities and Applications, Australia, and World Academy of Young Scientists, Hungary. His research papers have been published in national and international journals of repute.

Vijay Gupta is a professor at the Department of Mathematics at the Netaji Subhas University of Technology, New Delhi, India. He holds a Ph.D. from the Indian Institute of Technology Roorkee (formerly, the University of Roorkee), and his area of research is approximation theory, with a focus on linear positive operators. The author of 5 books, 15 book chapters, and over 300 research papers, he is actively involved in editing over 25 international scientific research journals.

Ana Maria Acu is a professor at the Department of Mathematics and Computer Science, Lucian Blaga University of Sibiu, Romania. She earned her Ph.D. in Mathematics from the Technical University of Cluj-Napoca, Romania. Professor Acu is an active member of various scientific organizations, editorial boards of scientific journals, and scientific committees, and her main research interest is approximation theory.

P. N. Agrawal is a professor at the Department of Mathematics, Indian Institute of Technology Roorkee, India. He received his Ph.D. degree from the Indian Institute of Technology Kanpur, India, in 1980. Having published his research papers in various journals of repute, Prof. Agarwal has delivered invited lectures and presented papers at a number of international conferences in India and abroad. His research interests include approximation theory, numerical methods, and complex analysis.

Expressions, Localization Results, and Voronovskaja Formulas for Generalized Durrmeyer Type Operators

Antonio-Jesús López-Moreno

Abstract We present a generalized sequence of Durrmeyer type operators that allows to summarize different formulas and results for different particular cases. We show for this sequence, several localization and Voronovskaja type results.

Keywords Durrmeyer type operators · Voronovskaja formula · Higher order derivative · Linear positive operators

1 Introduction

In 1967, Durrmeyer presented his now famous sequence of linear positive operators [12]. It immediately attracted the attention of the researchers due to their close connection to the classical and essential sequence of the Bernstein operators, their approximation properties, first studied by Derriennic [10], and their useful representation in terms of inner products and orthogonal polynomials. Since then, a never-ending list of modifications appeared in the literature. Different authors, with various purposes, have made use of all kinds of techniques to enlarge the, at this point, a wide class of what we could call Durrmeyer type operators. From the initial definition by Durrmeyer, namely

$$\mathbb{D}_n f(x) = (n+1)\sum_{i=0}^{n} p_{n,i}(x)\int_0^1 p_{n,i}(t)f(t)dt, \qquad p_{n,i}(x) = \binom{n}{i}x^i(1-x)^{n-i}, \tag{1}$$

for an integrable function $f : [0, 1] \to \mathbb{R}$ and $x \in [0, 1]$, different basis functions were considered in place of $p_{n,i}$ both inside and outside the integral, weighted inte-

This work is partially supported by Junta de Andalucía Research Project FQM-178, by Research Projects DGA (E-64), MTM2015-67006-P (Spain) and by FEDER founds.

A.-J. López-Moreno (✉)
Departamento de Matemáticas, Universidad de Jaén, 23071 Jaén, Spain
e-mail: ajlopez@ujaen.es

N. Deo et al. (eds.), *Mathematical Analysis I: Approximation Theory*,
Springer Proceedings in Mathematics & Statistics 306,
https://doi.org/10.1007/978-981-15-1153-0_1

grals and inner products instead of $\int_0^1 \cdot\, dt$ or part of the terms in the sum substituted with interpolatory ones. This all, apart from the huge list of operators obtained by composition or the extensions to the multivariate setting and to more general spaces. Especially, along the last 20 years, the production in this topic has been particularly intense and thus we have, a wide repertory of papers available where the approximation properties of many concrete Durrmeyer type modifications are studied. The structure of the analysis is usually, in each case, similar and after proposing a new version of Durrmeyer's sequence; the basic properties of the moments and some differentiation expressions for the modified operators are examined and a Voronovskaja type formula is established as a basic tool to derive direct or inverse results. We have to say that many times, the same or very similar arguments are repeated in different articles for every particular Durrmeyer variation. Although it is a fact that a small modification could make the computations to be completely different, it is also true that a more unified treatment is possible for some of the items involved in the analysis of this sort of operators. The aim of this paper is to show that this assertion is particularly true for the obtention of Voronovskaja type formulas and also for localization results of the sequences of Durrmeyer type operators.

For a sequence of operators $\{L_n\}_{n\in\mathbb{N}}$ which is an approximation method for certain space of functions, that is to say, $L_n f \to f$ for each f in that space, a Voronovskaja formula is an expression of the type

$$\lim_{n\to\infty} n\,(L_n f(x) - f(x)) = a_1(f,x). \tag{2}$$

These types of limits are particular cases (for $r = 1$) of an asymptotic expansion of order $r \in \mathbb{N}$ which is a representation for the convergence of the sequence of the form

$$L_n f(x) = f(x) + \sum_{i=1}^{r} \frac{1}{n^i} a_i(f,r,x) + o(n^{-r}),$$

for every x in the domain of the operators. Asymptotic expressions yield a deeper understanding of the approximation properties of the sequence and are the starting point to establish direct and inverse results or monotonicity and shape-preserving properties (elegant examples of how Voronovskaja formulas determine the direct/inverse results can be found in [5, 6]). Therefore, they are a necessary piece in the study of a sequence. Moreover, many of the classical operators present properties of simultaneous approximation, that is to say, the sequence will converge not only for the function but also for their derivatives. In that case, we need asymptotic expansions and Voronovskaja formulas also for those derivatives.

Connected with the asymptotic expansions, the localization properties of a sequence of operators are also a basic tool to analyze the convergence. It is well-known that, for a certain subinterval I, $f|_I = 0$ does not imply $L_n f|_I = 0$ but in general, we have a special behavior for the convergence at the points of I and for instance, since $a_1(f,x)$ usually is a differential operator, as a direct consequence of (2) we have that, for $x \in I$, at least,

$$L_n f(x) = o(n^{-1}).$$

This is what we call a localization result. A first simple instance can be found in [11] for the Bernstein operators but many other examples appear in the literature [24].

In this work, we are going to present a generalized Durrmeyer type sequence that summarizes many of the examples that we find in different papers. For this general sequence, we show a collection of formulas; we prove localization results and Voronovskaja type formulas. Some of them are already known for particular cases but the aim of this paper is to offer a unified approach for this class of operators at the same time that we also prove some new results and expressions.

The first section will be devoted to present the generalized Durrmeyer operators that we are going to study and a collection of basic formulas and theorems. In Sect. 2, we will prove localization results for the sequence of operators. In Sect. 3, we employ the results of Sects. 1 and 2 to derive Voronovskaja type formulas. Finally in the last section, we suggest several ideas to enlarge the class of operators for which our results are valid.

Notice that throughout the paper, t denotes the identity map $t : [0, \infty) \ni x \mapsto t(x) = x \in [0, \infty)$; meanwhile, x is a general fixed point of $[0, \infty)$. Therefore, we will use t to write functional expressions and x for pointwise formulas. Moreover, for any operator $L : E_1 \subseteq \mathbb{R}^{[0,\infty)} \to E_2 \subseteq \mathbb{R}^{[0,\infty)}$ and $f \in E_1$, $L(f)$ or Lf stand for the image function for f and $L(f)(x)$ or $Lf(x)$ is the evaluation of such a function at x. Moreover, we will use the following notation for ascending/descending factorial and generalized factorial numbers

$$x^{\underline{n}} = x(x-1)\cdots(x-n+1), \quad x^{\overline{n}} = x(x+1)\cdots(x+n-1),$$
$$x^{\underline{\alpha,n}} = x(x-\alpha)\cdots(x-(n-1)\alpha), \quad x^{\overline{\alpha,n}} = x(x+\alpha)\cdots(x+(n-1)\alpha),$$

for any $x, \alpha \in \mathbb{R}$ and $n \in \mathbb{N}_0 = \{0, 1, 2, \ldots\}$. Besides, for a sequence $\{a_n\}_{n\in\mathbb{N}}$, we write

$$a_n = o(n^{-\infty}) \qquad \text{whenever } a_n = o(n^{-r}),\ \forall r \in \mathbb{N}.$$

Finally for $r \in \mathbb{N}_0$, $\mathbb{P}_r$ will stand for the space of polynomials of degree r and D, D^k are the differential operators of order 1 and $k \in \mathbb{N}$, respectively.

2 Durrmeyer Type Operators

We can find in the literature, many generalizations of the Durrmeyer sequence of operators (1) [2, 4, 8, 13, 16, 18, 20, 23, 29]. We are going to consider here a definition wide enough to include several of the cases studied in the references about this topic. For this purpose, we will use the generalized sequence of Baskakov/Mastroianni in its Durrmeyer variant that we show in the following definition.

Definition 1 For $n, \alpha \in \mathbb{R}$ and the parameters $a \in \mathbb{R}$, $b \in \mathbb{Z}$, consider the functions

$$\phi_n^{[\alpha]}(x) = \begin{cases} (1+\alpha x)^{-\frac{n}{\alpha}}, & \text{if } \alpha \neq 0, \\ e^{-nx}, & \text{if } \alpha = 0 \end{cases} \quad \text{and} \quad H^{[\alpha]} = \begin{cases} [0, \infty), & \text{if } \alpha \geq 0, \\ [0, -\frac{1}{\alpha}], & \text{if } \alpha < 0. \end{cases}$$

Take also

$$\phi_{n,i}^{[\alpha]}(x) = \frac{(-1)^i}{i!} x^i D^i \phi_n^{[\alpha]}(x), \qquad C_n^{[\alpha]} = \int_H \phi_n^{[\alpha]}(t)dt = \frac{1}{n-\alpha}, \qquad N^{[\alpha]} = n + a - 2\alpha.$$

For $\alpha_1, \alpha_2 \in \mathbb{R}$ and a locally integrable function $f : H^{[\alpha_2]} \to \mathbb{R}$, we define the Baskakov generalized Durrmeyer operators as

$$\mathbb{D}_{n,a,b} f(x) = \frac{1}{C_{n+a}^{[\alpha_2]}} \sum_{i=\max\{0,-b\}}^{\infty} \phi_{n,i}^{[\alpha_1]}(x) \int_{H^{[\alpha_2]}} \phi_{n+a,i+b}^{[\alpha_2]}(t) f(t) dt. \tag{3}$$

As we see from the definition, we actually have $\mathbb{D}_{n,a,b} = \mathbb{D}_{n,a,b,\alpha_1,\alpha_2}$ but we will suppose α_1, α_2 to be fixed throughout the paper and for the sake of brevity, we will use the notation $\mathbb{D}_{n,a,b}$. Notice also that n can be any real number although it is usually taken as a natural one in the definitions for Durrmeyer type operators that we find in the literature. Many of the computations below are valid for $n \in \mathbb{R}$ (or at least for certain subinterval of $\mathbb{R}$) nevertheless, at some points, we will regard it as a natural number mainly in connection with the convergence properties that we will study later on. Moreover, inside the space of locally integrable functions on $H^{[\alpha_2]}$, we are going to consider the subclass of functions for which $\mathbb{D}_{n,a,b,\alpha_1,\alpha_2}$ is an approximation process and we denote it by $W_{\alpha_1,\alpha_2} = W_{\alpha_1,\alpha_2,a,b}$. We will assume some properties of the space W_{α_1,α_2} derived from the properties of the locally integrable functions and power series; for instance, the fact that $f, |f| \in W_{\alpha_1,\alpha_2}$ implies that $g \in W_{\alpha_1,\alpha_2}$ whenever $|g| \leq |f|$.

We have to take into account that $\mathbb{D}_{n,a,b,\alpha_1,\alpha_2}$ will be a linear positive operator only on $H^{[\alpha_1]}$. For this reason, it will also be convenient at certain points to use the notation

$$H^{[\alpha_1,\alpha_2]} = H^{[\alpha_1]} \cap H^{[\alpha_2]}.$$

It is important to take into account that this definition needs several technical considerations about the spaces of functions and intervals where it acts but we are not specially interested in these details and we propose this representation with the aim of summarizing formulas an expressions valid for several cases and therefore, at some points, it has to be seen only from a formal point of view. The fact is that by means of it we can offer some unified formulas that are valid for the particular cases; in particular, the expressions for the moments, central moments, and differentiation relations.

Of course, we could introduce even more parameters and we can find in several works some other proposals in this sense but with this definition and the modifications that we study in the following sections, we include an important part of the versions that we find along the last years in many papers. Therefore, the sequence that we have just introduced allows to analyze a wide enough class of operators.

Although in the definition we admit $b \in \mathbb{Z}$ to be negative, in this section, we will restrict the study to the case $b \in \mathbb{N}_0$. We will discuss some details of the case $b < 0$ in the last section since the properties of the sequence vary notably in that case. Therefore, throughout this section, we assume $b \in \mathbb{N}_0$.

Let us study the basic properties of this sequence. As we mentioned in the introduction, the usual path starts with the properties of the central moments and differentiation formulas. We summarize in the following results, the identities typically used to study this type of operators.

Theorem 2 *Let us suppose that $b \in \mathbb{N}_0$.*

(i) For $f \in W_{\alpha_1,\alpha_2}$ differentiable of order k on $H^{[\alpha_2]}$,

$$D^k \mathbb{D}_{n,a,b} f = \frac{n^{\overline{\alpha_1,k}}}{N^{\underline{\alpha_2,k}}} \mathbb{D}_{n+k\alpha_1,a-k(\alpha_1+\alpha_2),b+k}(D^k f).$$

(ii) For $f \in W_{\alpha_1,\alpha_2}$ differentiable enough,

$$\mathbb{D}_{n,a,b} f = \sum_{k=0}^{\infty} \frac{n^{\overline{\alpha_1,k}}}{N^{\underline{\alpha_2,k-1}}} \left(\int_{H^{[\alpha_2]}} \phi^{[\alpha_2]}_{n+a-k\alpha_2,b+k}(t) D^k f(t) dt \right) \frac{t^k}{k!}.$$

(iii) For $j \in \mathbb{N}_0$,

$$\mathbb{D}_{n,a,b}(t^j) = \frac{1}{N^{\underline{\alpha_2,j}}} \sum_{k=0}^{j} n^{\overline{\alpha_1,k}} \binom{j}{k} \frac{(b+j)!}{(b+k)!} t^k.$$

Where, from now on, for brevity we denote $N = N^{[\alpha_2]}$.

Proof For $\alpha \in \mathbb{R}$, some basic identities are

$$D\phi^{[\alpha]}_{n,i}(x) = n \left(\phi^{[\alpha]}_{n+\alpha,i-1}(x) - \phi^{[\alpha]}_{n+\alpha,i}(x) \right), \tag{4}$$

$$x(1+\alpha x) D\phi^{[\alpha]}_{n,i}(x) = \phi^{[\alpha]}_{n,i}(x)\,(i - nx), \tag{5}$$

$$\int_{H^{[\alpha]}} \phi^{[\alpha]}_{n,i}(t) t^j dt = \frac{1}{\alpha^{j+1}} \frac{(i+j)!}{i!} \frac{1}{\left(\frac{n}{\alpha} - j - 1\right)^{\underline{j+1}}}. \tag{6}$$

To obtain *(i)*, it is enough to prove the identity for $k = 1$ and to then iterate the formula. If we differentiate (3), by means of (4), after properly arranging the resulting sums and indexes, we obtain

$$D\mathbb{D}_{n,a,b}f(x)=\frac{n}{C^{[\alpha_2]}_{n+a}}\sum_{i=\max\{0,-b\}}^{\infty}\phi^{[\alpha_1]}_{n+\alpha_1,i}(x)\int_{H^{[\alpha_2]}}\underbrace{\left(\phi^{[\alpha_2]}_{n+a,i+b+1}(t)-\phi^{[\alpha_2]}_{n+a,i+b}(t)\right)}_{(*)}f(t)dt.$$

Now, inside the integral, we use again (4) to transform $(*)$ into $D\phi^{[\alpha_2]}_{n+a-\alpha_2,i+b+1}(t)$ and then we apply integration by parts. After adjusting the coefficients, we obtain the formula for the case $k=1$. Some little details about the limits of the sums and integrals depending on α_1, α_2 need only a basic analysis that is left to the reader.

(ii) can be obtained from a Taylor series for $\mathbb{D}_{n,a,b}f$ using *(i)* to compute the values of $D^k\mathbb{D}_{n,a,b}f(0)$.

Identity *(iii)* follows from *(ii)* and (6). □

Theorem 3 *Given $x\in H^{[\alpha_1]}$, let us denote*

$$V_{n,s}(x)=\mathbb{D}_{n,a,b}\left((t-x)^s\right)(x).$$

Then

$$\begin{aligned}(i)\qquad (N-\alpha_2 s)V_{n,s+1}(x)&=x(1+\alpha_1 x)DV_{n,s}(x)\\&\quad+sx(2+(\alpha_1+\alpha_2)x)V_{n,s-1}(x)\\&\quad-(-b+ax-(s+1)(1+2\alpha_2 x))\,V_{n,s}(x),\end{aligned}$$

$$(ii)\qquad V_{n,s}(x)=O(n^{-[\frac{s+1}{2}]}).$$

Proof By means of basic properties of the integration, we have that

$$DV_{n,s}(x)=\frac{1}{C^{[\alpha_2]}_{n+a}}\sum_{i=\max\{0,-b\}}^{\infty}D\phi^{[\alpha_1]}_{n,i}(x)\int_{H^{[\alpha]}}\phi^{[\alpha_2]}_{n+a,i+b}(t)f(t)dt-sV_{n,s-1}(x).$$

If we multiply both sides of the identity by $x(1+\alpha_1 x)$,

$$\begin{aligned}&x(1+\alpha_1 x)\left(DV_{n,s,0}(x)+sV_{n,s-1,0}(x)\right)\\&=\frac{1}{C^{[\alpha_2]}_{n+a}}\sum_{i=\max\{0,-b\}}^{\infty}x(1+\alpha_1 x)D\phi^{[\alpha_1]}_{n,i}(x)\int_{H^{[\alpha_2]}}\phi^{[\alpha_2]}_{n+a,i+b}(t)f(t)dt.\end{aligned}$$

Then we apply (6) inside the sum after which the coefficient $(i-nx)$ that appears can be moved inside the integral and written as

$$\underbrace{(i+b-(n+a)t)}_{(*)}+(n+a)(t-x)+(-b+ax)\,.$$

Now, for $(*)$ we again use (6) and write the resulting coefficient as $t(1+\alpha_2 t)=(1+2\alpha_2 x)(t-x)+\alpha_2(t-x)^2+x(1+\alpha_2 x)$ and we finish properly using the definition of $V_{n,i}(x)$, $i=s-1,s,s+1$.

Finally, *(ii)* can be proved from *(i)* by means of an standard induction argument. □

Here, we could stress one main difference between classical interpolatory and Durrmeyer type operators. If we considered the generalized Baskakov operators given, for certain $\alpha \in \mathbb{R}$, by

$$L_n f(x) = \sum_{i=0}^{\infty} \phi_{n,i}^{[\alpha]}(x) f\left(\frac{i}{n}\right),$$

we know that in a similar way, we can obtain also expressions for the moments and central moments. But in the case of L_n, the expressions can be written as a polynomial on n^{-1} in the form

$$L_n\left((t-x)^s\right)(x) = \sum_{i=[\frac{s+1}{2}]}^{s} A_i(x)\frac{1}{n^i}.$$

However, such a finite expression is not possible for Durrmeyer type operators and we need to consider a different non-polynomial basis to express the moments.

Theorem 4 *Consider the sequences* $\mathbf{n}_i = \{\frac{1}{N^{\underline{\alpha,i}}}\}_{n\in\mathbb{N}}$, $i = 1, 2, \ldots$, *and* $\mathbf{n}_0 = \{1\}_{n\in\mathbb{N}_0}$. *Then, for* $x \in H^{[\alpha_1]}$,

(i) $\mathbb{D}_{n,a,b}(t^j)(x) \in \operatorname{span}\{\mathbf{n}_0, \ldots, \mathbf{n}_j\}$,
(ii) $V_{n,s}(x) \in \operatorname{span}\{\mathbf{n}_{[\frac{s+1}{2}]}, \ldots, \mathbf{n}_s\}$.

Proof From Theorem 2-*(iii)*, we know that $N^{\underline{\alpha_2,j}}\mathbb{D}_{n,a,b}(t^j)(x)$ is an element of the space of polynomial on n of degree j for which $\{(N-\lambda\alpha_2)^{\underline{\alpha_2,j-\lambda}}\}_{\lambda=0,\ldots,j}$ is a base so that, for certain $A_0, \ldots, A_j \in \mathbb{R}$, we can write

$$N^{\underline{\alpha_2,j}}\mathbb{D}_{n,a,b}(t^j)(x) = \sum_{i=0}^{j} A_i (N-i\alpha_2)^{\underline{\alpha_2,j-i}} \Rightarrow \mathbb{D}_{n,a,b}(t^j)(x)$$
$$= \sum_{i=0}^{j} A_i \frac{(N-i\alpha_2)^{\underline{\alpha_2,j-i}}}{N^{\underline{\alpha_2,j}}} = \sum_{i=0}^{j} A_i \frac{1}{N^{\underline{\alpha_2,i}}}$$

which proves *(i)*. Now, for *(ii)*, we only need to consider Newton's binomial formula and Theorem 3-*(ii)*. □

3 Localization Results

As we know, the Durrmeyer type operators present good simultaneous approximation properties and we have convergence for the function and their derivatives. In that case, we can extend the localization properties presented in the introduction in the following sense. It is a simple fact (for instance, we can apply Theorem 2-*(i)*) that for $f \in \mathbb{P}_{k-1}$, we have that $D^k \mathbb{D}_{n,a,b} f = 0$ but we cannot assure the same conclusion when f is a polynomial only locally on an open subinterval $J \subseteq H^{[\alpha_2]}$ (that is to say, $f|_J = p \in \mathbb{P}_k$ or equivalently $D^k f|_J = 0$), but in that situation we can prove the following localization result.

Theorem 5 *Consider $f \in W_{\alpha_1,\alpha_2}$, of polynomial growth and differentiable of order k on $H^{[\alpha_2]}$ such that $D^k f|_J = 0$ for an open subinterval J of $H^{[\alpha_1,\alpha_2]}$. Then, for any $x \in J$,*

$$D^k \mathbb{D}_{n,a,b} f(x) = o(n^{-\infty}).$$

Proof For any even $r \in \mathbb{N}$ big enough, we can find $K_r > 0$ such that $\left|D^k f\right| \leq K_r (t-x)^r$. Now by means of Theorem 2-*(i)*, we have

$$\left|D^k \mathbb{D}_{n,a,b} f(x)\right| = \frac{n^{\overline{\alpha_1,k}}}{N^{\underline{\alpha_2,k}}} \left|\mathbb{D}_{n+k\alpha_1,a-k(\alpha_1+\alpha_2),b+k}(D^k f)(x)\right|$$

$$\leq \frac{n^{\overline{\alpha_1,k}}}{N^{\underline{\alpha_2,k}}} K_r \left|\mathbb{D}_{n+k\alpha_1,a-k(\alpha_1+\alpha_2),b+k}\left((t-x)^r\right)(x)\right| = O(n^{-\frac{r}{2}}),$$

where we have used that $\mathbb{D}_{n+k\alpha,a-2k\alpha,b+k}$ is positive and Theorem 3-*(ii)*. As r is arbitrarily large, we finish the proof. □

Nevertheless, this localization result is not suitable for many applications where a pointwise approach is required since it needs global conditions on the function f. For instance, the assumptions on f for a classical Voronovskaja type formula is to be twice differentiable at a point x and the global condition of the preceding theorem (f has to be differentiable of order two on the whole interval $H^{[\alpha_2]}$) is then to restrictive.

In order to improve Theorem 5, we need the alternative differentiation formula for the operators $\mathbb{D}_{n,a,b}$ that we obtain in the following result.

Theorem 6

$$D^s \mathbb{D}_{n,a,b} = n^{\overline{\alpha_1,s}} \Delta_1^s \left(\mathbb{D}_{n+s\alpha_1,a-s\alpha_1,\bullet}\right)(b),$$

where Δ_1^s denotes de forward difference of order s and step 1 given by

$$\Delta_1^s \left(\mathbb{D}_{n+s\alpha_1,a-s\alpha_1,\bullet}\right)(b) = \sum_{j=0}^{s} \binom{s}{j} (-1)^{s-j} \mathbb{D}_{n+s\alpha_1,a-s\alpha_1,b+j}.$$

Proof From (4) and for any function $f \in W_{\alpha_1,\alpha_2}$, we have

$$D\mathbb{D}_{n,a,b} f(x) = \frac{1}{C_{n+a}^{[\alpha_2]}} \sum_{i=\max\{0,-b\}}^{\infty} n \left(\phi_{n+\alpha_1,i-1}^{[\alpha_1]}(x) - \phi_{n+\alpha_1,i}^{[\alpha_1]}(x)\right) \int_{H^{[\alpha_2]}} \phi_{n+a,i+b}^{[\alpha_2]}(t) f(t) dt, \tag{7}$$

from which we only need to arrange the sums and parameters to obtain

$$D\mathbb{D}_{n,a,b} = n\left(\mathbb{D}_{n+\alpha_1,a-\alpha_1,b+1} - \mathbb{D}_{n+\alpha_1,a-\alpha_1,b}\right).$$

Finally, if we iterate this formula, we arrive at the expression of the theorem. □

We can find similar formulas in [2, Lemma 9] for the case $\alpha_1 = 1$, $\alpha_2 = 0$ or in the multivariate setting in [1, Lemma 5] for an extension of the original Durrmeyer operators ($\alpha_1 = \alpha_2 = -1$) to the d-dimensional simplex.

By means of the expressions given in the preceding result, it is possible to prove a pointwise localization result in the sense that we find in [24] where the following definition for pointwise degree is given.

Definition 7 Given a function $f : I \subseteq \mathbb{R} \to \mathbb{R}$ defined on certain subinterval I, differentiable of any order at $x \in I$, we define

$$\deg_x(f) = \min\{s \in \mathbb{N}_0 : D^i f(x) = 0, \forall i \geq s\} - 1,$$

where we assume the convention that $\min(\emptyset) = \infty$.

Theorem 8 *Given $f \in W_{\alpha_1,\alpha_2}$ of polynomial growth on $H^{[\alpha_2]}$ and $x \in H^{[\alpha_1,\alpha_2]}$ such that* $\deg_x(f) = k - 1$*, for certain $k \in \mathbb{N}$. Then,*

$$D^k\mathbb{D}_{n,a,b} f(x) = o(n^{-\infty}).$$

Proof Let us first prove the result for $k = -1$. For this purpose, let us consider a function f_0 in the conditions of the theorem for $k = -1$. Since f_0 is of polynomial growth, for even $r \in \mathbb{N}$ big enough we have that $|f_0| \leq 1 + Kt^r$. On the other hand, as $deg_x(f_0) = -1$, we know that $0 = f_0(x) = Df_0(x) = D^2 f_0(x) = \cdots$ and moreover, f_0 is differentiable of any order at x and in particular we can find $J = (x - \varepsilon, x + \varepsilon) \cap H^{[\alpha_2]}$ such that $f_0|_J$ is differentiable of order r. With this all at hand, we can find K_1 such that $|f_0| \leq K_1(t - x)^r$ and hence, by means of Theorems 6 and 3-*(ii)*,

$$\left|D^k\mathbb{D}_{n,a,b}(f_0)(x)\right| \leq \sum_{j=0}^{k}\binom{k}{j}\left|\mathbb{D}_{n+s\alpha_1,a-s\alpha_1,b+j}(f_0)(x)\right|$$

$$\leq \sum_{j=0}^{k}\binom{k}{j}K_1\left|\mathbb{D}_{n+s\alpha_1,a-s\alpha_1,b+j}\left((t-x)^r\right)(x)\right| = O(n^{-\frac{r}{2}}).$$

Since r is an arbitrary number, we finally deduce that $D^k\mathbb{D}_{n,a,b}(f_0)(x) = o(n^{-\infty})$.

Let us prove now the general case for f with $deg_x(f) = k - 1$. Now, for an even number $r \in \mathbb{N}$ big enough with $r > k$, we can find $J = (x - \varepsilon, x + \varepsilon) \cap H^{[\alpha_2]}$ such that $f|_J$ is differentiable of order r. Then we can also take $\varepsilon_1 < \varepsilon$ and $\tilde{f}$

differentiable of order r on $H^{[\alpha_2]}$ such that $\tilde{f}|_{J_1} = f|_{J_1}$ for $J_1 = (x - \varepsilon_1, x + \varepsilon_1) \cap H^{[\alpha_2]}$ and $\tilde{f}|_{H^{[\alpha_2]}-J} = 0$. Then, it is clear that $f - \tilde{f}$ is of polynomial growth with $\deg_x(f - \tilde{f}) = -1$ so that, by the case $k = -1$ proved above,

$$D^k \mathbb{D}_{n,a,b}(f - \tilde{f})(x) = o(n^{-\infty}). \tag{8}$$

It is also immediate that $D^k \tilde{f}$ is of polynomial growth (it has compact support) and besides $\deg_x(D^k \tilde{f}) = -1$. Now, as $\tilde{f}$ is differentiable of order k on $H^{[\alpha_2]}$, by Theorem 2-*(i)* and the already proved case $k = -1$ of the theorem, we have

$$D^k \mathbb{D}_{n,a,b} \tilde{f}(x) = \frac{n^{\overline{\alpha_1,k}}}{N^{\underline{\alpha_2,k}}} \mathbb{D}_{n+k\alpha_1, a-k(\alpha_1+\alpha_2), b+k}(D^k \tilde{f})(x) = o(n^{-\infty}),$$

and now with (8), we conclude the proof since the values for the derivatives of f and $\tilde{f}$ coincide at x. □

4 Voronovskaja Type Formulas

Sikkema's theorem [26] is the basic tool to compute the asymptotic expansion for a sequence of linear positive operators. We show here the version of this theorem that appears in [26].

Theorem 9 (Sikkema's Theorem [26]) *Let $H \subseteq \mathbb{R}$ be a subinterval, r be an even number and let $\{L_n : W \to C^\infty(H)\}_{n\in\mathbb{N}}$ be a sequence of linear positive operators defined in the linear subspace $W \subseteq \mathbb{R}^H$ such that $\mathbb{P}_r \subseteq W$. Let us suppose that for certain $x \in H$*

$$L_n\left((t - x)^{2s}\right)(x) = O(\phi(n)^{-s}), \quad s = \frac{r}{2}, \frac{r}{2} + 1,$$

where ϕ is an increasing strictly positive function such that $\lim_{n\to\infty} \phi(n) = +\infty$.

If f is of polynomial growth of degree r (there exists $p \in \mathbb{P}_r$ such that $|f| \le p$) and f is r times differentiable at x, then

$$L_n f(x) = \sum_{i=0}^{r} \frac{D^i f(x)}{i!} L_n\left((t - x)^i\right)(x) + o(\phi(n)^{-\frac{r}{2}}).$$

As a consequence of this result, we obtain the following Voronovskaja type formula for $\mathbb{D}_{n,a,b}$.

Theorem 10 *For $f \in W_{\alpha_1,\alpha_2}$ of polynomial growth on $H^{[\alpha_2]}$ twice differentiable at $x \in H^{[\alpha_1,\alpha_2]}$,*

$$\lim_{n\to\infty} n\left(\mathbb{D}_{n,a,b} f(x) - f(x)\right) = ((2\alpha_2 - a)x + b + 1)Df(x) + x\left(1 + \frac{\alpha_1 + \alpha_2}{2} x\right) D^2 f(x)$$

Proof From Theorem 2-*(iii)*, it is simple to compute

$$\lim_{n\to\infty} \mathbb{D}_{n,a,b}\left((t-x)\right)(x) = 1 + b + (2\alpha_2 - a)x,$$
$$\lim_{n\to\infty} \mathbb{D}_{n,a,b}\left((t-x)^2\right)(x) = x(2 + (\alpha_1 + \alpha_2)x),$$

which, since $\mathbb{D}_{n,a,b}(1) = 1$, together with Theorems 3 and 9 proves the formula. □

In the last proof, although in this case Theorem 9 can be applied for functions of polynomial growth of degree two, we can remove this restriction by means of Theorem 8 that allows considering any function of polynomial growth.

The differentiation formulas of Theorem 2 and the localization results of the preceding section make it possible to extend this Voronovskaja formula for higher order derivatives.

Theorem 11 *For $f \in W_{\alpha_1,\alpha_2}$ of polynomial growth on $H^{[\alpha_2]}$, $k+2$ times differentiable at $x \in H^{[\alpha_1,\alpha_2]}$,*

$$\lim_{n\to\infty} n \left(\frac{N^{\underline{\alpha_2,k}}}{n^{\overline{\alpha_1,k}}} D^k \mathbb{D}_{n,a,b} f(x) - D^k f(x) \right).$$
$$= ((2\alpha_2 - a + k(\alpha_1+\alpha_2))x + b + k + 1) D^{k+1} f(x) + x \left(1 + \frac{\alpha_1+\alpha_2}{2} x\right) D^{k+2} f(x).$$

Proof Since f is differentiable of order $k+2$ at x, we can find $J = (x-\varepsilon, x+\varepsilon) \cap H^{[\alpha_2]}$ and $\tilde{f}$ with compact support, differentiable of order k on $H^{[\alpha_2]}$ such that $f|_J = \tilde{f}|_J$. In that case, from Theorem 8, we know that

$$D^k \mathbb{D}_{n,a,b} \tilde{f}(x) = D^k \mathbb{D}_{n,a,b} f(x) + o(n^{-\infty}),$$

and then we can substitute f by $\tilde{f}$ along the rest of the proof. As $\tilde{f}$ is globally differentiable on $H^{[\alpha_2]}$, we can use for it Theorem 2-*(i)* and then

$$D^k \mathbb{D}_{n,a,b} \tilde{f}(x) - D^k \tilde{f}(x) = \frac{n^{\overline{\alpha_1,k}}}{N^{\underline{\alpha_2,k}}} \mathbb{D}_{n+k\alpha_1, a-k(\alpha_1+\alpha_2), b+k}(D^k \tilde{f})(x) - D^k \tilde{f}(x), \quad (9)$$

from which

$$\frac{N^{\underline{\alpha_2,k}}}{n^{\overline{\alpha_1,k}}} D^k \mathbb{D}_{n,a,b} \tilde{f}(x) - D^k \tilde{f}(x) = \mathbb{D}_{n+k\alpha_1, a-k(\alpha_1+\alpha_2), b+k}(D^k \tilde{f})(x) - D^k \tilde{f}(x).$$

Now we use Theorem 10 to compute the limit

$$\lim_{n\to\infty} n \left(\mathbb{D}_{n+k\alpha_1, a-k(\alpha_1+\alpha_2), b+k}(D^k \tilde{f})(x) - D^k \tilde{f}(x) \right)$$

to finish the proof. □

In [9, Theorems 3.1 and 4.1], Deo shows several examples of this type of Voronovskaja formulas for several values of α_1 and α_2.

We can also rewrite the preceding formula in the following terms.

Theorem 12 *For $f \in W_{\alpha_1,\alpha_2}$ of polynomial growth on $H^{[\alpha_2]}$, $k+2$ times differentiable at $x \in H^{[\alpha_1,\alpha_2]}$,*

$$\lim_{n\to\infty} n\left(D^k \mathbb{D}_{n,a,b} f(x) - D^k f(x)\right)$$
$$= D^k\left[(b - at + (\alpha_2-\alpha_1)t)\, Df\right](x) + D^{k+1}\left[t\left(1+\frac{\alpha_1+\alpha_2}{2}t\right)Df\right](x).$$

Proof Following the same arguments of the proof of Theorem 11, we can introduce the function $\tilde{f}$ for which from (9) we can write

$$D^k\mathbb{D}_{n,a,b}\tilde{f}(x) - D^k\tilde{f}(x) = \frac{n^{\overline{\alpha_1,k}}}{N^{\underline{\alpha_2,k}}}\left(\mathbb{D}_{n+k\alpha_1,a-k(\alpha_1+\alpha_2),b+k}(D^k\tilde{f})(x) - D^k\tilde{f}(x)\right.$$
$$\left. + \left(1 - \frac{N^{\underline{\alpha_2,k}}}{n^{\overline{\alpha_1,k}}}\right)D^k\tilde{f}(x)\right).$$

But

$$\frac{N^{\underline{\alpha_2,k}}}{n^{\overline{\alpha_1,k}}} = 1 + O(n^{-1}),$$
$$n\left(1 - \frac{N^{\underline{\alpha_2,k}}}{n^{\overline{\alpha_1,k}}}\right) = -ka + \frac{k(k-1)}{2}\alpha_1 + \frac{k(k+3)}{2}\alpha_2 + O(n^{-1}).$$

Now, this last identities along with Theorem 10 applied for $\mathbb{D}_{n+k\alpha_1,a-k(\alpha_1+\alpha_2)}$ yield, once we substitute again the derivatives of $\tilde{f}$ by the ones of f,

$$\lim_{n\to\infty} n\left(D^k\mathbb{D}_{n,a,b}f(x) - D^k f(x)\right)$$
$$= \left(-ka + \frac{k(k-1)}{2}\alpha_1 + \frac{k(k+3)}{2}\alpha_2\right)D^k f(x)$$
$$+ ((2\alpha_2 - a + k(\alpha_1+\alpha_2))x + b + k + 1)D^{k+1}f(x) + x\left(1 + \frac{\alpha_1+\alpha_2}{2}x\right)D^{k+2}f(x).$$

Finally, if we apply Leibnitz formula, it is simple to check that this identity is equivalent to the one that we want to prove. □

From this result, we conclude that the Voronovskaja formula of the operators $\mathbb{D}_{n,a,b}$ can be differentiated in the sense that the formula for the kth derivative of the operators is the kth derivative of the one for $\mathbb{D}_{n,a,b}$ (for more general result about this type of differentiation properties in the case of interpolatory operators see [25]).

When $\alpha_1 = \alpha_2 = -1$ and $a = b = 0$, this formula reduces to the one that we find in [22, Theorem 8] for the classical Durrmeyer operators. Later, Abel extended this expression obtaining the complete asymptotic expansion in [28].

5 Further Remarks

In this section, we include some additional comments an ideas that make it possible to extend the formulas shown before to some other classes of operators.

5.1 Mixed Summation Integration Operators

One important condition on the parameters a and b for Theorem 2 to hold is $b \geq 0$. As a matter of fact, for $b < 0$ we can define the sequence $\mathbb{D}_{n,a,b}$ but the differentiation formula given by Theorem 2-*(i)* is not true. For instance, if we want a differentiation formula for $b = -1$, we need to introduce a remainder term in the definition of $\mathbb{D}_{n,a,-1}$ of the type

$$\tilde{\mathbb{D}}_{n,a,-1} f(x) = \mathbb{D}_{n,a,-1} f(x) + \phi_{n,i}^{[\alpha_1]}(x) f(0)$$

and therefore we have operators of the form

$$\tilde{\mathbb{D}}_{n,a,-1} f(x) = \frac{1}{C_{n+a}^{[\alpha_2]}} \sum_{i=1}^{\infty} \phi_{n,i}^{[\alpha_1]}(x) \int_{H^{[\alpha_2]}} \phi_{n+a,i+b}^{[\alpha_2]}(t) f(t) dt + \phi_{n,i}^{[\alpha_1]}(x) f(0)$$

that fulfill a differentiation formula like

$$D\tilde{\mathbb{D}}_{n,a,-1} f = \mathbb{D}_{n,a,0} f \tag{10}$$

that connects these class of operators with the one presented in this paper.

In this manner, part of the summation–integration operators that we find in the literature [3, 7, 14, 15, 17, 19, 21, 27] can be seen as members of the family of operators that we present in this paper and we can derive formulas and results for them using the same arguments displayed in the preceding sections. In particular, from Theorem 12, as we indicated before, we know that for the operators $\mathbb{D}_{n,a,b}$, the Voronovskaja formula can be differentiated and then the formula for $D^k \mathbb{D}_{n,a,b}$ can be written as the kth derivative of

$$G = [(b - at + (\alpha_2 - \alpha_1)t)\, Df] + D\left[t\left(1 + \frac{\alpha_1 + \alpha_2}{2} t\right) Df\right].$$

In this case, from (10), if the Voronovskaja formula for $\tilde{\mathbb{D}}_{n,a,b}$ is

$$\lim_{n\to\infty} n\left(\tilde{\mathbb{D}}_{n,a,-1} f(x) - f(x)\right) = F(x),$$

we will have

$$DF = G$$

and the fórmula, F, for $\tilde{D}_{n,a,-1}$ is an antiderivative of the one for $D_{n,a,0}$. We find similar ideas in [3].

5.2 Multivariate Durrmeyer Operators

It is well-known that the generalized Baskakov operators can be extended to the multivariate case in several ways. For many of these extensions, part of the formulas that we presented is also valid. In the immediate case of the tensor product extensions, this is evident but let us check this for another nontrivial extension. Consider, for $m \in \mathbb{N}$,

$$\phi_n^{[\alpha]}(x) = \begin{cases} (1+\alpha|x|_1)^{-\frac{n}{\alpha}}, & \text{if } \alpha \neq 0 \\ e^{-n|x|_1}, & \text{if } \alpha = 0 \end{cases} \quad \text{and} \quad H^{[\alpha]} = \left\{ x \in [0,\infty)^m : |x|_1 \leq \begin{cases} \infty, & \text{if } \alpha \geq 0, \\ -\frac{1}{\alpha}, & \text{if } \alpha < 0 \end{cases} \right\},$$

where, for a vector $x = (x_1, x_2, \ldots, x_m)$, we denote $|x|_1 = x_0 + x_1 + \cdots + x_m$ and therefore, in this case, $\phi_n^{[\alpha]} : H^{[\alpha]} \subseteq \mathbb{R}^m \to \mathbb{R}$ is an m-variate function. For $k \in \mathbb{N}_0^m$, we also denote

$$\phi_{n,k}^{[\alpha]}(x) = \frac{(-1)^k}{k!} x^k D^k \phi_n^{[\alpha]}(0), \qquad C_n = \int_{H^{[\alpha]}} \phi_n^{[\alpha]}(t)dt,$$
$$\text{and} \qquad N = N^{[\alpha]} = n + a - (m+1)\alpha.$$

Here, we use the usual vectorial notation and for $j, k \in \mathbb{N}_0^m$,

$$k! = k_1! \cdots k_m!, \quad (x_1, \ldots, x_m)^{(k_1,\ldots,k_m)} = x_1^{k_1} \cdots x_m^{k_m}, \quad \binom{j}{k} = \frac{j!}{k!(j-k)!},$$

$$D^k = \frac{\partial^{|k|_1}}{\partial t_1^{k_1} \cdots \partial t_m^{k_m}} \quad \text{or, for } r \in \mathbb{R}, \quad r^k = r^{|k|_1}.$$

In terms of these functions, for the parameters $a \in \mathbb{R}$, $b \in \mathbb{Z}^m$, we can introduce the following multivariate version of Durrmeyer operators defined, for a locally integrable function, $f : H^{[\alpha]} \to \mathbb{R}$, by

$$\mathbb{D}_{n,a,b} f(x) = \frac{1}{C_{n+a}} \sum_{\substack{i=(i_1,\ldots,i_m)\in\mathbb{N}_0^m \\ i_1,\ldots,i_m \geq \max\{-b,0\}}} \phi_{n,i}^{[\alpha]}(x) \int_{H^{[\alpha]}} \phi_{n+a,i+b}^{[\alpha]}(t) f(t) dt.$$

Of course, it is possible to consider more general versions introducing also here two different basis functions inside and outside the integral as we did in the preceding sections but, to offer some sample formulas, this version of the multivariate operators is enough. In this way, for the operators that we have just defined, it is possible to prove the following version of the differentiation formula given in Theorem 2-*(i)*: for $f : H^{[\alpha]} \to \mathbb{R}$ differentiable enough, we have

$$D^k \mathbb{D}_{n,a,b} f = \frac{n^{\overline{\alpha,|k|_1}}}{N^{\underline{\alpha,|k|_1}}} \mathbb{D}_{n+|k|_1\alpha_1, a-2|k|_1\alpha, b+k}(D^k f)$$

from which we have the Taylor series expansion

$$\mathbb{D}_{n,a,b} f = \sum_{\substack{i=(i_1,\ldots,i_m)\in\mathbb{N}_0^m \\ i_1,\ldots,i_m \geq \max\{-b,0\}}} \frac{n^{\overline{\alpha,|k|_1}}}{N^{\underline{\alpha,|k|_1-1}}} \left(\int_{H^{[\alpha]}} \phi_{n+a-|k|_1\alpha, b+k}^{[\alpha]}(t) D^k f(t) dt \right) \frac{t^k}{k!}$$

that leads us to the following expression for the moment of exponent $j = (j_1, \ldots, j_m) \in \mathbb{N}_0^m$ of the operator

$$\mathbb{D}_{n,a,b}(t^j) = \frac{1}{N^{\underline{\alpha,|j|_1}}} \sum_{\substack{k=(k_1,\ldots,k_m)\in\mathbb{N}_0^m \\ \max\{-b,0\}\leq k_s \leq j_s, \\ s=1,\ldots,m}} n^{\overline{\alpha,|k|_1}} \binom{j}{k} \frac{(b+j)!}{(b+k)!} t^k,$$

where $t = (t_1, \ldots, t_m) : \mathbb{R}^m \to \mathbb{R}^m$ is the identity map.

As we can see by means of these examples, many of the formulas and results can be extended to the multivariate case.

References

1. U. Abel, E.E. Berdysheva, Complete asymptotic expansion for multivariate Bernstein–Durrmeyer operators and quasi-interpolants. J. Approx. Theory **162**, 201–220 (2010)
2. U. Abel, V. Gupta, M. Ivan, Asymptotic approximation of functions and their derivatives by generalized Baskakov-Szász-Durrmeyer operators. Anal. Theory Appl. **21**(1), 15–26 (2005)
3. U. Abel, V. Gupta, R.N. Mohapatra, Local approximation by a variant of Bernstein-Durrmeyer operators. Nonlinear Anal. **68**, 3372–3381 (2008)
4. T. Acar, V. Gupta, A. Aral, Rate of convergence for generalized Szász operators. Bull. Math. Sci. **1**, 99–113 (2011)
5. J. Bustamante, J.M. Quesada, A property of Ditzian-Totik second order moduli. Appl. Math. Lett. **23**(5), 576–580 (2010)

6. D. Cárdenas-Morales, P. Garrancho, Local saturation of conservative operators. Acta Math. Hung. **100**(1–2), 83–95 (2003)
7. D. Cárdenas-Morales, V. Gupta, Two families of Bernstein-Durrmeyer type operators. Appl. Math. Comput. **248**, 342–353 (2014)
8. D. Cárdenas-Morales, P. Garrancho, I. Rasa, Approximation properties of Bernstein-Durrmeyer type operators. Appl. Math. Comput. **232**, 1–8 (2014)
9. N. Deo, Direct result on exponential-type operators. Appl. Math. Comput. **204**(1), 109–115 (2008)
10. M.M. Derriennic, Sur l'approximation des fonctions integrables sur [0, 1] par des polynomes de Bernstein modifies. J. Approx. Theory **31**, 325–343 (1981)
11. R.A. DeVore, G.G. Lorentz, *Constructive Approximation*. A Series of Comprehensive Studies in Mathematics, vol. 303 (Springer, Berlin, 1993)
12. J.L. Durrmeyer, Une formule d'inversion de la transformqeé de Laplace: applications à la théorie des moments, Thése de 3e cycle, Faculté des Sciences de l'Université de Paris (1967)
13. V. Gupta, P.N. Agrawal, A.R. Gairola, On the integrated Baskakov type operators. Appl. Math. Comput. **213**(2), 419–425 (2009)
14. V. Gupta, M.K. Gupta, Rate of convergence for certain families of summation-integral type operators. J. Math. Anal. Appl. **296**, 608–618 (2004)
15. V. Gupta, M.K. Gupta, V. Vasishtha, Simultaneous approximation by summation-integral type operators. Nonlinear Funct. Anal. Appl. **8**(3), 399–412 (2003)
16. V. Gupta, G.S. Srivastava, A. Sahai, On simultaneous approximation by Szász-beta operators. Soochow J. Math. **21**(1), 1–11 (1995)
17. V. Gupta, Rate of approximation by a new sequence of linear positive operators. Comput. Math. Appl. **45**(12), 1895–1904 (2003)
18. V. Gupta, R. Yadav, Rate of convergence for generalized Baskakov operators. Arab J. Math. Sci. **18**(1), 39–50 (2012)
19. V. Gupta, R.N. Mohapatra, Z. Finta, A certain family of mixed summation-integral type operators. Math. Comput. Model. **42**(1–2), 181–191 (2005)
20. V. Gupta, A.-J. López-Moreno, J.-M. Latorre-Palacios, On simultaneous approximation of the Bernstein Durrmeyer operators. Appl. Math. Comput. **213**(1), 112–120 (2009)
21. V. Gupta E. Erkus, On a hybrid family of summation integral type operators. J. Inequal. Pure Appl. Math. **7**(1) (2006), Article 23, 11 pp
22. M. Heilmann, M.W. Müller, Direct and converse results on simultaneous approximation by the method of Bernstein-Durrmeyer operators, in *Algorithms for Approximation I*, ed. by J.C. Mason, M.G. Cox (Chapman & Hall, London, 1989), pp. 107–116
23. M. Heilmann, M.W. Müller, On simultaneous approximation by the method of Baskakov-Durrmeyer operators. Numer. Funct. Anal. Optim. **10**(1–2), 127–138 (1989)
24. A.-J. López-Moreno, J.-M. Latorre-Palacios, Localization results for generalized Baskakov/Mastroianni and composite operators. J. Math. Anal. Appl. **380**(2), 425–439 (2011)
25. A.-J. López-Moreno, F.-J. Muñoz-Delgado, Asymptotic expansion of multivariate conservative linear operators. J. Comput. Appl. Math. **150**(2), 219–251 (2003)
26. P.C. Sikkema, On some linear positive operators. Indag. Math. (Proc.) **73**, 327–337 (1970)
27. H.M. Srivastava, V. Gupta, A certain family of summation-integral type operators. Math. Comput. Model. **37**(12–13), 1307–1315 (2003)
28. A. Ulrich, Asymptotic approximation by Bernstein-Durrmeyer operators and their derivatives. Approx. Theory Appl. **16**(2), 1–12 (2000)
29. G. Ulusoy, E. Deniz, A. Aral, Simultaneous approximation with generalized Durrmeyer operators. Appl. Math. Comput. **260**, 126–134 (2015)

Lupaş–Kantorovich Type Operators for Functions of Two Variables

P. N. Agrawal and Abhishek Kumar

Abstract Agratini [1] introduced the Lupas–Kantorovich type operators. Manav and Ispir [18] defined a Durrmeyer variant of the operators proposed by Lupas and studied some of their approximation properties. Later, they [17] considered the bivariate case of these operators and studied the degree of approximation by means of the complete and partial moduli of continuity and the order of convergence by using Peetre's K-functional. The associated GBS (Generalized Boolean Sum) operators were also investigated in the same paper. Our goal is to define the bivariate Chlodowsky Lupas–Kantorovich type operators and study their degree of approximation. We also introduce the associated GBS operators and investigate the rate of convergence of these operators for Bögel continuous and Bögel differentiable functions with the aid of mixed modulus of smoothness.

Keywords Peetre's K-functional · Bögel continuous · Bögel differentiable · Mixed modulus of smoothness

1 Introduction

For $f \in C(S),\ S := [0, \infty)$, the space of all real-valued continuous functions on S, Agratini [1] considered the operators

$$L_n(f; x) = 2^{-nx} \sum_{k=0}^{\infty} \frac{(nx)_k}{2^k (k!)} f\left(\frac{k}{n}\right), \qquad x \geq 0 \tag{1.1}$$

proposed by Lupas [16] and having a form similar with the Szász–Mirakyan operators and investigated the degree of approximation by the operators (1.1) in terms of the modulus of continuity and also established a Voronovskaja type asymptotic

P. N. Agrawal (✉) · A. Kumar
Department of Mathematics, IIT Roorkee, Roorkee, India
e-mail: pnappfma@gmail.com

A. Kumar
e-mail: anikk6887@gmail.com

N. Deo et al. (eds.), *Mathematical Analysis I: Approximation Theory*,
Springer Proceedings in Mathematics & Statistics 306,
https://doi.org/10.1007/978-981-15-1153-0_2

theorem for these operators. Erencin and Tasdelen [9] discussed a generalization of the operators (1.1) defined as

$$L_n^*(f;x) = 2^{-a_n x}\sum_{k=0}^{\infty} l_{n,k}(x) f\left(\frac{k}{b_n}\right), \quad for \ \ x \in S, \ n \in \{1,2,\ldots\}, \tag{1.2}$$

where $l_{n,k}(x) = \frac{(a_n x)_k}{2^k (k!)}$ and $\{a_n\}$, $\{b_n\}$ are increasing and unbounded sequences of positive real numbers such that

$$\frac{1}{b_n} \to 0, \ \ as \ n \to \infty \quad and \quad \frac{a_n}{b_n} = 1 + O\left(\frac{1}{b_n}\right), \quad as \ \ n \to \infty$$

and studied some approximation properties of the operators given by (1.2) in weighted approximation. Subsequently, the same authors [9] considered a Kantorovich type variant of these operators as follows:

$$K_n(f;x) = 2^{-a_n x} b_n \sum_{k=0}^{\infty} l_{n,k}(x) \int_{\frac{k}{b_n}}^{\frac{(k+1)}{b_n}} f(t)dt \tag{1.3}$$

and established some local direct results with the aid of the modulus of continuity and Peetre's K-functional.

Ispir and Manav [18] introduced a sequence of Durrmeyer type summation integral operators based on Lupas–Szasz basis functions and studied some approximation properties. Manav and Ispir [17] defined a bivariate extension of the operators considered in [18] and investigated the rate of convergence by means of the total and partial moduli of continuity and the Peetre's K-functional and also studied the associated GBS operators. For some other significant contributions in this direction, we refer the reader to the papers [11–15].

The goal of the present study is to introduce the bivariate extension of the operators defined by (1.3) as

$$K_{n,m}^*(f(t,s);x,y) = 2^{-a_n x} 2^{-d_m y} b_n c_m \sum_{j=0}^{\infty}\sum_{k=0}^{\infty} l_{n,m}^{k,j}(x,y) \int_{\frac{k}{b_n}}^{\frac{(k+1)}{b_n}} \int_{\frac{j}{c_m}}^{\frac{(j+1)}{c_m}} f(t,s)dt\,ds, \tag{1.4}$$

where

$$l_{n,m}^{k,j}(x,y) = \frac{(a_n x)_k}{2^k (k!)} \cdot \frac{(d_m y)_j}{2^j (j!)},$$

$$\lim_{n\to\infty} \frac{a_n}{b_n} = 1, \ \ \lim_{n\to\infty} \frac{1}{b_n} = 0,$$

and

$$\lim_{n\to\infty} \frac{d_m}{c_m} = 1, \ \ \lim_{n\to\infty} \frac{1}{c_m} = 0,$$

investigate the rate of convergence of the operators defined by (1.4) by means of the total and partial moduli of continuity and the order of convergence with the aid of the Peetre's K-functional.The degree of approximation of functions belonging to a weighted space is determined in terms of the weighted modulus of continuity. Further, we introduce the GBS operators associated with the operators defined by (1.4) and study the rate of approximation by means of the mixed modulus of smoothness for Bögel continuous and Bögel differentiable functions.

2 Basic Results

Lemma 2.1 ([9]) *For the operators L_n^* defined by (1.2), there hold the following identities*

(i) $L_n^*(1;x) = 1;$
(ii) $L_n^*(t;x) = \frac{a_n}{b_n}x;$
(iii) $L_n^*(t^2;x) = \frac{a_n^2}{b_n^2}x^2 + 2\frac{a_n}{b_n^2}x;$
(iv) $L_n^*(t^3;x) = \frac{a_n^3}{b_n^3}x^3 + 6\frac{a_n^2}{b_n^3}x^2 + 6\frac{a_n}{b_n^3}x;$
(v) $L_n^*(t^4;x) = \frac{a_n^4}{b_n^4}x^4 + 12\frac{a_n^3}{b_n^4}x^3 + 36\frac{a_n^2}{b_n^4}x^2 + 26\frac{a_n}{b_n^4}x.$

Consequently, we have

Lemma 2.2 ([9]) *The operators defined by (1.3) verify*

(i) $K_n(1;x) = 1;$
(ii) $K_n(t;x) = \frac{a_n}{b_n}x + \frac{1}{2b_n};$
(iii) $K_n(t^2;x) = \frac{a_n^2}{b_n^2}x^2 + 3\frac{a_n}{b_n^2}x + \frac{1}{3b_n^2}.$
Further, using Lemma 2.1, by a simple calculation it follows that
(iv) $K_n(t^3;x) = \frac{a_n^3}{b_n^3}x^3 + (\frac{12}{15})\frac{a_n^2}{b_n^3}x^2 + 10\frac{a_n}{b_n^3}x + \frac{1}{4b_n^3};$
(v) $K_n(t^4;x) = \frac{a_n^4}{b_n^4}x^4 + 14\frac{a_n^3}{b_n^4}x^3 + 50\frac{a_n^2}{b_n^4}x^2 + 43\frac{a_n}{b_n^4}x + \frac{1}{5b_n^4}.$

Let $e_{i,j}(t,s) = t^i s^j,\ (i,j) \in \mathbb{N}_0 \times \mathbb{N}_0$, *with* $i+j \leq 4$ *and* $\mathbb{N}_0 = \mathbb{N} \cup \{0\}$.

Lemma 2.3 *For the operators $K_{n,m}^*$ defined by (1.4), there hold the following identities*

(i) $K_{n,m}^*(e_{0,0};x,y) = 1;$
(ii) $K_{n,m}^*(e_{1,0};x,y) = \frac{a_n}{b_n}x + \frac{1}{2b_n};$
(iii) $K_{n,m}^*(e_{0,1};x,y) = \frac{d_m}{c_m}y + \frac{1}{2c_m};$
(iv) $K_{n,m}^*(e_{2,0};x,y) = \frac{a_n^2}{b_n^2}x^2 + 3\frac{a_n}{b_n^2}x + \frac{1}{3b_n^2};$
(v) $K_{n,m}^*(e_{0,2};x,y) = \frac{d_m^2}{c_m^2}y^2 + 3\frac{d_m}{c_m^2}y + \frac{1}{3c_m^2};$
(vi) $K_{n,m}^*(e_{3,0};x,y) = \frac{a_n^3}{b_n^3}x^3 + (\frac{12}{15})\frac{a_n^2}{b_n^3}x^2 + 10\frac{a_n}{b_n^3}x + \frac{1}{4b_n^3};$

(vii) $K^*_{n,m}(e_{0,3};x,y)=\frac{d_m^3}{c_m^3}y^3+(\frac{12}{15})\frac{d_m^2}{c_m^3}y^2+10\frac{d_m}{c_m^3}y+\frac{1}{4d_m^3}$;
(viii) $K^*_{n,m}(e_{4,0};x,y)=\frac{a_n^4}{b_n^4}x^4+14\frac{a_n^3}{b_n^4}x^3+50\frac{a_n^2}{b_n^4}x^2+43\frac{a_n}{b_n^4}x+\frac{1}{5b_n^4}$;
(ix) $K^*_{n,m}(e_{0,4};x,y)=\frac{d_m^4}{c_m^4}y^4+14\frac{d_m^3}{c_m^4}y^3+50\frac{d_m^2}{c_m^4}y^2+43\frac{d_m}{c_m^4}y+\frac{1}{5c_m^4}$.

Lemma 2.4 *As a consequence of Lemma 2.3, we obtain*

(i) $K^*_{n,m}(t-x;x,y)=\left(\frac{a_n}{b_n}-1\right)x+\frac{1}{2b_n}$;

(ii) $K^*_{n,m}(s-y;x,y)=\left(\frac{d_m}{c_m}-1\right)y+\frac{1}{2c_m}$;

(iii) $K^*_{n,m}((t-x)^2;x,y)=\left(\frac{a_n}{b_n}-1\right)^2x^2+\left(3\frac{a_n}{b_n^2}-\frac{1}{b_n}\right)x+\frac{1}{3b_n^2}$;

(iv) $K^*_{n,m}((s-y)^2;x,y)=\left(\frac{d_m}{c_m}-1\right)^2y^2+\left(3\frac{d_m}{c_m^2}-\frac{1}{c_m}\right)y+\frac{1}{3c_m^2}$;

(v) $K^*_{n,m}((t-x)^4;x,y)=\left(\frac{a_n^4}{b_n^4}-\frac{a_n^3}{b_n^3}+6\frac{a_n^2}{b_n^2}-4\frac{a_n}{b_n}+1\right)x^4+\left(14\frac{a_n^3}{b_n^4}-30\frac{a_n^2}{b_n^3}+\right.$
$\left.18\frac{a_n}{b_n^2}-\frac{2}{b_n}\right)x^3+\left(50\frac{a_n^2}{b_n^4}-40\frac{a_n}{b_n^3}+\frac{2}{b_n^2}\right)x^2+\left(43\frac{a_n}{b_n^4}-\frac{1}{b_n^3}\right)x+\frac{1}{5b_n^4}$;

(vi) $K^*_{n,m}((s-y)^4;x,y)=\left(\frac{d_m^4}{c_m^4}-\frac{d_m^3}{c_m^3}+6\frac{d_m^2}{c_m^2}-4\frac{d_m}{c_m}+1\right)y^4+\left(14\frac{d_m^3}{c_m^4}-30\frac{d_m^2}{c_m^3}+\right.$
$\left.18\frac{d_m}{c_m^2}-\frac{2}{c_m}\right)y^3+\left(50\frac{d_m^2}{c_m^4}-40\frac{d_m}{c_m^3}+\frac{2}{c_m^2}\right)y^2+\left(43\frac{d_m}{c_m^4}-\frac{1}{c_m^3}\right)y+\frac{1}{5c_m^4}$.

Remark 1 For the operators $K^*_{n,m}$, we have

(i) $K^*_{n,m}((t-x);x,y)=O(\frac{1}{b_n})(1+x)$, as $n\to\infty$;
(ii) $K^*_{n,m}((s-y);x,y)=O(\frac{1}{c_m})(1+y)$, as $m\to\infty$;
(iii) $K^*_{n,m}((t-x)^2;x,y)=O(\frac{1}{b_n})(1+x+x^2)$, as $n\to\infty$;
(iv) $K^*_{n,m}((s-y)^2;x,y)=O(\frac{1}{c_m})(1+y+y^2)$, as $m\to\infty$;
(iv) $K^*_{n,m}((t-x)^4;x,y))=O(\frac{1}{b_n})(1+x+x^2+x^3+x^4)$, as $n\to\infty$;
(v) $K^*_{n,m}((s-y)^4;x,y))=O(\frac{1}{c_m})(1+y+y^2+y^3+y^4)$, as $m\to\infty$.

Theorem 1 *Let* $f\in C(S^2)$, *then*

$$\lim_{n,m\to\infty}K^*_{n,m}(f;x,y)=f(x,y)$$

holds uniformly on each compact subset of S^2.

Proof From Lemma 2.4, for every $(x,y)\in S^2$, we obtain

$$\lim_{n,m\to\infty}K^*_{n,m}(e_{i,j};x,y)=e_{i,j}(x,y),\quad i,j\in\{0,1,2,3,4\},\ with\ i+j\le 4.$$

uniformly on every compact subset of S^2. Therefore, by Bohman Korovkin theorem, the required result is immediate.

3 Main Results

Rate of Approximation for the Operators $K^*_{n,m}$

Let $C_B(S^2)$ denote the space of all bounded and uniformly continuous functions on S^2 endowed with the norm $\|f\| = \sup_{(x,y)\in S^2} |f(x,y)|$.

Following [10], the complete modulus of continuity for the bivariate case is defined as follows:

$$\overline{\omega}(f;\gamma) = \sup\{|f(t,s) - f(x,y)| : \sqrt{(t-x)^2 + (s-y)^2} \le \gamma\} \tag{3.1}$$

for every $(t,s), (x,y) \in S^2$. Further, the partial moduli of continuity with respect to x and y is given by

$$\omega_1(f,\gamma) = \sup\{|f(x_1,y) - f(x_2,y)| : y \in S \ \ and \ \ |x_1 - x_2| \le \gamma\} \tag{3.2}$$

and

$$\omega_2(f,\gamma) = \sup\{|f(x,y_1) - f(x,y_2)| : x \in S \ \ and \ \ |y_1 - y_2| \le \gamma\}. \tag{3.3}$$

They satisfy the properties of the usual modulus of continuity.

Now, we discuss the rate of convergence of the sequence of operators $K^*_{n,m}$ to the function $f \in C_B(S^2)$.

Theorem 2 *For $f \in C_B(S^2)$, the inequalities*

$$\left| K^*_{n,m}(f;x,y) - f(x,y) \right| \le 2\overline{\omega}(f;\gamma_{n,m}(x,y))$$

and

$$\left| K^*_{n,m}(f;x,y) - f(x,y) \right| \le 2\left\{ \omega_1(f;\mu_{n,2}(x)) + \omega_2(f;\nu_{m,2}(y)) \right\} \quad hold,$$

where

$$\gamma_{n,m}(x,y) = \left\{ \left(\frac{a_n}{b_n} - 1\right)^2 x^2 + \left(3\frac{a_n}{b_n^2} - \frac{1}{b_n}\right)x + \frac{1}{3b_n^2} + \left(\frac{d_m}{c_m} - 1\right)^2 y^2 + \left(3\frac{d_m}{c_m^2} - \frac{1}{c_m}\right)y + \frac{1}{3c_m^2} \right\}^{\frac{1}{2}},$$

$$\mu_{n,2}(x) = \left\{ \left(\frac{a_n}{b_n} - 1\right)^2 x^2 + \left(3\frac{a_n}{b_n^2} - \frac{1}{b_n}\right)x + \frac{1}{3b_n^2} \right\}^{\frac{1}{2}},$$

and

$$\nu_{m,2}(y) = \left\{ \left(\frac{d_m}{c_m} - 1 \right)^2 y^2 + \left(3\frac{d_m}{c_m^2} - \frac{1}{c_m} \right) y + \frac{1}{3c_m^2} \right\}^{\frac{1}{2}}.$$

Proof Using definition (3.1), we get

$$\begin{aligned}
&\left| K_{n,m}^*(f; x, y) - f(x, y) \right| \\
&\quad \leq K_{n,m}^*\Big(|f(t, s) - f(x, y)|; x, y \Big) \\
&\quad \leq K_{n,m}^*\Big(\overline{\omega}(f; \sqrt{(t-x)^2 + (s-y)^2}); x, y \Big) \\
&\quad \leq K_{n,m}^*\left(\overline{\omega}(f, \gamma) \left\{ 1 + \frac{\sqrt{(t-x)^2 + (s-y)^2}}{\gamma} \right\}; x, y \right) \\
&\quad \leq \overline{\omega}(f, \gamma) \left[1 + \frac{1}{\gamma} K_{n,m}^*\Big(\sqrt{(t-x)^2 + (s-y)^2}); x, y \Big) \right].
\end{aligned}$$

Now, applying the Cauchy–Schwarz inequality, we have

$$\begin{aligned}
&\left| K_{n,m}^*(f; x, y) - f(x, y) \right| \\
&\quad \leq \overline{\omega}(f, \gamma) \left[1 + \frac{1}{\gamma} \left\{ K_{n,m}^*\Big((t-x)^2 + (s-y)^2; x, y \Big) \right\}^{\frac{1}{2}} \right] \\
&\quad \leq \overline{\omega}(f, \gamma) \left[1 + \frac{1}{\gamma} \left\{ K_{n,m}^*\Big((t-x)^2; x \Big) + K_{n,m}^*\Big((s-y)^2; y \Big) \right\}^{\frac{1}{2}} \right] \\
&\quad \leq \overline{\omega}(f, \gamma) \left[1 + \frac{1}{\gamma} \left\{ \left(\frac{a_n}{b_n} - 1 \right)^2 x^2 + \left(3\frac{a_n}{b_n^2} - \frac{1}{b_n} \right) x \right.\right. \\
&\qquad\qquad + \frac{1}{3b_n^2} + \left(\frac{d_m}{c_m} - 1 \right)^2 y^2 \\
&\qquad\qquad \left.\left. + \left(3\frac{d_m}{c_m^2} - \frac{1}{c_m} \right) y + \frac{1}{3c_m^2} \right\}^{\frac{1}{2}} \right].
\end{aligned}$$

Choosing $\gamma = \gamma_{n,m}(x, y)$, we obtain the first assertion of the theorem.

Now, using the properties of the partial moduli of continuity (3.2), (3.3) and Cauchy–Schwarz inequality, for any $\delta_1, \delta_2 > 0$, we have

$$\begin{aligned}
&\left| K_{n,m}^*(f; x, y) - f(x, y) \right| \\
&\quad \leq K_{n,m}^*\Big(|f(t, s) - f(x, y)|; x, y \Big)
\end{aligned}$$

$$= K^*_{n,m}\Big(|f(t,s) - f(t,y) + f(t,y) - f(x,y)|; x, y \Big)$$

$$\leq K^*_{n,m}\Big(|f(t,y) - f(x,y)|; x, y \Big) + K^*_{n,m}\Big(|f(t,s) - f(t,y)|; x, y \Big)$$

$$\leq \omega_1(f, \delta_1)\left\{1 + \frac{1}{\delta_1}\Big(K^*_{n,m}((t-x)^2; x, y) \Big)^{\frac{1}{2}} \right\}$$

$$+ \omega_2(f, \delta_2)\left\{1 + \frac{1}{\delta_2}\Big(K^*_{n,m}((s-y)^2; x, y) \Big)^{\frac{1}{2}} \right\}$$

$$\leq \omega_1(f, \delta_1)\left\{1 + \frac{1}{\delta_1}\left(\left(\frac{a_n}{b_n} - 1\right)^2 x^2 + \left(3\frac{a_n}{b_n^2} - \frac{1}{b_n}\right)x + \frac{1}{3b_n^2}\right)^{\frac{1}{2}} \right\}$$

$$+ \omega_2(f, \delta_2)\left\{1 + \frac{1}{\delta_2}\left(\left(\frac{d_m}{c_m} - 1\right)^2 y^2 + \left(3\frac{d_m}{c_m^2} - \frac{1}{c_m}\right)y + \frac{1}{3c_m^2}\right)^{\frac{1}{2}} \right\}.$$

Choosing $\delta_1 = \mu_{n,2}(x)$ and $\delta_2 = \nu_{m,2}(y)$, the second assertion of the theorem is proved.

Now, we find the order of approximation of the operators $K^*_{n,m}(g(t,s); x, y)$ to $g(x, y) \in C_B(S^2)$ by means of the Peetre's K-functional.

Let $C^2_B(S^2)$ be the space of all functions $g \in C_B(S^2)$ such that $\frac{\partial^i g}{\partial x^i}, \frac{\partial^i g}{\partial y^i}$ for $i = 1, 2$, belong to $C_B(S^2)$. The norm on the space $C^2_B(S^2)$ is defined as

$$\|g\|_{C^2_B(S^2)} = \|g\| + \sum_{i=1}^{2}\left(\left\|\frac{\partial^i g}{\partial x^i}\right\| + \left\|\frac{\partial^i g}{\partial y^i}\right\| \right).$$

Following [8], the Petree's K-functional of the function $g \in C_B(S^2)$ is defined as

$$K(g, \gamma) = \inf_{f \in C^2_B(S^2)} \left\{ \|g - f\| + \gamma \|f\|_{C^2_B(S^2)} \right\}, \tag{3.4}$$

for $\gamma > 0$.

It is also known that

$$K(g, \gamma) \leq C\left\{ \overline{\omega}_2(g, \sqrt{\gamma}) + \min(1, \gamma)\|g\| \right\} \tag{3.5}$$

holds for all $\gamma > 0$.

The constant C in the above inequality is independent of γ and g and $\overline{\omega}_2(g, \gamma)$ is the second-order modulus of continuity for the bivariate case.

Theorem 3 *For the function $f \in C_B(S^2)$, the following inequality*

$$\left|K^*_{n,m}(f;x,y)-f(x,y)\right|$$
$$\leq C\left\{\overline{\omega}_2(f;\sqrt{C_{n,m}(x,y)})+\min(1,C_{n,m}(x,y))\|f\|\right\}$$
$$+\overline{\omega}\left(f;\sqrt{\left(\frac{2x(a_n-b_n)+1}{2b_n}\right)^2+\left(\frac{2y(d_m-c_m)+1}{2c_m}\right)^2}\right)\ holds,$$

where C is a constant independent of f and

$$C_{n,m}(x,y)=\mu_{n,2}(x)+\mu^2_{n,1}(x)+\nu_{m,2}(y)+\nu^2_{m,1}(y),$$
$$\mu_{n,1}(x)=\frac{2x(a_n-b_n)+1}{2b_n},$$
$$\nu_{m,1}(y)=\frac{2y(d_m-c_m)+1}{2c_m}\ and\ \mu_{n,2}(x),$$
$$\nu_{m,2}(y)\ are\ defined\ as\ in\ Theorem\ 2.$$

Proof We define the auxiliary operators as follows:

$$\overline{K}^*_{n,m}(f;x,y)=K^*_{n,m}(f;x,y)+f(x,y)-f\left(\frac{2a_nx+1}{2b_n},\frac{2d_my+1}{2c_m}\right). \tag{3.6}$$

By using Lemmas 2.2 and 2.3, we have $\overline{K}^*_{n,m}(1;x,y)=1$ and

$$\overline{K}^*_{n,m}((t-x);x,y)=0,\ \overline{K}^*_{n,m}((s-y);x,y)=0. \tag{3.7}$$

Let $g\in C^2_B(S^2)$ and $(t,s)\in S^2$. Using the Taylor's theorem, we get

$$\begin{aligned}g(t,s)-g(x,y)&=g(t,y)-g(x,y)+g(t,s)-g(t,y)\\&=(t-x)\frac{\partial}{\partial x}g(x,y)+\int_x^t(t-u)\frac{\partial^2}{\partial u^2}g(u,y)du\\&\quad+(s-y)\frac{\partial}{\partial y}g(x,y)\\&\quad+\int_y^s(s-v)\frac{\partial^2}{\partial v^2}g(x,v)dv.\end{aligned} \tag{3.8}$$

Applying $\overline{K}^*_{n,m}$ on both sides of the equality (3.8) and using (3.7), we have

$$\begin{aligned}\overline{K}^*_{n,m}(g;x,y)-g(x,y) &= K^*_{n,m}\left(\int_x^t (t-u)\frac{\partial^2}{\partial u^2}g(u,y)du;x,y\right)\\ &\quad -\int_x^{\frac{2a_nx+1}{2b_n}}\left(\frac{2a_nx+1}{2b_n}-u\right)\frac{\partial^2}{\partial u^2}g(u,y)du\\ &\quad + K^*_{n,m}\left(\int_y^s (s-v)\frac{\partial^2}{\partial v^2}g(x,v)dv;x,y\right)\\ &\quad -\int_y^{\frac{2d_my+1}{2c_m}}\left(\frac{2d_my+1}{2c_m}-v\right)\frac{\partial^2}{\partial v^2}g(x,v)dv.\end{aligned}$$

Hence,

$$\begin{aligned}&\left|\overline{K}^*_{n,m}(g;x,y)-g(x,y)\right|\\ &\quad\le K^*_{n,m}\left(\left|\int_x^t |t-u|\left|\frac{\partial^2}{\partial u^2}g(u,y)\right|du\right|;x,y\right)\\ &\qquad+\left|\int_x^{\frac{2a_nx+1}{2b_n}}\left|\frac{2a_nx+1}{2b_n}-u\right|\left|\frac{\partial^2}{\partial u^2}g(u,y)\right|du\right|\\ &\qquad+K^*_{n,m}\left(\left|\int_y^s |s-v|\left|\frac{\partial^2}{\partial v^2}g(x,v)\right|dv\right|;x,y\right)\\ &\qquad+\left|\int_y^{\frac{2d_my+1}{2c_m}}\left|\frac{2d_my+1}{2c_m}-v\right|\left|\frac{\partial^2}{\partial v^2}g(x,v)\right|dv\right|\\ &\quad\le\left\{K^*_{n,m}\left((t-x)^2;x,y\right)+\left(\frac{2a_nx+1}{2b_n}-x\right)^2\right\}\|g\|_{C^2_B(S^2)}\\ &\qquad+\left\{K^*_{n,m}\left((s-y)^2;x,y\right)+\left(\frac{2d_my+1}{2c_m}-y\right)^2\right\}\|g\|_{C^2_B(S^2)}.\end{aligned}$$

By using the value of $\mu_{n,2}(x)$ and $\nu_{m,2}(y)$ as in Theorem 2 and taking $\mu_{n,1}(x)=\frac{2x(a_n-b_n)+1}{2b_n}$ and $\nu_{m,1}(y)=\frac{2y(d_m-c_m)+1}{2c_m}$, we have

$$\begin{aligned}&|\overline{K}^*_{n,m}(g;x,y)-g(x,y)|\\ &\quad\le\left\{\mu_{n,2}(x)+\mu^2_{n,1}(x)+\nu_{m,2}(y)+\nu^2_{m,1}(y)\right\}\|g\|_{C^2_B(S^2)}\\ &\quad= C_{n,m}(x,y)\|g\|_{C^2_B(S^2)}.\end{aligned}\tag{3.9}$$

For $f\in C_B(S^2)$, we have

$$\left|\overline{K}^*_{n,m}(f;x,y)\right|$$
$$\leq \left|K^*_{n,m}(f;x,y)\right| + |f(x,y)| + f\left(\frac{2a_nx+1}{2b_n}, \frac{2d_my+1}{2c_m}\right) \leq 3\|f\|. \qquad (3.10)$$

Hence, from Eqs. (3.6), (3.9) and (3.10), we have

$$\begin{aligned}|K^*_{n,m}&(f;x,y) - f(x,y)|\\ &\leq |\overline{K}^*_{n,m}((f-g);x,y)| + |\overline{K}^*_{n,m}(g;x,y) - g(x,y)|\\ &\quad + |g(x,y) - f(x,y)|\\ &\quad + \left|f\left(\frac{2a_nx+1}{2b_n}, \frac{2d_my+1}{2c_m}\right) - f(x,y)\right|\\ &\leq 4\|f-g\|_{C_B(S^2)} + C_{n,m}(x,y)\|g\|_{C^2_B(S^2)}\\ &\quad + \overline{\omega}\left(f;\sqrt{\left(\frac{2a_nx+1}{2b_n}-x\right)^2 + \left(\frac{2d_my+1}{2c_m}-y\right)^2}\right).\end{aligned}$$

Taking infimum on the right-hand side over all $g \in C^2_B(S^2)$ and using (3.5), we have

$$\begin{aligned}\Big|K^*_{n,m}&(f;x,y) - f(x,y)\Big|\\ &\leq 4K(f, C_{n,m}(x,y))\\ &\quad + \overline{\omega}\left(f;\sqrt{\left(\frac{2a_nx+1}{2b_n}-x\right)^2 + \left(\frac{2d_my+1}{2c_m}-y\right)^2}\right)\\ &\leq C\left\{\overline{\omega}_2(f;\sqrt{C_{n,m}(x,y)}) + \min(1, C_{n,m}(x,y))\|f\|\right\}\\ &\quad + \overline{\omega}\left(f;\sqrt{\left(\frac{2x(a_n-b_n)+1}{2b_n}\right)^2 + \left(\frac{2y(d_m-c_m)+1}{2c_m}\right)^2}\right).\end{aligned}$$

Hence, the theorem is proved.

Following [2], for $0 < \lambda_1 \leq 1$ and $0 < \lambda_2 \leq 1$, we define the Lipschitz class $Lip_M(\lambda_1, \lambda_2)$ for the bivariate case as follows:

$$|f(t,s) - f(x,y)| \leq M\ |t-x|^{\lambda_1}\ |s-y|^{\lambda_2},$$

where $(t,s),\ (x,y) \in S^2$ are arbitrary.

Theorem 4 *For $f \in Lip_M(\lambda_1, \lambda_2)$, we have*

$$|K^*_{n,m}(f;x,y) - f(x,y)| \leq M\ (\mu_{n,2}(x))^{\frac{\lambda_2}{2}}\ (\nu_{m,2}(y))^{\frac{\lambda_2}{2}},$$

where $\mu_{n,2}(x)$ and $\nu_{m,2}(y)$ are defined as in Theorem 2.

Proof By our hypothesis, we can write

$$\begin{aligned}|K^*_{n,m}(f;x,y)-f(x,y)| &\leq K^*_{n,m}(|f(t,s)-f(x,y)|;x,y)\\ &\leq M\ K^*_{n,m}(|t-x|^{\lambda_1}\ |s-y|^{\lambda_2};x,y)\\ &\leq M\ K^*_{n,m}(|t-x|^{\lambda_1};x,y)\ K^*_{n,m}(|s-y|^{\lambda_2};x,y).\end{aligned}$$

Now, using Lemma 2.2 and Hölder's inequality with $p_1=\frac{2}{\lambda_1}$, $q_1=\frac{2}{2-\lambda_1}$ and $p_2=\frac{2}{\lambda_2}$, $q_2=\frac{2}{2-\lambda_2}$, we get

$$\begin{aligned}|K^*_{n,m}(f;x,y)-f(x,y)| &\leq M\left(K^*_{n,m}((t-x)^2;x,y)\right)^{\frac{\lambda_1}{2}}\left(K^*_{n,m}(1;x,y)\right)^{\frac{2-\lambda_1}{2}}\\ &\quad\times\left(K^*_{n,m}((s-y)^2;x,y)\right)^{\frac{\lambda_2}{2}}\left(K^*_{n,m}(1;x,y)\right)^{\frac{2-\lambda_2}{2}}\\ &\leq M\ (\mu_{n,2}(x))^{\frac{\lambda_1}{2}}\ (\nu_{m,2}(y))^{\frac{\lambda_2}{2}},\end{aligned}$$

which implies the desired result.

Now, we discuss the weighted estimate of the degree of approximation of a function defined on a weighted space of functions of two variables, by the linear positive operator $K^*_{n,m}$. Let

$$\begin{aligned}C_\rho := \{f\in C(S^2) : |f(x,y)| &\leq M_f\rho(x,y),\\ &M_f \textit{ is a positive constant depending on } f \textit{ only}\},\end{aligned}$$

where $\rho(x,y)$ is a weight function.

Following [11], for all $f\in C^0_\rho$, the weighted modulus of continuity is defined as

$$\omega_\rho(f;\gamma_1,\gamma_2)=\sup_{(x,y)\in S^2}\ \sup_{|h_1|\leq\gamma_1,\ |h_2|\leq\gamma_2}\left(\frac{f(x+h_1,y+h_2)-f(x,y)}{\rho(x,y)\rho(h_1,h_2)}\right),\quad (3.11)$$

where C^0_ρ is the subspace of all functions $f\in C_\rho$ such that $\lim_{x\to\infty}\frac{|f(x,y)|}{\rho(x,y)}$ exists finitely.

In our next result, let us assume $\rho(x,y)=1+x^2+y^2$.

Theorem 5 *If $f \in C_\rho^0$, then the inequality*

$$\sup_{(x,y)\in S^2} \frac{|K_{n,m}^*(f;x,y)-f(x,y)|}{(1+x^2+y^2)^4} \leq K\,\omega_\rho\left(f;\left(\frac{1}{b_n}\right)^{\frac{1}{2}},\left(\frac{1}{c_m}\right)^{\frac{1}{2}}\right),$$

holds for sufficiently large n and m, where K is a constant independent of $n,\ m$.

Proof From [15],

$$\begin{aligned}
&|f(t,s)-f(x,y)|\\
&\quad\leq 8(1+x^2+y^2)\omega_\rho(f;\gamma_n,\delta_m)\\
&\quad\left(1+\frac{|t-x|}{\gamma_n}\right)\left(1+\frac{|s-y|}{\delta_m}\right)(1+(t-x)^2)(1+(s-y)^2).
\end{aligned}$$

Now, applying the operator $K_{n,m}^*$ on both sides of the above equation,

$$\begin{aligned}
&|K_{n,m}^*(f;x,y)-f(x,y)|\\
&\quad\leq 8(1+x^2+y^2)\omega_\rho(f;\gamma_n,\delta_m)b_n 2^{-a_n x}\sum_{k=0}^{\infty} l_{n,k}(x)\\
&\quad\int_{\frac{k}{b_n}}^{\frac{(k+1)}{b_n}}\left\{1+(t-x)^2+\frac{|t-x|}{\gamma_n}+\frac{1}{\gamma_n}|t-x|(t-x)^2\right\}dt\\
&\quad\times c_m 2^{-d_m y}\sum_{j=0}^{\infty} l_{m,j}(y)\\
&\quad\int_{\frac{j}{c_m}}^{\frac{(j+1)}{c_m}}\left\{1+(s-y)^2+\frac{|s-y|}{\delta_m}+\frac{1}{\delta_m}|s-y|(s-y)^2\right\}ds.
\end{aligned}$$

By simple calculations and applying Cauchy–Schwarz inequality, we have

$$\begin{aligned}
&|K_{n,m}^*(f;x,y)-f(x,y)|\\
&\quad\leq 8(1+x^2+y^2)\omega_\rho(f;\gamma_n,\delta_m)\times\Bigg[1+K_{n,m}^*((e_{1,0}-x)^2;x,y)\\
&\quad+\frac{1}{\gamma_n}\sqrt{K_{n,m}^*((e_{1,0}-x)^2;x,y)}\\
&\quad+\frac{1}{\gamma_n}\sqrt{K_{n,m}^*((e_{1,0}-x)^2;x,y)K_{n,m}^*((e_{1,0}-x)^4;x,y)}\Bigg]\\
&\quad\times\Bigg[1+K_{n,m}^*((e_{0,1}-y)^2;x,y)+\frac{1}{\delta_m}\sqrt{K_{n,m}^*((e_{0,1}-y)^2;x,y)}\\
&\quad+\frac{1}{\delta_m}\sqrt{K_{n,m}^*((e_{0,1}-y)^2;x,y)K_{n,m}^*((e_{0,1}-y)^4;x,y)}\Bigg].
\end{aligned}$$

By using Remark 1, we get

$$
\begin{aligned}
&|K^*_{n,m}(f;x,y)-f(x,y)| \\
&\le M(1+x^2+y^2)\omega_\rho(f;\gamma_n,\delta_m)\Bigg[1+\frac{1}{b_n}(1+x+x^2) \\
&+\frac{1}{\gamma_n}\sqrt{\frac{1}{b_n}(1+x+x^2)} \\
&+\frac{1}{\gamma_n}\sqrt{\frac{1}{b_n}(1+x+x^2)\ \frac{1}{b_n}(1+x+x^2+x^3+x^4)}\Bigg] \\
&\times\Bigg[1+\frac{1}{c_m}(1+y+y^2)+\frac{1}{\delta_m}\sqrt{\frac{1}{c_m}(1+y+y^2)} \\
&+\frac{1}{\delta_m}\sqrt{\frac{1}{c_m}(1+y+y^2)\left(\frac{1}{c_m}\right)(1+y+y^2+y^3+y^4)}\Bigg],
\end{aligned}
$$

where $M>0$, is some positive constant.

Taking $\gamma_n=(\frac{1}{b_n})^{\frac{1}{2}}$ *and* $\delta_m=(\frac{1}{c_m})^{\frac{1}{2}}$, we have

$$
\begin{aligned}
&|K^*_{n,m}(f;x,y)-f(x,y)| \\
&\le M(1+x^2+y^2)\omega_\rho\left(f;\left(\frac{1}{b_n}\right)^{\frac{1}{2}},\left(\frac{1}{c_m}\right)^{\frac{1}{2}}\right) \\
&\Bigg[1+(1+x+x^2)+\sqrt{(1+x+x^2)} \\
&+\sqrt{(1+x+x^2)(1+x+x^2+x^3+x^4)}\Bigg] \\
&\times\Bigg[1+(1+y+y^2)+\sqrt{(1+y+y^2)} \\
&+\sqrt{(1+y+y^2)(1+y+y^2+y^3+y^4)}\Bigg].
\end{aligned}
$$

Hence for the sufficiently large n, m, we obtain

$$
\sup_{(x,y)\in S^2}\frac{|K^*_{n,m}(f;x,y)-f(x,y)|}{(1+x^2+y^2)^4}\le K\ \omega_\rho\left(f;\left(\frac{1}{b_n}\right)^{\frac{1}{2}},\left(\frac{1}{c_m}\right)^{\frac{1}{2}}\right),
$$

which yields the desired result.

Construction of GBS (Generalized Boolean Sum) Operators

In this section, we shall give a generalization of the operator (1.4) for the B-continuous functions. For this, we shall introduce a GBS operator associated with the bivariate operator (1.4) and investigate some of its approximation properties.

The concepts of B-continuous and B-differentiable function were introduced by Bögel [6, 7]. In approximation theory, the well-known Korovkin type theorem for B-continuous functions is developed by Badea et al. in [3, 4] using the Boolean sum approach.

The approximation properties of the bivariate Bernstein type operators and corresponding generalized Boolean sum operators were investigated in [5].

Let X and Y be compact subsets of $\mathbb{R}$. A function $f : X \times Y \to \mathbb{R}$ is called B-continuous (Bögel continuous) at a point $(x, y) \in X \times Y$ if

$$\lim_{(t,s)\to(x,y)} \Delta f[(t, s); (x, y)] = 0,$$

where $\Delta f[(t, s); (x, y)] = f(x, y) - f(x, s) - f(t, y) + f(t, s)$ denotes the mixed difference.

A function $f : X \times Y \to \mathbb{R}$ is called B-bounded on $X \times Y$ if there exists $M > 0$ such that

$$|\Delta f[(t, s); (x, y)]| \leq M,$$

for every $(x, y), (t, s) \in X \times Y$. Since $X \times Y$ is a compact subset of $\mathbb{R}^2$, each B-continuous function is a B-bounded function on $X \times Y \to \mathbb{R}$.

Let $B_b(X \times Y)$ denote all B-bounded functions on $X \times Y \to \mathbb{R}$, equipped with the norm

$$\|f\|_B = \sup_{(x,y),\ (t,s)\in X\times Y} |\Delta f[(t, s); (x, y)]|.$$

We denote by $C_b(X \times Y)$, the space of all B-continuous function on $X \times Y$. Let $B(X \times Y)$ and $C(X \times Y)$ denote the space of all bounded functions and the space of all continuous functions on $X \times Y$, respectively, endowed with the sup-norm $\|.\|_\infty$. It is known that $C(X \times Y) \subset C_b(X \times Y)$ [6]. A function $f : X \times Y \to \mathbb{R}$ is called a B-differentiable (Bögel differentiable) function at $(x, y) \in X \times Y$ if the limit

$$\lim_{(t,s)\to(x,y)} \frac{\Delta f[(t, s); (x, y)]}{(t - x)(s - y)},$$

exists and is finite.

The limit is said to be the B-differential of f at the point (x, y) and is denoted by $D_B(f; x, y)$ and the space of all B-differentiable functions is denoted by $D_b(X \times Y)$.

The mixed modulus of smoothness of $f \in C_b(X \times Y)$ is defined as

$$\omega_B(f, \gamma_1, \gamma_2) = \sup\{|\Delta f[(t, s); (x, y)]| : |x - t| < \gamma_1, \\ |y - s| < \gamma_2\},$$

for all (x, y), $(t, s) \in X \times Y$, and for any $(\gamma_1, \gamma_2) \in (0, \infty) \times (0, \infty)$ with $\omega_B : [0, \infty) \times [0, \infty) \to \mathbb{R}$. The basic properties of ω_B were obtained by Badea et al. in [3, 6] which are similar to the properties of the usual modulus of continuity.

For any $f \in C_b(I_{cd})$, $I_{cd} := [0, c] \times [0, d]$, and $m, n \in \mathbb{N}$, we define the GBS operators of the operators $K^*_{n,m}$ given by (1.4), as

$$\overline{G}_{n,m}(f(t,s); x, y) = K^*_{n,m}(f(t,y) + f(x,s) - f(t,s); x, y), \tag{3.12}$$

for all $(x, y) \in I_{cd}$.

Hence for any $f \in C_b(I_{cd})$, the GBS operator associated with the operator $K^*_{n,m}$ is defined as follows:

$$\begin{aligned}
&\overline{G}_{n,m}(f; x, y)\\
&= b_n 2^{-a_n x} c_m 2^{-d_m y} \sum_{j=0}^{\infty}\sum_{k=0}^{\infty} l^{k,j}_{n,m}(x, y)\\
&\int_{\frac{k}{b_n}}^{\frac{(k+1)}{b_n}} \int_{\frac{j}{c_m}}^{\frac{(j+1)}{c_m}} \Big(f(t, y) + f(x, s) - f(t, s)\Big) dsdt,
\end{aligned}$$

where the operator $\overline{G}_{n,m}$ is well defined from the space $C_b(I_{cd})$ into $C(I_{cd})$. It is clear that $\overline{G}_{n,m}$ is a linear operator.

Theorem 6 *For every $f \in C_b(I_{cd})$, at each point $(x, y) \in I_{cd}$, the operator $\overline{G}_{n,m}(f; x, y)$ verifies the following inequality*

$$|\overline{G}_{n,m}(f; x, y) - f(x, y)| \leq 4\omega_B(f; \sqrt{\mu_{n,2}(x)}, \sqrt{\nu_{m,2}(y)}),$$

where $\mu_{n,2}(x)$ and $\nu_{m,2}(y)$ are defined as in Theorem 2.

Proof By the property

$$\omega_B(f; \lambda_1\delta_1, \lambda_2\delta_2) \leq (1 + \lambda_1)(1 + \lambda_2)\omega_B(f; \delta_1, \delta_2); \quad \lambda_1, \lambda_2 > 0,$$

we can write

$$\begin{aligned}
|\Delta f[(t, s); (x, y)]| &\leq \omega_B(f; |t - x|, |s - y|)\\
&\leq \left(1 + \frac{|t - x|}{\delta_1}\right)\left(1 + \frac{|s - y|}{\delta_2}\right)\omega_B(f; \delta_1, \delta_2),
\end{aligned} \tag{3.13}$$

for every $(t, s) \in I_{cd}$ and any $\delta_1, \delta_2 > 0$. From (3.12) and the definition of the mixed difference $\Delta f[(t, s); (x, y)]$, on applying Lemma 2.3 and the inequality (3.13), we get

$$
\begin{aligned}
&|\overline{G}_{n,m}(f; x, y) - f(x, y)| \\
&\leq K^*_{n,m}(|\Delta f[(t, s); (x, y)]|; x, y) \\
&\leq \Bigg(K^*_{n,m}(1; x, y) + \frac{1}{\sqrt{\mu_{n,2}(x)}} K^*_{n,m}(|t - x|; x, y) \\
&\quad + \frac{1}{\sqrt{\mu_{n,2}(x)}\,\sqrt{\nu_{m,2}(y)}} K^*_{n,m}(|t - x|; x, y)\, K^*_{n,m}(|s - y|; x, y) \\
&\quad + \frac{1}{\sqrt{\nu_{m,2}(y)}} K^*_{n,m}(|s - y|; x, y) \Bigg) \omega_B\Big(f; \sqrt{\mu_{n,2}(x)}, \sqrt{\nu_{m,2}(y)} \Big).
\end{aligned}
$$

Now, applying the Cauchy–Schwarz inequality,

$$
\begin{aligned}
&|\overline{G}_{n,m}(f; x, y) - f(x, y)| \\
&\leq \Bigg(K^*_{n,m}(e_{0,0}; x, y) + \frac{1}{\sqrt{\mu_{n,2}(x)}} \sqrt{K^*_{n,m}((t - x)^2; x, y)} \\
&\quad + \frac{1}{\sqrt{\mu_{n,2}(x)}\,\sqrt{\nu_{m,2}(y)}} \sqrt{K^*_{n,m}((t - x)^2; x, y)}\, \sqrt{K^*_{n,m}((s - y)^2; x, y)} \\
&\quad + \frac{1}{\sqrt{\nu_{m,2}(y)}} \sqrt{K^*_{n,m}((s - y)^2; x, y)} \Bigg) \omega_B\Big(f; \sqrt{\mu_{n,2}(x)}, \sqrt{\nu_{m,2}(y)} \Big) \\
&= 4\, \omega_B\Big(f; \sqrt{\mu_{n,2}(x)}, \sqrt{\nu_{m,2}(y)} \Big).
\end{aligned}
$$

This completes the proof.

Following ([7], p. 382), for $f \in C_b(I_{cd})$, the Lipschitz class $Lip^*_M(\lambda_1, \lambda_2)$ with $\lambda_1, \lambda_2 \in (0, 1]$ is defined as follows: $Lip^*_M(\lambda_1, \lambda_2) = \{f \in C_b(I_{cd}) : |\Delta f[(t, s); (x, y)]| \leq M|t - x|^{\lambda_1} |s - y|^{\lambda_2}\}$, for $M > 0$ and $(t, s), (x, y) \in I_{cd}$.

In our next result, we determine the degree of approximation for the operators $\overline{G}_{n,m}$ by means of the class $Lip_M(\lambda_1, \lambda_2)$ class of Bögel continuous functions.

Theorem 7 *For $f \in Lip^*_M(\lambda_1, \lambda_2)$, we have*

$$
|\overline{G}_{n,m}(f; x, y) - f(x, y)| \leq M\ (\mu_{n,2}(x))^{\frac{\lambda_1}{2}}\ (\nu_{m,2}(y))^{\frac{\lambda_2}{2}},
$$

for $M > 0,\ \lambda_1, \lambda_2 \in (0, 1]$ and $\mu_{n,2}(x)$ and $\nu_{m,2}(x)$ are defined as in Theorem 2.

Proof From the definition of the mixed difference $\Delta f[(t, s); (x, y)]$, (3.12) and by our hypothesis, we may write

$$
\begin{aligned}
|\overline{G}_{n,m}(f; x, y) - f(x, y)| &\leq K^*_{n,m}(|\Delta f[(t, s); (x, y)]|; x, y) \\
&\leq M K^*_{n,m}(|t - x|^{\lambda_1} |s - y|^{\lambda_2}; x, y) \\
&= M K^*_{n,m}(|t - x|^{\lambda_1}; x, y) K^*_{n,m}(|s - y|^{\lambda_2}; x, y).
\end{aligned}
$$

Now, using the Hölder's inequality with $p_1 = \frac{2}{\lambda_1}$, $q_1 = \frac{2}{2-\lambda_1}$ and $p_2 = \frac{2}{\lambda_2}$, $q_2 = \frac{2}{2-\lambda_2}$, we are led to

$$\begin{aligned}
&|\overline{G}_{n,m}(f; x, y) - f(x, y)| \\
&\leq M(K^*_{n,m}((t-x)^2; x, y))^{\frac{\lambda_1}{2}} (K^*_{n,m}(e_{0,0}; x, y))^{\frac{2-\lambda_1}{2}} \\
&\quad \times (K^*_{n,m}((s-y)^2; x, y))^{\frac{\lambda_2}{2}} (K^*_{n,m}(e_{0,0}; x, y))^{\frac{2-\lambda_2}{2}}.
\end{aligned}$$

In view of Lemma 2.2, the desired result is immediate.

Theorem 8 *For $f \in D_b(I_{cd})$ with $D_B f \in B(I_{cd})$ and each $(x, y) \in I_{cd}$, we have*

$$\begin{aligned}
&|\overline{G}_{n,m}(f; x, y) - f(x, y)| \\
&\leq \|D_B f\|_\infty \frac{1}{\sqrt{b_n}} \frac{1}{\sqrt{c_m}} \sqrt{\xi_1(c)\eta_1(d)} \\
&\quad + \left(\frac{1}{\sqrt{b_n}} \frac{1}{\sqrt{c_m}} \sqrt{\xi_1(c)\eta_1(d)} + \frac{1}{\sqrt{c_m}} \sqrt{\xi_2(c)\eta_1(d)} \right. \\
&\quad + \frac{1}{\sqrt{b_n}} \sqrt{\xi_1(c)\eta_2(d)} \\
&\quad \left. + \frac{1}{\sqrt{b_n}} \frac{1}{\sqrt{c_m}} \xi_1(c)\eta_1(d) \right) \omega_B \left(D_B f; \frac{1}{\sqrt{b_n}}, \frac{1}{\sqrt{c_m}} \right).
\end{aligned}$$

Proof Since $f \in D_b(I_{cd})$ and $D_B f \in B(I_{cd})$,

$$\begin{aligned}
D_B f(x, y) &= \lim_{(t,s)\to(x,y)} \frac{\Delta f[(t, s); (x, y)]}{(t-x)(s-y)} \\
\Rightarrow \qquad \Delta f[(t, s); (x, y)] &= (t-x)(s-y) D_B f(\alpha_1, \alpha_2),
\end{aligned}$$

where α_1, α_2 lie between t and x, s, and y, respectively.

Using the following relation

$$\begin{aligned}
D_B f(\alpha_1, \alpha_2) = \Delta D_B f(\alpha_1, \alpha_2) + D_B f(\alpha_1, y) \\
+ D_B f(x, \alpha_2) - D_B f(x, y),
\end{aligned}$$

we obtain

$$
\begin{aligned}
&|K_{n,m}^{*}(\Delta f[(t,s);(x,y)];x,y)| \\
&\quad = |K_{n,m}^{*}((t-x)(s-y)D_B f(\alpha_1,\alpha_2);x,y)| \\
&\quad \le K_{n,m}^{*}(|t-x||s-y||\Delta D_B f(\alpha_1,\alpha_2)|;x,y) \\
&\qquad + K_{n,m}^{*}(|t-x||s-y|(|D_B f(\alpha_1,y)| \\
&\qquad + |D_B f(x,\alpha_2)| + |D_B f(x,y)|);x,y) \\
&\quad \le K_{n,m}^{*}(|t-x||s-y|\omega_B(D_B f;|\alpha_1-x|,|\alpha_2-y|);x,y) \\
&\qquad + 3\|D_B f\|_\infty K_{n,m}^{*}(|t-x||s-y|;x,y).
\end{aligned}
$$

Hence taking into account

$$
\begin{aligned}
&\omega_B(D_B f;|\alpha_1-x|,|\alpha_2-y|) \\
&\quad \le \omega_B(D_B f;|t-x|,|s-y|) \\
&\quad \le \left(1+\frac{1}{\delta_1}|t-x|\right)\left(1+\frac{1}{\delta_2}|s-y|\right)\omega_B(D_B f;\delta_1,\delta_2),
\end{aligned}
$$

for any $\delta_1, \delta_2 > 0$ and applying the Cauchy–Schwarz inequality, we obtain

$$
\begin{aligned}
&|\overline{G}_{n,m}(f;x,y) - f(x,y)| \\
&\quad = |K_{n,m}^{*}(\Delta f[(t,s);(x,y)];x,y)| \\
&\quad \le 3\|D_B f\|_\infty \sqrt{K_{n,m}^{*}((t-x)^2(s-y)^2;x,y)} \\
&\qquad + \Bigg(\sqrt{K_{n,m}^{*}((t-x)^2(s-y)^2;x,y)} \\
&\qquad + \frac{1}{\delta_1}\sqrt{K_{n,m}^{*}((t-x)^4(s-y)^2;x,y)} \\
&\qquad + \frac{1}{\delta_2}\sqrt{K_{n,m}^{*}((t-x)^2(s-y)^4;x,y)} \\
&\qquad + \frac{1}{\delta_1\delta_2}K_{n,m}^{*}((t-x)^2(s-y)^2;x,y)\Bigg)\omega_B(D_B f;\delta_1,\delta_2).
\end{aligned}
$$

From Lemma 2.4, we observe that for $(x,y),\ (t,s) \in I_{cd}$, we have $K_n^{*}((t-x)^2; x) \le (\frac{1}{b_n})\xi_1(c)$, $K_m^{*}((s-y)^2;y) \le (\frac{1}{c_m})\eta_1(d)$, and $K_n^{*}((t-x)^4;x) \le (\frac{1}{b_n})\xi_2(c)$, $K_m^{*}((s-y)^4;y) \le (\frac{1}{c_m})\eta_2(d)$, where $\xi_i(c)$ and $\eta_j(d)$, $i,j = 1,2$ are some constants depending on c and d, respectively.

Since

$$
\begin{aligned}
&K_{n,m}^{*}((t-x)^{2i}(s-y)^{2j};x,y) \\
&\quad = K_{n,m}^{*}((t-x)^{2i};x,y)K_{n,m}^{*}((t-x)^{2j};x,y), \quad i,j = 1,2,
\end{aligned}
$$

we get

$$
\begin{aligned}
&|\overline{G}_{n,m}(f;x,y)-f(x,y)|\\
&\le 3\|D_B f\|_\infty\sqrt{\left(\frac{1}{b_n}\right)\xi_1(c)\left(\frac{1}{c_m}\right)\eta_1(d)}\\
&+\Bigg[\sqrt{\left(\frac{1}{b_n}\right)\xi_1(c)\left(\frac{1}{c_m}\right)\eta_1(d)}+\frac{1}{\delta_1}\sqrt{\left(\frac{1}{b_n}\right)\xi_2(c)\left(\frac{1}{c_m}\right)\eta_1(d)}\\
&+\frac{1}{\delta_2}\sqrt{\left(\frac{1}{b_n}\right)\xi_1(c)\left(\frac{1}{c_m}\right)\eta_2(d)}\\
&+\frac{1}{\delta_1\delta_2}\left(\frac{1}{b_n}\right)\xi_1(c)\left(\frac{1}{c_m}\right)\eta_1(d)\Bigg]\omega_B(D_B f;\delta_1,\delta_2).
\end{aligned}
$$

Choosing $\delta_1=\sqrt{(\frac{1}{b_n})}$ and $\delta_2=\sqrt{(\frac{1}{c_m})}$, we get the desired result.

References

1. O. Agratini, On a sequence of linear and positive operators. Facta Univ. (NIŠ) Ser. Math. Inform. **14**, 41–48 (1999)
2. P.N. Agrawal, N. Ispir, Degree of approximation for bivariate Cholodowsky-Szasz-Charlier type operators. Results Math. **69**(3–4), 369–385 (2016)
3. C. Badea, C. Cottin, Korovkin-type theorems for generalized Boolean sum operators, *Colloquia Mathematica Societatis Janos Bolyai, Approximation Theory, Kecskemĕt (Hungary)*, vol. 58 (1990), pp. 51–67
4. C. Badea, I. Badea, H.H. Gonska, A test function theorem and approximation by pseudo polynomials. Bull. Aust. Math. Soc. **34**, 53–64 (1986)
5. D. Barbosu, GBS operators of Schurer-Stancu type. Ann. Univ. Craiova Math. Comput. Sci. Ser. **30**(2), 34–39 (2003)
6. K. Bögel, Über die mehrdimensionale differentiation. Jahresber. Dtsch. Math.-Ver. **65**, 45–71 (1962)
7. K. Bögel, Mehrdimensionale Differentiation von Funktionen mehrerer reeller Vernderlichen. J. Reine Angew. Math. **170**, 197–217 (2009)
8. P. L. Butzer, H. Berens, *Semi-groups of Operators and Approximation, Computational Science and Engineering*. Grundlehren der mathematischen Wissenschaften, vol. 145 (Springer, New York, 1967)
9. A. Erençin, F. Taşdelen, On certain Kantorovich type operators. Fasc. Math. **41**, 65–71 (2009)
10. A.K. Gazanfer, I. Büyükyazici, Approximation by certain linear positive operators of two variables. Abstr. Appl. Anal. **2014**, Article ID 782080, 6 pp
11. N.K. Govil, V. Gupta, D. Soybaş, Certain new classes of Durrmeyer type operators. Appl. Math. Comput. **225**, 195–203 (2013)
12. V. Gupta, A new class of Durrmeyer operators. Adv. Stud. Contemp. Math. **23**(2), 219–224 (2013)
13. V. Gupta, R. Yadav, On approximation of certain integral operators. Acta Math. Vietnam. **39**, 193–203 (2014)
14. V. Gupta, T.M. Rassias, R. Yadav, Approximation by Lupaş-Beta integral operators. Appl. Math. Comput. **236**, 19–26 (2014)

15. N. Ispir, C. Atakut, Approximation by modified Szasz-Mirakjan operators on weighted spaces. Proc. Indian Acad. Sci. (Math. Sci.) **112**(4), 571–578 (2002)
16. A. Lupaş, The approximation by some positive linear operators, in *Approximation Theory (Proceedings of the International Dortmound Meeting on Approximation Theory)*, ed. by M.W. Muller et al. (Akademie Verlag, Berlin, 1991), pp. 201–229
17. N. Manav, N. Ispir, Approximation by blending type operators based on Szasz-Lupaş basis functions. Gen. Math. **24**(1–2), 105–119 (2016)
18. N. Manav, N. Ispir, Approximation by the summation integral type operators based on Lupaş-Szasz basis functions. J. Sci. Arts **18**(4), 853–868 (2018)

Bernstein Polynomial Multiwavelets Operational Matrix for Solution of Differential Equation

Shweta Pandey, Sandeep Dixit and S. R. Verma

Abstract A new application of the Bernstein polynomials based multiwavelets approach for the numerical solution of differential equations is given. In the proposed method, Bernstein polynomial multiwavelets are obtained by using orthonormality of Bernstein polynomials. We present the operational matrix of integration of Bernstein polynomial based multiwavelets basis which diminishes the taken differential equation into the system of algebraic equations for less demanding calculations. High accuracy of these results even in the case of a small number of polynomials is observed. The convergence and exactness are described by comparing the ascertained approximated solution and the known analytical solution. The error estimates of the approximate solution are given and also some comparative examples with figures are given to confirm the reliability and accuracy of the proposed method. Some physical problems that lead to the differential equations are examined by the proposed method.

Keywords Bernstein polynomials · Bernstein polynomial multiwavelets operational matrix · Bernstein polynomial multiwavelets · Differential equation

---

S. Pandey (✉) · S. R. Verma
Department of Mathematics and Statistics, Gurukula Kangri Vishwavidyalaya, Haridwar 249404, India
e-mail: shwetapandey154@gmail.com

S. R. Verma
e-mail: drsrverma@gkv.ac.in

S. Dixit
Department of Mathematics, University of Petroleum and Energy Studies, Dehradun 248007, India
e-mail: SDIXIT@ddn.upes.ac.in

N. Deo et al. (eds.), *Mathematical Analysis I: Approximation Theory*, Springer Proceedings in Mathematics & Statistics 306,
https://doi.org/10.1007/978-981-15-1153-0_3

1 Introduction

The mathematical equation that shows a relationship between some function (generally symbolize physical quantities in case of a physical model) with its derivatives (correspond to their rates of change) is called a differential equation (DE). The problems of mathematical physics such as mechanics, elasticity and linearised theory of water waves, steady-state heat conduction, and radioactive heat transfer problems can be easily derived using differential equations. In economics and biology, DE is used to represent the behavior of complex systems. In recent decades, orthogonal functions and polynomial series have received enormous consideration in dealing with different problems of dynamic systems. Lepik [1] gave a solution of DE based on the Haar wavelet. Different operational matrices to get the solution of DE are known till now [2–4]. In this paper, a wavelet-based numerical method to solve DE based on Bernstein polynomial multiwavelets is given. We solved an example based on the LC circuit to show the application of the proposed method. An LC circuit is an electric circuit comprising of a capacitor (C) and an inductor (L) together as shown in Fig. 1.

The resonance effect of the LC circuit has major applications in communications systems and signal processing such as when we tune a radio to a specific radio channel, the LC circuits are set at resonance for that particular carrier frequency and also it is used for picking out a signal at a specific frequency from a more complex signal or generating signals at a specific frequency; these are the key components in many applications such as Mixers, Oscillators, Filters, and Tuners. A parallel and series resonant circuit gives current and voltage magnification, respectively. The DE governing the current flow in a series LC circuit when subject to a sinusoidal applied voltage $v(p) = v_0 \sin(\omega p)$ is

$$L\frac{d^2 i}{dp^2} + \frac{1}{C}i = \omega V_0 \cos \omega p; \tag{1}$$

First, we integrate the given DE, then remove the integral operators by approximating the given function with the help of Bernstein polynomial multiwavelets operational matrix of integration which converts the above DE to a system of algebraic equations for easier computations. One more example is given to better understand the proposed method.

Fig. 1 LC circuit diagram

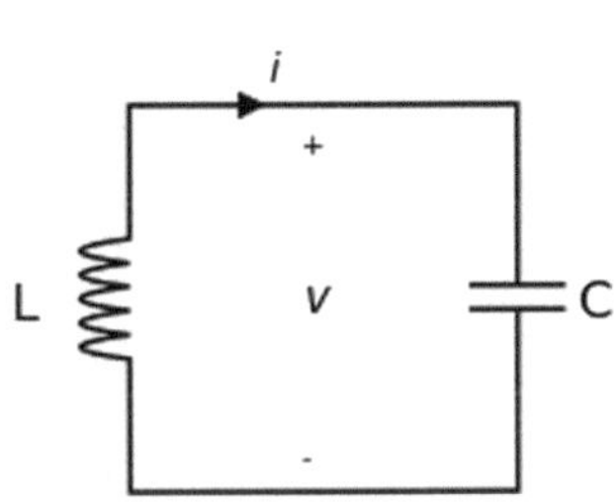

2 Wavelets and Bernstein Polynomial Multiwavelets

Wavelets used to compress the given information in a manner so that the resulting signal is a better representation of the information than the given signal which was in the original form. Wavelets has achieved the present growth as a result of various numerical investigation [5–10]. The continuous variation of the translation and dilation parameters u and v gives the following continuous wavelet form [5]

$$\psi_{u,v}(p) = |u|^{-1/2}\,\psi\left(\frac{p-v}{u}\right), \quad u, v \in R, \quad u \neq 0 \tag{2}$$

when these two parameters u and v are regulated to $u = 2^{-d}, v = n\,2^{-d}$, then new discrete wavelets family is obtained from the above equation as $\psi_{d,i}(p) = 2^{d/2}\psi\left(2^d p - i\right), \; d, i \in Z$, and $\int_R \psi(p)dp = 0$. Bernstein polynomials of order n characterized over the interval [0, 1] are given as

$$B_{m,n}(p) = \binom{n}{m} p^m (1-p)^{n-m}, \quad \forall\, m = 0, 1, 2, \ldots, n \tag{3}$$

Six orthonormal Bernstein polynomials of order 5 which are obtained from Bernstein polynomials $B_{m,n}(p)$ given in Eq. 3 by utilizing Gram–Schmidt process are given below:

$$\begin{aligned}
b_0(p) &= \sqrt{15}\,(1-p)^5 \\
b_1(p) &= 3\,(p-1)^4\,(-1+11p) \\
b_2(p) &= -\sqrt{7}\,(p-1)^3\left(1-20p+55\,p^2\right) \\
b_3(p) &= \sqrt{5}\,(p-1)^2\left(-1+27p-135p^2+165p^3\right) \\
b_4(p) &= \sqrt{3}\left(1-33p+248p^2-696p^3+810p^4-330p^5\right) \\
b_5(p) &= \;-1+35p-280p^2+840p^3-1050p^4+462p^5\,.
\end{aligned}$$

Bernstein polynomials orthonormal multiwavelets $\psi_{i,j}(p) = \psi(d, i, j, p)$ where d is dilation parameter assumes any positive integer, $j = 0, 1, 2, \ldots, N$ is Bernstein polynomial degree, translation parameter $i = 0, 1, 2, \ldots, 2^d - 1$, and normalized time p. They are characterized on the interval [0, 1] as [11]

$$\psi_{i,j}(p) = \begin{cases} 2^{d/2} b_j(2^d p - i) & \frac{i}{2^d} \le p < \frac{i+1}{2^d} \\ 0 & otherwise, \end{cases} \tag{4}$$

where $b_j(p)$ represents an order j orthonormal Bernstein polynomial. On taking $N = 5$ and dilation parameter $d = 0$ with help of orthonormal Bernstein polynomials, six Bernstein polynomials orthonormal multiwavelets can be obtained as

$$\psi_{0,0}(p) = \begin{cases} b_0(p), & 0 \le p < 1 \\ 0, & otherwise \end{cases},$$

$$\psi_{0,1}(p) = \begin{cases} b_1(p), & 0 \le p < 1 \\ 0, & otherwise \end{cases},$$

$$\vdots$$

$$\psi_{0,5}(p) = \begin{cases} b_5(p), & 0 \le p < 1 \\ 0, & otherwise \end{cases}.$$

Similarly, we can get $\psi_{i,j}(p)$ of different order j and dilation parameter d.

3 Function Approximation

As $f(p) \in L^2[0, 1]$, we may expand $f(p)$ as follows

$$f(p) = \sum_{i=0}^{\infty}\sum_{j=0}^{\infty} e_{ij}\, \psi_{_{ij}}(p) \tag{5}$$

where $e_{ij} = \langle f(p),\ \psi_{ij}(p)\rangle$ and $\langle .,\ .\rangle$ denote the inner product on the Hilbert space $L^2(R)$. In Eq. 5, the infinite series is truncated at levels $i = 2^d - 1$ and $j = N$, we obtain an approximate representation for $f(p)$ as

$$f\ (p) \approx \sum_{i=1}^{2^d-1}\sum_{j=0}^{N} e_{ij}\, \psi_{_{ij}}\ (p) = \ E^T\ \Psi\ (p) \tag{6}$$

where E and Ψ are $2^d(N+1) \times 1$ order matrices given by

$$E = [e_{00},\ \ldots,\ e_{0N};\ e_{10}, \ldots, e_{1N}; \ldots; e_{(2^d-1)0}, \ldots, e_{(2^d-1)N}]^T$$
$$\Psi(p) = \ [\psi_{00}(p), \ldots, \psi_{0N}(p); \psi_{10}(p), \ldots, \psi_{1N}(p); \quad \ldots; \psi_{(2^d-1)0}(p), \ldots, \psi_{(2^d-1)N}(p)]^T.$$

4 Bernstein Polynomial Multiwavelets Operational Matrix of Integration

In this section, the operational matrix of integration P is derived. First, we find the matrix P. The six basis functions are given by the approach based on transforming the DE to integral equations by integration, then truncating orthogonal series and approximating various signals involved in the equation by using the operational matrix of integration to remove the integral operations. The elements are the basis functions, orthogonal on a certain interval [a, b]. Gu and Jiang [11] derived the Haar wavelets based operational matrix of integration. In this paper, the integration of $\Psi(p)$ is approximated by Bernstein polynomials orthonormal multiwavelets series by Bernstein polynomials orthonormal multiwavelets coefficient matrix P

$$\int_0^p \Psi(p)\,dp = P_{2^d(N+1)\times 2^d(N+1)}\Psi(p), \tag{7}$$

where P is $2^d(N+1)$ order Bernstein polynomials orthonormal multiwavelets based operational matrix of integration.

5 Method of Solution

In this section, solution of DE

$$L\frac{d^2i}{dp^2} + \frac{1}{C}i = \omega V_0 \cos\omega p$$

governing the flow of current in a series LC circuit is given subject to an applied voltage $v(p) = v_0 \sin(\omega p)$ with boundary conditions $i(p) = 0$ and $i'(p) = 0$ whose exact solution is given by $C_1 cos\frac{p}{\sqrt{LC}} + C_2 sin\frac{p}{\sqrt{LC}} + \frac{C_3\omega V_0}{1-\omega^2 LC}\cos\omega p$. On integrating Eq. 1 with respect to p, we get

$$L\int_0^p \frac{d^2i}{dp^2}dp + \frac{1}{C}\int_0^p i\,dp = \int_0^p D^T\,\psi(p)dp. \tag{8}$$

Taking $i(p) = E^T\psi(p), \quad \omega V_0\cos\omega p = D^T\psi(p)$

$$L\int_0^p \frac{d^2i}{dp^2}dp + \frac{1}{C}\int_0^p E^T\psi(p)dp = \int_0^p D^T\,\psi(p)dp. \tag{9}$$

Using Eqs. 7 and 9 with boundary conditions, Eq. 10 becomes

$$L\frac{di}{dp} + \frac{1}{C} \times E^T P\,\psi(p) = D^T P\,\psi(p) \tag{10}$$

integrating the above equation with respect to p

$$L\int_0^p \frac{di}{dp}dp + \frac{1}{C}E^T P\int_0^p \psi(p)\,dp = D^T P\int_0^p \psi(p)\,dp \tag{11}$$

Again, using boundary conditions and Eqs. 7 and 9

$$L\frac{di}{dp}+\frac{C^T}{E}P^2\psi(p)=D^T P^2\psi(p)\,(13)L\,E^T+\frac{1}{C}E^T P^2=D^T P^2 \tag{12}$$

$$XE=Z \tag{13}$$

where $X=L\,I+\frac{1}{C}\left(P^2\right)^T$ and $Z=\left(P^2\right)^T D$ and I is the identity matrix. Equation 15 is a set of algebraic equations which can be solved for E where P is $2^d(N+1)$ order square matrix, called Bernstein multiwavelets based operational matrix. On putting the values of E into Eq. 9, we get the desired solution.

6 Illustrative Examples

The subsequent examples are illustrated to show the effectiveness and steadiness of our algorithm. Note that in first example, the series in Eq. 6 is truncated at level $j=5$. The exactness of the proposed method is shown by manipulating the absolute error, $\Delta\xi(p_k)$, which is given by

$$E(p)=\left|\xi(p_k)-\tilde{\xi}(p_k)\right|, \tag{14}$$

where $\tilde{\xi}(p_k)$ is the approximate solution calculated at point p_k and the exact solution at the corresponding point is $\xi(p_k)$.

6.1 Example 1

Consider the problem

$$0.25y'(t)+y(t)=u(t) \tag{15}$$

with $y(0)=0$ where $u(t)$ is the unit function. The analytic solution of Eq. 15 is $y(t)=1-\exp(-4t)$. Gu and Jiang [11] solved this problem by using Haar wavelets with six and ten basis functions. Razzaghi and Yousefi [12] solved this problem using Legendre wavelets, with $M=3$ and $K=2$; here, we solve this by Bernstein polynomial multiwavelets, with $j=5$; $d=0$ and $j=5$; $d=1$. We assume the unknown function $y(t)$ is given by $y(t)=C^T\psi(t)$, to get the solution of DE; take $y(t)=C^T\psi(t)$, $u(t)=D^T\psi(t)$. Integrating Eq. 15, we get

$$0.25C^T\psi(t)=-\int_0^t C^T\psi(t)dt+\int_0^t D^T\psi(t)\,dt$$

$$0.25\,C^T\psi(t)=-C^T P\psi(t)+D^T P\psi(t)$$

$$0.25\,C^T + \kappa C^T P = D^T P$$

$$LC = M \tag{16}$$

where $L = 0.25I + P^T$ and $M = P^T D$ and I is the identity matrix. Equation 16 is a set of algebraic equations which can be solved for C where P is $2^d(N+1)$ order square matrix, called Bernstein multiwavelets based operational matrix. After getting the value of C, we get the desired solution (Table 1 and Fig. 2).

6.2 Example 2

Consider the LC circuit given in Eq. 1 with $L = 2\,\mathrm{H}$, $C = 0.5\,F$, and the source voltage $v(p) = v_0 \sin(\omega p)$ where $V_0 = 1V$ with boundary conditions $i(p) = 0$ and $i'(p) = 0$ whose exact solution is given by $i(p) = \frac{1}{3}(\cos(t) - \cos(2t))$ using Bern-

Table 1 Approximate and exact solution of Example 1

p	Exact solutions $\xi 1(p)$	Approximate solutions $\xi 2(p)$ (For $d = 0$)	Approximate solutions $\xi 2(p)$ (For $d = 1$)
0.0	0.000000	0.000006	0.000995
0.2	0.550671	0.550663	0.550898
0.4	0.798103	0.798078	0.797948
0.6	0.909282	0.909288	0.909294
0.8	0.959238	0.959208	0.959326
0.9	0.981684	0.981975	0.982680

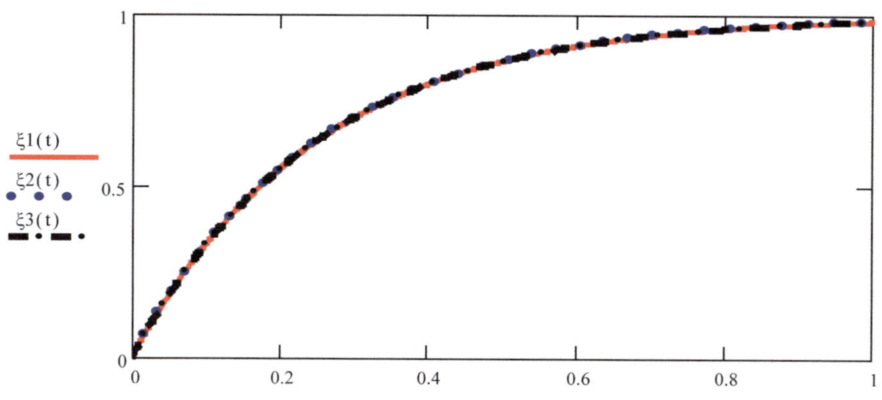

Fig. 2 Comparison of solutions: exact solution $\xi 1(t)$, approximated solution $\xi 2(t)$ (with dilation parameter $d = 0$), and approximated solution $\xi 3(t)$ (with dilation parameter $d = 1$)

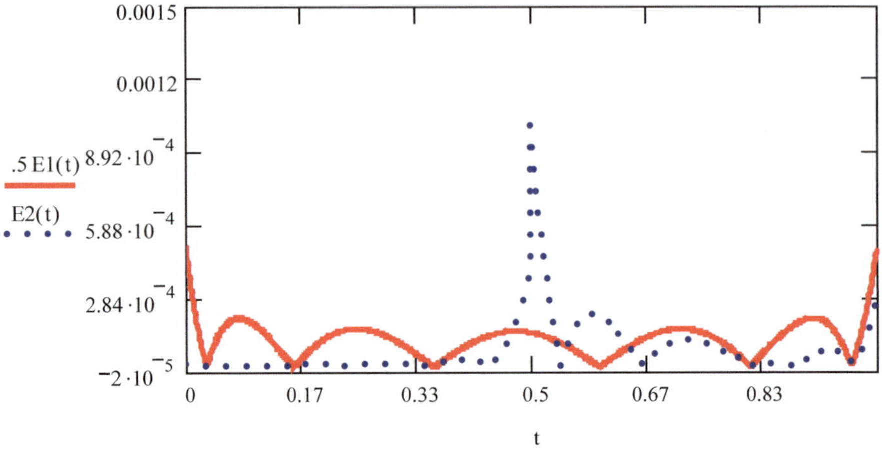

Fig. 3 Comparison of absolute errors: $E1(t)$ and $E2(t)$ for dilation parameter $d = 0$ and $d = 1$

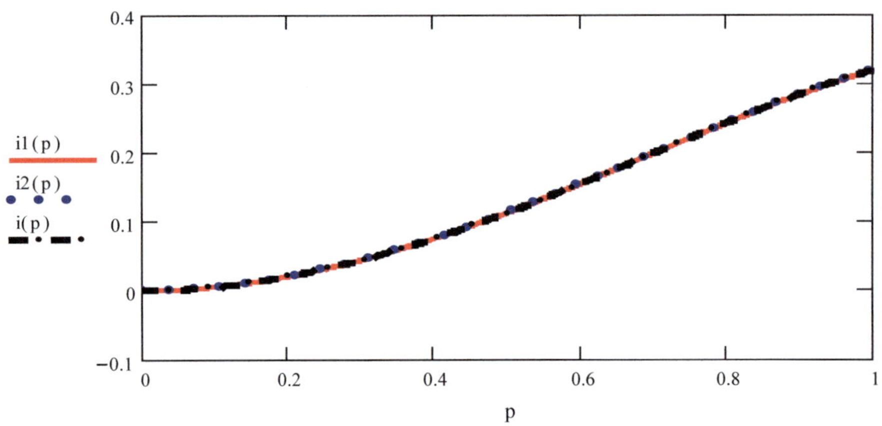

Fig. 4 Comparison of solutions: exact solution $i(p)$, approximated solution $i1(p)$ (with dilation parameter $d = 0$), and $i2(p)$ (with dilation parameter $d = 1$)

stein polynomial multiwavelets operational matrix of integration we solve the above LC circuit problem and calculate the approximated solution $i1(p)$ and respectively $i2(p)$ for $d = 0$ and $d = 1$ Bernstein polynomial multiwavelets of order by utilizing the above algorithm from Eqs. 8 to 1. Figures representing the exact and approximate solution and absolute errors $E1(p) = |i(p) - i1(p_k)|$ and $E2(p) = |i(p) - i2(p_k)|$ are given in (Figs. 3, 4, 5).

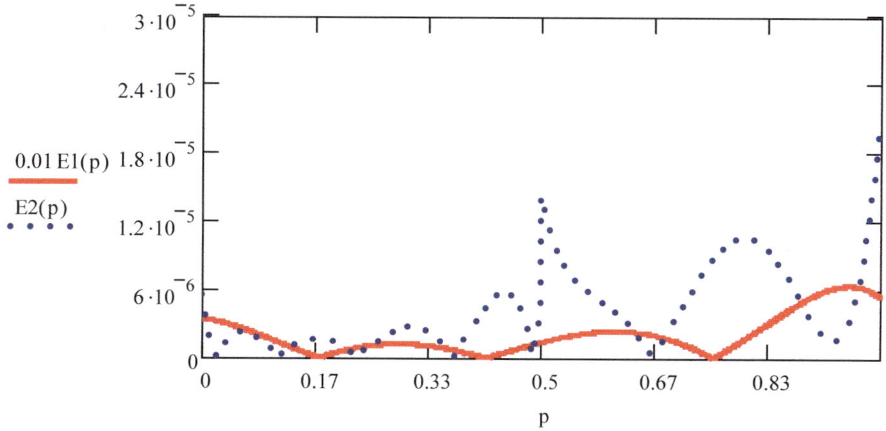

Fig. 5 Comparison of absolute errors: $E1(p)$ and $E2(p)$ for dilation parameter $d = 0$ and $d = 1$

7 Conclusion

We have used Bernstein polynomial multiwavelets to construct the operational matrix of integration with two dilation parameters $d = 0$ and $d = 1$, which diminishes the given differential equation to the system of algebraic equation for easy computations. Here, we have used the Bernstein multiwavelets operation matrix of integration approach to finding the numerical solutions of the differential equations. Furthermore, our technique demonstrates the comparison between the solution for two different dilation parameters. The selection of only six orthonormal polynomials of degree 5 makes the method easy and straight forward to use.

References

1. Ü. Lepik, Numerical solution of differential equations using Haar wavelets. Math. Comput. Simul. **682**, 127–143 (2005)
2. L. Shi, X. Ding, Z. Chen, Q. Ma, A new class of operational matrices method for solving fractional neutral pantograph differential equations. Adv. Differ. Equ. **1**, 94 (2018)
3. Y. Öztürk, Numerical solution of systems of differential equations using operational matrix method with chebyshev polynomials. J. Taibah Univ. Sci. **122**, 155–162 (2018)
4. M.H.T. Alshbool, A.S. Bataineh, I. Hashim, O.R. Isik, Solution of fractional-order differential equations based on the operational matrices of new fractional Bernstein functions. J. King Saud Univ.-Sci. **291**, 1–18 (2017)
5. J.S. Gu, W.S. Jiang, The Haar wavelets operational matrix of integration. Int. J. Syst. Sci. **27**, 623–628 (1996)
6. J.O. Stromberg, in *Proceedings of Harmonic Analysis* (University of Chicago, 1981), pp. 475–494
7. A. Grossmann, J. Morlet, SIAM J. Math. Anal. **15**, 723–736 (1984)

8. Y. Meyer, *Analysis at Urbana I.: Analysis in Function Spaces* (Cambridge University Press, Cambridge, 1989)
9. S.G. Mallat, Trans. Am. Math. Soc. **315**, 69 (1989)
10. I. Daubechies, Commun. Pure Appl. Math. **41**, 909 (1988)
11. S.A. Yousefi, B-polynomial multiwavelets approach for the solution of Abel's integral equation. Int. J. Comput. Math. **872**, 310–316 (2010)
12. M. Razzaghi, S. Yousefi, The legendre wavelets operational matrix of integration. Int. J. Syst. Sci. **324**, 495–502 (2001)

Convergence Estimates of Certain Exponential Type Operators

Vijay Gupta

Abstract The present paper deals with the approximation properties of certain exponential type operators, which is one of the operators proposed by Ismail-May (1978). We calculate the moments and obtain a direct result and an error estimation.

Keywords Exponential type operators · Lorentz type result · Moment estimates · Convergence estimates · Ismail-May operators

1 Introduction

In continuation of the remarkable work by May [15], Ismail-May [13] considered some more exponential type operators and studied the approximation properties involving direct results. The exponential type operators satisfy the following partial differential equation:

$$\frac{\partial}{\partial x}W(n,x,t)=\frac{n(t-x)}{p(x)}W(n,x,t),$$

where $W(n,x,t)$ is the kernel of exponential type operators $S_n(f,x)=\int_{-\infty}^{\infty} W(n,x,t)f(t)dt$, Ismail and May [13] recovered some known operators and constructed some new operators. One of the operators proposed in [13] is defined as

$$T_n(f,x)=e^{-n\sqrt{x}}\left\{n\int_0^\infty e^{-nt/\sqrt{x}}t^{-1/2}I_1(2n\sqrt{t})f(t)dt+f(0)\right\}, \tag{1}$$

V. Gupta (✉)
Department of Mathematics, Netaji Subhas University of Technology, Sector 3 Dwarka, New Delhi 110078, India
e-mail: vijaygupta2001@hotmail.com

N. Deo et al. (eds.), *Mathematical Analysis I: Approximation Theory*, Springer Proceedings in Mathematics & Statistics 306,
https://doi.org/10.1007/978-981-15-1153-0_4

where I_1 is modified Bessel's function of first kind given by

$$I_n(z) = \sum_{k=0}^{\infty} \frac{\left(\frac{z}{2}\right)^{n+2k}}{k!\Gamma(n+k+1)}.$$

Also the Lorentz type result for the operators (1) satisfy

$$2x^{3/2}\frac{\partial}{\partial x}[e^{-n\sqrt{x}}e^{-nt/\sqrt{x}}] = n(t-x)[e^{-n\sqrt{x}}e^{-nt/\sqrt{x}}]. \tag{2}$$

In the past several years convergence estimates are an active area of research. Singh and Sharma in [16] generalized a result of Garrett-Stanojevic [9] concerning convergence of certain cosine sums. In the case of linear positive operators, many operators have been constructed in past seven decades and many approximation results have been established. It is difficult here to refer many papers, some of the results in real and complex domain are due to Agarwal–Gupta [1], Deo and collaborators (see [3, 4]), Gal–Gupta [8], Lorentz [14], Ditzian–Totik [6], Gupta–Tachev [10] Gupta et al. [11], etc.

As per our knowledge, the operators T_n specifically were not studied independently by researchers, although some researchers have considered exponential type operators. The behavior of these operators is different, the present article deals with some of the approximation properties of Ismail-May operators T_n, we obtain some convergence estimates for such operators.

2 Moment Estimation

In the sequel, we calculate here the moment estimates of the operators T_n.

Lemma 1 *For the operators defined by (1), if* $e_r = t^r, r = 0, 1, 2, \ldots,$ *then we have*

$$T_n(e_0, x) = 1, \;\; T_n(e_1, x) = x, \;\; T_n(e_2, x) = x^2 + \frac{2x^{3/2}}{n}$$

$$T_n(e_3, x) = x^3 + \frac{6x^{5/2}}{n} + \frac{6x^2}{n^2}$$

$$T_n(e_4, x) = x^4 + \frac{12x^{7/2}}{n} + \frac{36x^3}{n^2} + \frac{24x^{5/2}}{n^3}.$$

Proof First by definition of T_n, we have

$$
\begin{aligned}
T_n(e_0,x) &= e^{-n\sqrt{x}}\left\{n\int_0^\infty e^{-nt/\sqrt{x}}t^{-1/2}\sum_{k=0}^\infty \frac{\left(n\sqrt{t}\right)^{1+2k}}{k!(k+1)!}dt+1\right\}\\
&= e^{-n\sqrt{x}}\left\{\sum_{k=0}^\infty \frac{n^{2k+2}}{k!(k+1)!}\int_0^\infty e^{-nt/\sqrt{x}}t^k dt+1\right\}\\
&= e^{-n\sqrt{x}}\left\{\sum_{k=0}^\infty \frac{n^{2k+2}}{k!(k+1)!}\int_0^\infty e^{-u}u^k\frac{x^{(k+1)/2}}{n^{k+1}}du+1\right\}\\
&= e^{-n\sqrt{x}}\left\{\sum_{k=0}^\infty \frac{n^{k+1}}{(k+1)!}x^{(k+1)/2}+1\right\}\\
&= e^{-n\sqrt{x}}\left\{\sum_{k=1}^\infty \frac{n^{k}}{k!}x^{k/2}+1\right\}\\
&= e^{-n\sqrt{x}}\left\{\sum_{k=0}^\infty \frac{n^{k}}{k!}x^{k/2}\right\}=e^{-n\sqrt{x}}e^{n\sqrt{x}}=1.
\end{aligned}
$$

Next

$$
\begin{aligned}
T_n(e_r,x) &= e^{-n\sqrt{x}}\left\{n\int_0^\infty e^{-nt/\sqrt{x}}t^{-1/2}\sum_{k=0}^\infty \frac{\left(n\sqrt{t}\right)^{1+2k}}{k!(k+1)!}t^r dt\right\}\\
&= e^{-n\sqrt{x}}\left\{\sum_{k=0}^\infty \frac{n^{2k+2}}{k!(k+1)!}\int_0^\infty e^{-nt/\sqrt{x}}t^{k+r}dt\right\}\\
&= e^{-n\sqrt{x}}\left\{\sum_{k=0}^\infty \frac{n^{2k+2}}{k!(k+1)!}\int_0^\infty e^{-u}u^{k+r}\frac{x^{(k+r+1)/2}}{n^{k+r+1}}du\right\}\\
&= e^{-n\sqrt{x}}\left\{\sum_{k=0}^\infty \frac{n^{k-r+1}}{k!(k+1)!}\Gamma(k+r+1).x^{(k+r+1)/2}\right\}\\
&= e^{-n\sqrt{x}}\left\{x^{(r+1)/2}n^{-r+1}\sum_{k=0}^\infty \frac{n^{k}}{k!}x^{k/2}\frac{(k+r)!}{(k+1)!}\right\}.
\end{aligned}
$$

Substituting $r=1,2,3,4,\dots$ the moments are given as

$$
\begin{aligned}
T_n(e_1,x) &= e^{-n\sqrt{x}}\left\{x^{(r+1)/2}n^{-r+1}\sum_{k=0}^\infty \frac{n^{k}}{k!}x^{k/2}\frac{(k+r)!}{(k+1)!}\right\}\\
&= e^{-n\sqrt{x}}\left\{x\sum_{k=0}^\infty \frac{n^{k}}{k!}x^{k/2}\right\}=x.
\end{aligned}
$$

$$\begin{aligned}T_n(e_2,x) &= e^{-n\sqrt{x}}\left\{x^{(r+1)/2}n^{-r+1}\sum_{k=0}^{\infty}\frac{n^k}{k!}x^{k/2}\frac{(k+r)!}{(k+1)!}\right\}\\ &= e^{-n\sqrt{x}}\left\{\frac{x^{3/2}}{n}\sum_{k=0}^{\infty}\frac{n^k}{k!}x^{k/2}(k+2)\right\}\\ &= e^{-n\sqrt{x}}\left\{\frac{x^{3/2}}{n}\sum_{k=1}^{\infty}\frac{n^k}{(k-1)!}x^{k/2}+\frac{2x^{3/2}}{n}\sum_{k=0}^{\infty}\frac{n^k}{k!}x^{k/2}\right\}\\ &= e^{-n\sqrt{x}}\left\{x^2\sum_{k=0}^{\infty}\frac{n^k}{k!}x^{k/2}+\frac{2x^{3/2}}{n}\sum_{k=0}^{\infty}\frac{n^k}{k!}x^{k/2}\right\}=x^2+\frac{2x^{3/2}}{n}.\end{aligned}$$

Similarly, we obtain the other two moments, we omit the details.

Lemma 2 *If the central moments are defined by* $\mu_{n,m}(x)=T_n((t-x)^m,x), m=0,1,2,\ldots,$ *then we have*

$$\begin{aligned}\mu_{n,0}(x) &= 1\\ \mu_{n,1}(x) &= 0\\ \mu_{n,2}(x) &= \frac{2x^{3/2}}{n}\\ \mu_{n,3}(x) &= \frac{6x^2}{n^2}\\ \mu_{n,4}(x) &= \left(\frac{24x^{5/2}}{n^3}+\frac{12x^3}{n^2}\right).\end{aligned}$$

The proof of above lemma follows using Lemma 1, just we have to apply the linearity property of the operators.

Remark 1 By definition of T_n, we have

$$\begin{aligned}T_n(e^{At},x) &= e^{-n\sqrt{x}}\left\{n\int_0^{\infty}e^{-nt/\sqrt{x}}t^{-1/2}\sum_{k=0}^{\infty}\frac{\left(n\sqrt{t}\right)^{1+2k}}{k!(k+1)!}e^{At}dt+1\right\}\\ &= e^{-n\sqrt{x}}\left\{\sum_{k=0}^{\infty}\frac{n^{2k+2}}{k!(k+1)!}\int_0^{\infty}e^{-t(n/\sqrt{x}-A)}t^k dt+1\right\}\\ &= e^{-n\sqrt{x}}\left\{\sum_{k=0}^{\infty}\frac{n^{2k+2}}{k!(k+1)!}\int_0^{\infty}e^{-u}u^k\frac{x^{(k+1)/2}}{(n-A\sqrt{x})^{k+1}}du+1\right\}\\ &= e^{-n\sqrt{x}}\left\{\sum_{k=0}^{\infty}\frac{n^{2k+2}}{k!(k+1)!}\Gamma(k+1).\frac{x^{(k+1)/2}}{(n-A\sqrt{x})^{k+1}}+1\right\}\end{aligned}$$

$$= e^{-n\sqrt{x}} \left\{ \sum_{k=0}^{\infty} \frac{\left(\frac{n^2\sqrt{x}}{n-A\sqrt{x}}\right)^{k+1}}{(k+1)!} \right\} = e^{\frac{nAx}{(n-A\sqrt{x})}}.$$

3 Convergence Estimates

Suppose $C_B[0,\infty)$ be the space of all continuous and bounded functions on $[0,\infty)$ endowed with the norm $\|f\| = \sup\{|f(x)| : x \in [0,\infty)\}$. Further let us consider the following K functional:

$$K_2(f,\delta) = \inf_{g \in C_B^2[0,\infty)} \{\|f-g\| + \delta\|g''\|\},$$

where $\delta > 0$ and $C_B^2[0,\infty) = \{g \in C_B[0,\infty) : g', g'' \in C_B[0,\infty)\}$. Following [5], one has the inequality

$$K_2(f,\delta) \leq C\omega_2(f,\sqrt{\delta}), \delta > 0. \tag{3}$$

Theorem 1 *Let $f \in C_B[0,\infty)$, then we have*

$$|T_n(f,x) - f(x)| \leq C\omega_2\left(f, \frac{x^{3/4}}{\sqrt{n}}\right).$$

Proof Let $g \in C_B^2[0,\infty)$ and $x, t \in [0,\infty)$, by Taylor's formula, we have

$$g(t) = g(x) + (t-x)g'(x) + \int_x^t (t-u)g''(u)du.$$

Then using Lemma 1, we have

$$\begin{aligned} |T_n(g,x) - g(x)| &= \left| T_n\left(\int_x^t (t-u)g''(u)du, x\right)\right| \\ &\leq \frac{1}{2}\mu_{n,2}(x)\ \|g''\| = \frac{x^{3/2}}{n}\|g''\|. \end{aligned} \tag{4}$$

Also, we have

$$|T_n(f,x)| \leq \|f\|. \tag{5}$$

Therefore using (4) and (5), we have

$$\begin{aligned}|T_n(f,x)-f(x)| &\le |T_n(f-g,x)-(f-g)(x)|+|T_n(g,x)-g(x)|\\ &\le 2\|f-g\|+\frac{x^{3/2}}{n}\|g''\|.\end{aligned} \tag{6}$$

Taking infimum over all $g \in C_B^2[0,\infty)$, and using (3), we get the desired result.

Theorem 2 *Let f be bounded and integrable function on the interval $[0,\infty)$, possessing a second derivative of f at a fixed point $x \in [0,\infty)$, then*

$$\lim_{n\to\infty} n\,(T_n(f,x)-f(x)) = x^{3/2} f''(x).$$

Proof By the Taylor's formula, we have

$$f(t) = f(x) + f'(x)(t-x) + \frac{1}{2} f''(x)(t-x)^2 + \lambda(t,x)(t-x)^2, \tag{7}$$

where $\lambda(t,x)$ is the remainder term and $\lim_{n\to\infty} \lambda(t,x) = 0$. Applying T_n to the Eq. (7), we obtain

$$\begin{aligned}T_n(f,x)-f(x) &= T_n(t-x,x) f'(x) + T_n\left((t-x)^2,x\right)\frac{f''(x)}{2}\\ &\quad + T_n\left(\lambda(t,x)(t-x)^2,x\right)\end{aligned}$$

Next using Cauchy–Schwarz inequality, we have

$$T_n\left(\lambda(t,x)(t-x)^2,x\right) \le \sqrt{T_n\left(\lambda^2(t,x),x\right)}\sqrt{T_n\left((t-x)^4,x\right)}. \tag{8}$$

As $\lambda^2(x,x)=0$, we have

$$\lim_{n\to\infty} T_n\left(\lambda^2(t,x),x\right) = \lambda^2(x,x) = 0 \tag{9}$$

uniformly with respect to $x \in [0,A]$. Now from (8), (9) and Lemma 2, we get

$$\lim_{n\to\infty} T_n\left(\lambda(t,x)(t-x)^2,x\right) = 0.$$

Then using Lemma 2, we obtain

$$\begin{aligned}&\lim_{n\to\infty} (T_n(f,x)-f(x))\\ &= \lim_{n\to\infty} f'(x)\mu_{n,1}(x) + \frac{1}{2} f''(x)\mu_{n,2}(x)\\ &\quad + T_n\left(\lambda(t,x)(t-x)^2;x\right)\\ &= x^{3/2} f''(x).\end{aligned}$$

Next result provides an estimate in weighted approximation, we give the following result.

Let us consider $C^*[0, \infty)$ the space of all continuous functions satisfying the condition $\lim_{x\to\infty} \frac{f(x)}{1+x^{3/2}}$ is finite and belonging to the class $B[0, \infty)$, where $B[0, \infty) = \{f :$ for every $x \in [0, \infty), |f(x)| \leq C(1 + x^{3/2})$, C being certain constant depending on $f\}$. The norm on $C^*[0, \infty)$ is defined by

$$\|f\| = \sup_{x\in[0,\ \infty)} \frac{|f\ (x)|}{1 + x^{3/2}}.$$

Theorem 3 *For each $f \in C^*[0, \infty)$, we have*

$$\lim_{n\to\infty} \|T_n(f) - f\| = 0.$$

Proof Using Theorem in [7], in order to prove the theorem, it is sufficient to show that

$$\lim_{n\to\infty} \|T_n(e_m, x) - e_m\| = 0, m = 0, 1, 2.$$

Obviously the operators T_n preserve constant and linear function, we only have to show the validity at $m = 2$. We can write

$$\left\|T_n\ (e_2, x) - x^2\right\| \leq \sup_{x\in[0,\ \infty)} \frac{2x^{3/2}}{n(1 + x^{3/2})}.$$

which implies that

$$\lim_{n\to\infty} \left\|T_n\ (e_2, x) - x^2\right\| = 0$$

This completes the proof of the theorem.

Let $C^*[0, \infty)$ denotes the Banach space of all real-valued continuous functions on $[0, \infty)$ with the property that $\lim_{x\to\infty} f(x)$ exists and is finite, endowed with the uniform norm.

Also, for every $\delta \geq 0$ the modulus of continuity (see [12]) is given by

$$\omega^*(f, \delta) = \sup_{\substack{x,t\geq 0 \\ |e^{-x}-e^{-t}|\leq\delta}} |f(x) - f(t)|.$$

Theorem 4 *Let $f, f'' \in C^*[0, \infty)$, then, for $x \in [0, \infty)$, the following inequality holds:*

$$\left|n\left[T_n(f,x)-f(x)\right]-x^{3/2}f''(x)\right|$$
$$\leq 2\omega^*(f'',n^{-1/2})\left[2x^{3/2}+\left(\frac{24x^{5/2}}{n}+12x^3\right)^{1/2}\left[n^2T_n\left((e^{-x}-e^{-t})^4,x\right)\right]^{1/2}\right].$$

Proof By the Taylor's expansion, we have

$$f(t)=\sum_{i=0}^{2}(t-x)^i\,\frac{f^{(i)}(x)}{i!}+h(t,x)\;(t-x)^2,$$

where

$$h(t,x):=\frac{f''(\xi)-f''(x)}{2},$$

with ξ lying between x and t. Applying the operator T_n to above equality, we can write that

$$\left|T_n(f,x)-\mu_{n,0}(x)f(x)-\mu_{n,1}(x)\,f'(x)-\tfrac{1}{2}\,\mu_{n,2}(x)f''(x)\right|$$
$$=\left|T_n\left(h(t,x)\;(t-x)^2,x\right)\right|.$$

Thus, using Lemma 2, we get

$$\left|n\left[T_n(f,x)-f(x)\right]-x^{3/2}f''(x)\right|=\left|n\;T_n\left(h(t,x)\;(t-x)^2,x\right)\right|.$$

Using the methods as given in [2, Th. 2], we can write

$$|h(t,x)|\leq 2\left(1+\frac{\left(e^{-x}-e^{-t}\right)^2}{\delta^2}\right)\omega^*(f'',\delta).$$

Hence, after applying Cauchy–Schwarz inequality we get

$$n\;T_n\left(|h(t,x)|\,(t-x)^2,x\right)\leq 2\,n\;\omega^*(f'',\delta)\;\mu_{n,2}(x)$$
$$+\frac{2n}{\delta^2}\;\omega^*(f'',\delta)\;\sqrt{T_n\left((e^{-x}-e^{-t})^4,x\right)}\sqrt{T_n\left((t-x)^4,x\right)}.$$

Considering $\delta=n^{-1/2}$, we obtain

$$nT_n\left(|h(t,x)|\,(t-x)^2,x\right)\leq 2\omega^*\left(f'',\frac{1}{\sqrt{n}}\right)\left[n\mu_{n,2}(x)+\sqrt{n^2T_n\left((e^{-x}-e^{-t})^4,x\right)}\sqrt{n^2\mu_{n,4}(x)}\right].$$

Finally using Lemma 2, we get the required result.

Remark 2 The convergence of the Ismail-May operators T_n in the above theorem takes place for n sufficiently large. Using Remark 1, for $A=-1,-2,-3,-4$ and by the mathematical software, we find that

$$
\begin{aligned}
&\lim_{n\to\infty} n^2\, T_n\left(\left(e^{-x}-e^{-t}\right)^4, x\right)\\
&= \lim_{n\to\infty} n^2 \left(T_n(e^{-4t},x) - 4e^{-x}T_n(e^{-3t},x) + 6e^{-2x}T_n(e^{-2t},x) - 4e^{-3x}T_n(e^{-t},x) + e^{-4x}\right)\\
&= \lim_{n\to\infty} n^2 \left[e^{\frac{-4nx}{(n+4\sqrt{x})}} - 4e^{-x}e^{\frac{-3nx}{(n+3\sqrt{x})}} + 6e^{-2x}e^{\frac{-2nx}{(n+2\sqrt{x})}} - 4e^{-3x}e^{\frac{-nx}{(n+1\sqrt{x})}} + e^{-4x}\right]\\
&= \lim_{n\to\infty} n^2 \left[e^{\frac{-4nx}{(n+4\sqrt{x})}} - 4e^{\frac{-3x\sqrt{x}-4nx}{(n+3\sqrt{x})}} + 6e^{\frac{-4x\sqrt{x}-4nx}{(n+2\sqrt{x})}} - 4e^{\frac{-3x\sqrt{x}-4nx}{(n+\sqrt{x})}} + e^{-4x}\right]\\
&= 12e^{-4x}x^3.
\end{aligned}
$$

References

1. R.P. Agarwal, V. Gupta, On q-analogue of a complex summation-integral type operators in compact disks. J. Inequal Appl. (1), 111 (2012). https://doi.org/10.1186/1029-242X-2012-111
2. E. Deniz, A. Aral, V. Gupta, Note on Szász–Mirakyan–Durrmeyer operators preserving $e^{2ax}, a>0$. Numer. Funct. Anal. Optim. **39**(2), 201–207 (2017)
3. N. Deo, Faster rate of convergence on Srivastava-Gupta operators. Appl. Math. Comput. **218**(21), 10486–10491 (2012)
4. N. Deo, M.A. Noor, M.A. Siddiqui, On approximation by a class of new Bernstein type operators. Appl. Math. Comput. **201**(1–2), 604–612 (2008)
5. R.A. DeVore, G.G. Lorentz, *Constructive Approximation* (Springer, Berlin, 1993)
6. Z. Ditzian, V. Totik, *Moduli of Smoothness* (Springer, New York, 1987)
7. A.D. Gadzhiev, Theorems of the type of P.P. Korovkin type theorems. Math. Zametki **20**(5), 781–786 (1976); English translation, Math. Notes **20**(5–6), 996–998 (1976)
8. S.G. Gal, V. Gupta, Quantitative estimates for a new complex Durrmeyer operator in compact disks. Appl. Math. Comput. **218**(6), 2944–2951 (2011)
9. J.W. Garrett, C.V. Stanojevic, On L' convergence of certain cosine sums. Proc. Am. Math. Soc. **54**, 101–105 (1976)
10. V. Gupta, G. Tachev, *Approximation with Positive Linear Operators and Linear Combinations, Series: Developments in Mathematics*, vol. 50 (Springer, Cham, 2017)
11. V. Gupta, T.M. Rassias, P.N. Agrawal, A.M. Acu, *Recent Advances in Constructive Approximation Theory* (Springer, Cham, 2018)
12. A. Holhoş, The rate of approximation of functions in an infinite interval by positive linear operators. Stud. Univ. Babes-Bolyai Math. **2**, 133–142 (2010)
13. M. Ismail, C.P. May, On a family of approximation operators. J. Math. Anal. Appl. **63**, 446–462 (1978)
14. G.G. Lorentz, *Bernstein Polynomials* (University of Toronto Press, Toronto, 1953)
15. C.P. May, Saturation and inverse theorems for combinations of a class of exponential-type operators. Can. J. Math. **28**, 1224–1250 (1976)
16. N. Singh, K.K. Sharma, Convergence of certain cosine sums in a metric space-L. Proc. Am. Math. Soc. **72**(1), 117–120 (1978)

A Better Error Estimation on Generalized Positive Linear Operators Based on PED and IPED

Neha Bhardwaj

Abstract In this paper, we consider King type modification of generalized positive linear operators based on Pólya-Eggenberger Distribution (PED) as well as inverse Pólya-Eggenberger Distribution (IPED). We investigate the rate of convergence of these operators with the aid of the Peetre's K_2 functional and study the order of approximation for functions in Lipschitz type space.

Keywords Pólya-Eggenberger distribution · Lipschitz-type space · Modulus of continuity · Peetre's K_2-functional · Voronovskaya result

2010 AMS Subject Classification 41A10 · 41A25 · 41A30 · 41A63 · 26A15

1 Introduction

The nth Bernstein polynomial of real-valued function f on the closed unit interval $[0, 1]$ is defined as

$$B_n(f)(x) = \sum_{k=0}^{n} \binom{n}{k} x^k (1-x)^{n-k} f\left(\frac{k}{n}\right), \qquad x \in [0, 1]. \tag{1.1}$$

It is well known that the sequence $\{B_n(f)\}_{n\in N}$ converges uniformly to f on $[0, 1]$ and the Bernstein polynomials and their generalization as well as modification have an important role in approximation theory (see, for instance, [1–5, 8, 10, 12, 13, 15, 17]).

In 1968, Stancu [19] introduced a new class of positive linear operators based on Pólya-Eggenberger Distribution (PED) and associated with a real-valued function on $[0, 1]$ as:

N. Bhardwaj (✉)
Department of Applied Mathematics, Amity Institute of Applied Sciences, Amity University, Noida 201303, India
e-mail: neha_bhr@yahoo.co.in

N. Deo et al. (eds.), *Mathematical Analysis I: Approximation Theory*, Springer Proceedings in Mathematics & Statistics 306,
https://doi.org/10.1007/978-981-15-1153-0_5

$$B_n^{(\alpha)}(f;x) = \sum_{k=0}^{n} \binom{n}{k} \frac{x^{[k,-\alpha]}(1-x)^{[n-k,-\alpha]}}{1^{[n,-\alpha]}} f\left(\frac{k}{n}\right) \tag{1.2}$$

where α is a nonnegative parameter which may depend only on the natural number n and $m^{[n,h]} = m\,(m-h)\,(m-2h)\cdots\left(m-\overline{n-1}h\right)$, $m^{[0,h]} = 1$ represents the factorial power of m with increment h.

In view of these concernments in 1970, Stancu [20] introduced a generalized form of Baskakov operators based on inverse Pólya-Eggenberger Distribution (IPED) for a real-valued function bounded on $[0, \infty)$, given by

$$V_n^{(\alpha)}(f;x) = \sum_{k=0}^{n} \binom{n+k-1}{k} \frac{1^{[n,-\alpha]}x^{[k,-\alpha]}}{(1+x)^{[n+k,-\alpha]}} f\left(\frac{k}{n}\right) \tag{1.3}$$

In 2017, Deo and Dhamija [7] considered new positive linear operators $L_n^{(\alpha)}$, for each f, real-valued function bounded on interval $[0, \infty)$, as

$$L_{n,\lambda}^{(\alpha)}(f;x) = \sum_{k} w_{n,k}^{(\alpha)}(x)\, f\left(\frac{k}{n}\right), x \in [0,1], n = 1, 2, \ldots, \tag{1.4}$$

where $\alpha = \alpha\,(n) \to 0$ when $n \to \infty$, p and k are nonnegative integers and for $\lambda = -1, 0$, we have

$$w_{n,k}^{(\alpha)}(x) = \frac{n+p}{n+p+\overline{\lambda+1}k} \binom{n+p+\overline{\lambda+1}k}{k} \frac{x^{[k,-\alpha]}(1+\lambda x)^{[n+p+\lambda k,-\alpha]}}{\left(1+\overline{\lambda+1}x\right)^{\left[n+p+\overline{\lambda+1}k,-\alpha\right]}},$$

using the notation $\overline{t-r}\alpha = (t-r)\,\alpha$. Equation (1.4) is the generalized form of two operators (1.2) and (1.3) and associated with PED and IPED (Eggenberger and Pólya).

Deo et al. [9] also studied local approximation theorem, weighted approximation, and estimation of rate of convergence for absolutely continuous function having derivatives of bounded variation for generalized positive linear Kantorovich operators associated to PED as well as IPED.

Let $f \in C_B[0, \infty)$ be the space of all real-valued bounded and uniformly continuous functions on $[0, \infty)$, equipped with the norm $\|f\| = \sup_{x \in [0,\infty)} |f(t)|$. The classical Peetre's K_2-functional and the second modulus of smoothness of a function $f \in C_B[0, \infty)$ are defined, respectively, by

$$K_2(f, \delta) = \inf\left\{\|f-g\| + \delta\left\|g''\right\| : g \in W_\infty^2\right\}, \quad \delta > 0$$

where $W_\infty^2 = \left\{g \in C_B[0, \infty) : g', g'' \in C_B[0, \infty)\right\}$. From [10], there exists a positive constant C such that

$$K_2(f,\delta) \le C\omega_2\left(f,\sqrt{\delta}\right) \tag{1.5}$$

and

$$\omega_2(f,\sqrt{\delta}) = \sup_{0<h\le\delta}\ \sup_{x\in[0,\infty)} \left|f(x+2h)-2f(x+h)+f(x)\right|.$$

In this paper, we study a King type modification of generalized linear positive operators (1.4). We establish the rate of convergence of these operators in terms of second-order modulus of continuity via the approach of Peetre's K-functional and also determine the rate of approximation for functions in a Lipschitz-type space.

2 Basic Results

Lemma 2.1 ([7]) *For the generalized positive linear operators defined by (1.4), there hold the identities:*

$$L_{n,\lambda}^{(\alpha)}(1;x) = 1$$

$$L_{n,\lambda}^{(\alpha)}(t;x) = \left(\frac{n+p}{n}\right)\frac{x}{\left(1-\overline{\lambda+1}\alpha\right)}$$

$$L_{n,\lambda}^{(\alpha)}\left(t^2;x\right) = \left(\frac{n+p}{n^2}\right)\frac{1}{(1-\lambda\alpha)\left(1-\overline{\lambda+1}\alpha\right)}\left[\frac{(n+p+\lambda+1)\,x\,(x+\alpha)}{1-2\overline{\lambda+1}\alpha}+x\,(1+\lambda x)\right]$$

$$L_{n,\lambda}^{(\alpha)}\left(t^3;x\right) = \left[\frac{(n+p+2\lambda+1)\left(n+p+2\overline{2\lambda+1}\right)(x+\alpha)(x+2\alpha)}{\left(1-\overline{3\lambda+2}\alpha\right)\left(1-\overline{5\lambda+3}\alpha\right)}\right.$$
$$\left.+\frac{3\,(n+p+2\lambda+1)\,(x+\alpha)}{\left(1-\overline{3\lambda+2}\alpha\right)}+1\right]$$

$$L_{n,\lambda}^{(\alpha)}\left(t^4;x\right) = \frac{(n+p)\,x}{n^4\left(1-\overline{\lambda+1}\alpha\right)}$$
$$\left[\frac{(n+p+2\lambda+1)\left(n+p+2\overline{2\lambda+1}\right)\left(n+p+3\overline{2\lambda+1}\right)(x+\alpha)(x+2\alpha)(x+3\alpha)}{\left(1-\overline{3\lambda+2}\alpha\right)\left(1-\overline{5\lambda+3}\alpha\right)\left(1-\overline{7\lambda+4}\alpha\right)}\right.$$
$$+\frac{6\,(n+p+2\lambda+1)\left(n+p+2\overline{2\lambda+1}\right)(x+\alpha)(x+2\alpha)}{\left(1-\overline{3\lambda+2}\alpha\right)\left(1-\overline{5\lambda+3}\alpha\right)}$$
$$\left.+\frac{7\,(n+p+2\lambda+1)\,(x+\alpha)}{\left(1-\overline{3\lambda+2}\alpha\right)}+1\right]$$

Consequently, we have

Lemma 2.2 ([7]) *The generalized linear positive operators (1.4) satisfy:*

$$L_{n,\lambda}^{\alpha}(t-x;x)=\frac{\left(p+n\overline{\lambda+1}\alpha\right)x}{n\left(1-\overline{\lambda+1}\alpha\right)}$$

$$L_{n,\lambda}^{\alpha}\left((t-x)^2;x\right)=\frac{n+p}{n(1-\lambda\alpha)\left(1-\overline{\lambda+1}\alpha\right)}\left[(1-\lambda\alpha)\left(1-\overline{\lambda+1}\alpha\right)\frac{nx^2}{n+p}\right.$$
$$\left.+\frac{(n+p+\lambda+1)x(x+\alpha)}{n\left(1-2\overline{\lambda+1}\alpha\right)}+\frac{x(1+\lambda x)}{n}-2(1-\lambda\alpha)x^2\right]$$

$$L_{n,\lambda}^{\alpha}\left((t-x)^3;x\right)=\frac{(n+p)x(x+\alpha)}{n^2\left(1-\overline{\lambda+1}\alpha\right)}$$
$$\left[\frac{(n+p+2\lambda+1)\left(n+p+2\overline{2\lambda+1}\right)(x+2\alpha)}{n\left(1-\overline{3\lambda+2}\alpha\right)\left(1-\overline{5\lambda+3}\alpha\right)}+\frac{3(n+p+2\lambda+1)}{n\left(1-\overline{3\lambda+2}\alpha\right)}\right.$$
$$\left.-\frac{3(n+p+\lambda+1)x}{(1-\lambda\alpha)\left(1-\overline{2(\lambda+1)}\alpha\right)}\right]+\frac{(n+p)x}{n(1-\overline{\lambda+1}\alpha)}$$
$$\left[\frac{1}{n^2}-3x\left(\frac{px+\overline{\lambda+1}\alpha}{(n+p)}\right)-\frac{3}{(1-\lambda\alpha)}\frac{x(1+\lambda x)}{n}\right]$$
$$-2x^3\left[2-\frac{3(n+p)}{n\left(1-\overline{\lambda+1}\alpha\right)}\right]$$

$$L_{n,\lambda}^{(\alpha)}((t-x)^4;x)=\frac{(n+p)x(x+\alpha)}{n^2\left(1-\overline{\lambda+1}\alpha\right)}$$
$$\left[\frac{(n+p+2\lambda+1)\left(n+p+2\overline{2\lambda+1}\right)\left(n+p+3\overline{2\lambda+1}\right)(x+2\alpha)(x+3\alpha)}{n^2\left(1-\overline{3\lambda+2}\alpha\right)\left(1-\overline{5\lambda+3}\alpha\right)\left(1-\overline{7\lambda+4}\alpha\right)}\right.$$
$$+\frac{2(3-2nx)(n+p+2\lambda+1)\left(n+p+2\overline{2\lambda+1}\right)(x+2\alpha)}{n^2\left(1-\overline{3\lambda+2}\alpha\right)\left(1-\overline{5\lambda+3}\alpha\right)}$$
$$\left.+\frac{(7-12nx)(n+p+2\lambda+1)}{n^2\left(1-\overline{3\lambda+2}\alpha\right)}+\frac{6(n+p+2\lambda+1)x^2}{(1-\lambda x)\left(1-2\left(\overline{\lambda+1}\right)\alpha\right)}\right]$$
$$+\frac{(n+p)x}{n\left(1-\overline{\lambda+1}\alpha\right)}\left[\frac{1-4nx}{n^3}+8x^2\left(\frac{px+\overline{\lambda+1}\alpha}{(n+p)}\right)\frac{6x^2(1+\lambda\alpha)}{n(1-\lambda\alpha)}\right]$$
$$+3x^4\left[3-\frac{4(n+p)}{n\left(1-\overline{\lambda+1}\alpha\right)}\right]$$

Many researchers have studied King type modification for different sequences of linear positive operators (see, for instance [6, 11, 14, 16, 18]). Now we consider a similar type of modification for the operators given by (1.4).

We assume that $\{u_n(x)\}$ is a sequence of real-valued continuous functions defined on $[0,\infty)$ with $0\le u_n(x)<\infty$, for $x\in[0,\infty)$; then we have

$$\hat{L}_{n,\lambda}^{(\alpha)}(f;x)=\sum_{k} w_{n,k}^{(\alpha)}(u_n(x))\, f\left(\frac{k}{n}\right), \tag{2.1}$$

where $w_{n,k}^{(\alpha)}(u_n(x))=\frac{n+p}{n+p+\overline{\lambda+1}k}\binom{n+p+\overline{\lambda+1}k}{k}\frac{u_n(x)^{[k,-\alpha]}(1+\lambda u_n(x))^{[n+p+\lambda k,-\alpha]}}{(1+\overline{\lambda+1}u_n(x))^{\left[n+p+\overline{\lambda+1}k,-\alpha\right]}}$

and $u_n(x)=\frac{n}{n+p}\left(1-\overline{\lambda+1}\alpha\right)x,\ n,\ p\in\mathbb{N}.$

By a simple computation, we obtain the following result.

Lemma 2.3 *For the modified generalized positive linear operators (2.1), there hold the following equalities:*

$$\hat{L}_{n,\lambda}^{(\alpha)}(1;x)=1$$

$$\hat{L}_{n,\lambda}^{(\alpha)}(t;x)=x$$

$$\hat{L}_{n,\lambda}^{(\alpha)}(t^2;x)=\frac{1}{n(1-\lambda\alpha)}\left[\frac{n}{n+p}(1-\overline{\lambda+1}\alpha)\left[\frac{n+p+\lambda+1}{1-2\overline{\lambda+1}}\right]x+\left(1+\frac{n+p+\lambda+1}{1-2\overline{\lambda+1}\alpha}\alpha\right)\right]x$$

$$\begin{aligned}\hat{L}_{n,\lambda}^{(\alpha)}(t^3;x)=\frac{x}{n^2}\Bigg[&\frac{(n+p+2\lambda+1)\left(n+p+2\overline{2\lambda+1}\right)}{\left(1-\overline{3\lambda+2\alpha}\right)\left(1-\overline{5\lambda+3\alpha}\right)}\left(\frac{n\left(1-\overline{\lambda+1}\alpha\right)x}{n+p}+\alpha\right)\\&\left(\frac{n\left(1-\overline{\lambda+1}\alpha\right)x}{n+p}+2\alpha\right)+\frac{3(n+p+2\lambda+1)}{\left(1-\overline{3\lambda+2\alpha}\right)}\left(\frac{n\left(1-\overline{\lambda+1}\alpha\right)x}{n+p}+\alpha\right)+1\Bigg]\end{aligned}$$

$$\begin{aligned}\hat{L}_{n,\lambda}^{(\alpha)}(t^4;x)=\frac{x}{n^3}\Bigg[&\frac{(n+p+2\lambda+1)\left(n+p+2\overline{2\lambda+1}\right)\left(n+p+3\overline{2\lambda+1}\right)}{\left(1-\overline{3\lambda+2\alpha}\right)\left(1-\overline{5\lambda+3\alpha}\right)\left(1-\overline{7\lambda+4\alpha}\right)}\\&\left[\left(\frac{n\left(1-\overline{\lambda+1}\alpha\right)x}{n+p}+\alpha\right)\left(\frac{n\left(1-\overline{\lambda+1}\alpha\right)x}{n+p}+2\alpha\right)\left(\frac{n\left(1-\overline{\lambda+1}\alpha\right)x}{n+p}+3\alpha\right)\right]\\&+\frac{6(n+p+2\lambda+1)\left(n+p+2\overline{2\lambda+1}\right)}{\left(1-\overline{3\lambda+2\alpha}\right)\left(1-\overline{5\lambda+3\alpha}\right)}\left(\frac{n\left(1-\overline{\lambda+1}\alpha\right)x}{n+p}+\alpha\right)\\&\left(\frac{n\left(1-\overline{\lambda+1}\alpha\right)x}{n+p}+2\alpha\right)+\frac{7(n+p+2\lambda+1)}{\left(1-\overline{3\lambda+2\alpha}\right)}\left(\frac{n\left(1-\overline{\lambda+1}\alpha\right)x}{n+p}+\alpha\right)+1\Bigg]\end{aligned}$$

Lemma 2.4 *The modified generalized linear positive operators (2.1) satisfy:*

$$\hat{L}_{n,\lambda}^{(\alpha)}(t-x;x)=0$$

$$\begin{aligned}\hat{L}_{n,\lambda}^{(\alpha)}((t-x)^2;x)=\frac{x}{n(1-\lambda\alpha)}\Bigg[&\frac{n}{n+p}\left(1-\overline{\lambda+1}\alpha\right)\left(\frac{n+p+\lambda+1}{1-2\overline{\lambda+1}\alpha}+\alpha\right)x\\&+\left(1+\frac{n+p+\lambda+1}{1-2\overline{\lambda+1}\alpha}\alpha\right)\Bigg]-x^2\end{aligned}$$

$$\hat{L}_{n,\lambda}^{(\alpha)}((t-x)^3;x)=\frac{x}{n^2}\Bigg[\frac{(n+p+2\lambda+1)\left(n+p+2\overline{2\lambda+1}\right)}{\left(1-\overline{3\lambda+2\alpha}\right)\left(1-\overline{5\lambda+3\alpha}\right)}\left(\frac{n\left(1-\overline{\lambda+1}\alpha\right)x}{n+p}+\alpha\right)$$

$$\left(\frac{n\left(1-\overline{\lambda+1}\alpha\right)x}{n+p}+2\alpha\right)+\frac{3\,(n+p+2\lambda+1)}{\left(1-\overline{3\lambda+2\alpha}\right)}$$
$$\times\left(\frac{n\left(1-\overline{\lambda+1}\alpha\right)x}{n+p}+\alpha\right)+1\Bigg]$$
$$-x^3-3x\left[\frac{x}{n(1-\lambda\alpha)}\left[\frac{n}{n+p}\left(1-\overline{\lambda+1}\alpha\right)\left(\frac{n+p+\lambda+1}{1-2\overline{\lambda+1}\alpha}+\alpha\right)x\right.\right.$$
$$\left.\left.+\left(1+\frac{n+p+\lambda+1}{1-2\overline{\lambda+1}\alpha}\alpha\right)\right]-x^2\right]$$

$$\hat{L}^{(\alpha)}_{n,\lambda}((t-x)^4;x)=\frac{x}{n^3}\Bigg[\frac{(n+p+2\lambda+1)\left(n+p+2\overline{2\lambda+1}\right)\left(n+p+3\overline{2\lambda+1}\right)}{\left(1-\overline{3\lambda+2\alpha}\right)\left(1-\overline{5\lambda+3\alpha}\right)\left(1-\overline{7\lambda+4\alpha}\right)}$$
$$\Bigg[\left(\frac{n\left(1-\overline{\lambda+1}\alpha\right)x}{n+p}+\alpha\right)\left(\frac{n\left(1-\overline{\lambda+1}\alpha\right)x}{n+p}+2\alpha\right)$$
$$\times\left(\frac{n\left(1-\overline{\lambda+1}\alpha\right)x}{n+p}+3\alpha\right)\Bigg]+\frac{6\,(n+p+2\lambda+1)\left(n+p+2\overline{2\lambda+1}\right)}{\left(1-\overline{3\lambda+2\alpha}\right)\left(1-\overline{5\lambda+3\alpha}\right)}$$
$$\times\left(\frac{n\left(1-\overline{\lambda+1}\alpha\right)x}{n+p}+\alpha\right)\left(\frac{n\left(1-\overline{\lambda+1}\alpha\right)x}{n+p}+2\alpha\right)+\frac{7\,(n+p+2\lambda+1)}{\left(1-\overline{3\lambda+2\alpha}\right)}$$
$$\times\left(\frac{n\left(1-\overline{\lambda+1}\alpha\right)x}{n+p}+\alpha\right)+1\Bigg]$$
$$-x^4-\frac{4x^2}{n^2}\Bigg[\frac{(n+p+2\lambda+1)\left(n+p+2\overline{2\lambda+1}\right)}{\left(1-\overline{3\lambda+2\alpha}\right)\left(1-\overline{5\lambda+3\alpha}\right)}\left(\frac{n\left(1-\overline{\lambda+1}\alpha\right)x}{n+p}+\alpha\right)$$
$$\left(\frac{n\left(1-\overline{\lambda+1}\alpha\right)x}{n+p}+2\alpha\right)+\frac{3\,(n+p+2\lambda+1)}{\left(1-\overline{3\lambda+2\alpha}\right)}$$
$$\left(\frac{n\left(1-\overline{\lambda+1}\alpha\right)x}{n+p}+\alpha\right)+1\Bigg]$$
$$-x^3-3x\left[\frac{x}{n(1-\lambda\alpha)}\left[\frac{n}{n+p}\left(1-\overline{\lambda+1}\alpha\right)\left(\frac{n+p+\lambda+1}{1-2\overline{\lambda+1}\alpha}+\alpha\right)x\right.\right.$$
$$\left.\left.+\left(1+\frac{n+p+\lambda+1}{1-2\overline{\lambda+1}\alpha}\alpha\right)\right]-x^2\right]+\frac{6x^3}{n(1-\lambda\alpha)}\left[\frac{n}{n+p}\left(1-\overline{\lambda+1}\alpha\right)\right.$$
$$\left.\times\left(\frac{n+p+\lambda+1}{1-2\overline{\lambda+1}\alpha}+\alpha\right)x+\left(1+\frac{n+p+\lambda+1}{1-2\overline{\lambda+1}\alpha}\alpha\right)\right]-x^2$$

3 Voronovskaya Type Results

Theorem 3.1 *For $f \in C_B[0,\infty)$, we have*

$$\left|\hat{L}_{n,\lambda}^{(\alpha)}(f;x) - f(x)\right| \leq C\omega_2\left(f, \frac{\sqrt{\mu_{n,\lambda}^{(\alpha)}(x)}}{2}\right), \tag{3.1}$$

where C is a positive constant and

$$\mu_{n,\lambda}^{(\alpha)}(x) = \frac{x}{n(1-\lambda\alpha)}\left[\frac{n}{n+p}\left(1-\overline{\lambda+1}\alpha\right)\left(\frac{(n+p+\lambda+1)}{\left(1-2\overline{\lambda+1}\alpha\right)}+\alpha\right)x\right]$$
$$+\left(1+\frac{(n+p+\lambda+1)}{\left(1-2\overline{\lambda+1}\alpha\right)}\alpha\right).$$

Proof Let $g \in W_\infty^2$. Using Taylor's expansion, we get

$$g(y) = g(x) + g'(x)(y-x) + \int_x^y (y-u)g''(u)du$$

From Lemma 2.4, we have

$$\left(\hat{L}_{n,\lambda}^{(\alpha)}g\right)(x) - g(x) = \left(\hat{L}_{n,\lambda}^{(\alpha)}\int_x^y (y-u)g''(u)du\right)(x).$$

We know that

$$\left|\int_x^y (y-u)g''(u)du\right| \leq (y-x)^2\left\|g''\right\|.$$

Therefore

$$\left|\left(\hat{L}_{n,\lambda}^{(\alpha)}g\right)(x) - g(x)\right| \leq \left(\hat{L}_{n,\lambda}^{(\alpha)}(y-x)^2\right)(x)\left\|g''\right\| = \mu_{n,\lambda}^{(\alpha)}(x)\left\|g''\right\|.$$

By Lemma 2.3, we have

$$\left|\left(\hat{L}_{n,\lambda}^{(\alpha)}f\right)(x)\right| \leq \left|\sum_k w_{n,k}^{(\alpha)}(u_n(x))\, f\left(\frac{k}{n}\right)\right| \leq \|f\|.$$

Hence

$$\left|\left(\hat{L}_{n,\lambda}^{(\alpha)}f\right)(x)-f(x)\right|\leq\left|\left(\hat{L}_{n,\lambda}^{(\alpha)}(f-g)\right)(x)-(f-g)(x)\right|+\left|\left(\hat{L}_{n,\lambda}^{(\alpha)}g\right)(x)-g(x)\right|$$
$$\leq 2\left\|f-g\right\|+\mu_{n,\lambda}^{(\alpha)}(x)\left\|g''\right\|$$

where
$\mu_{n,\lambda}^{(\alpha)}(x)=\frac{x}{n(1-\lambda\alpha)}\left[\frac{n}{n+p}\left(1-\overline{\lambda+1}\alpha\right)\left(\frac{n+p+\lambda+1}{1-2\overline{\lambda+1}\alpha}+\alpha\right)x+\left(1+\frac{n+p+\lambda+1}{1-2\overline{\lambda+1}\alpha}\alpha\right)\right]-x^2$
taking the infimum on the right side over all $g\in W_{\infty}^{2}$ and using (1.5), we get the required result. □

Remark 3.2 ([8]) Under the same conditions of Theorem 3.1, we obtain

$$\left|L_{n,\lambda}^{(\alpha)}(f;x)-f(x)\right|\leq\omega\left(f,\frac{px+nx(\lambda+1)\alpha}{n\left(1-(\lambda+1)\alpha\right)}\right)+A\omega_2\left(f,\frac{\sqrt{\psi_{n,\lambda}^{(\alpha)}(x)}}{2}\right)$$

where A is a positive constant and

$$\psi_{n,\lambda}^{(\alpha)}(x)=L_{n,\lambda}^{(\alpha)}((t-x)^2;x)+\left\{\frac{px+nx\left(\lambda+1\right)\alpha}{n\left(1-\left(\lambda+1\right)\alpha\right)}\right\}^2$$

Now, we compute the rate of convergence of these operators by means of Lipschitz class $Lip_M^*(\beta)$, $0<\beta\leq 1$. Consider the following Lipshitz type space

$$Lip_M^*(\beta):=\left\{f\in C_B[0,\infty):\left|f(y)-f(x)\right|\leq M\frac{|y-x|^{\beta}}{(x+y)^{\beta/2}};x,y\in(0,\infty)\right\}$$

where M is a positive constant.

Theorem 3.3 *For all* $x\in[0,\infty)$ *and* $f\in Lip_M^*(\beta)$, $0<\beta\leq 1$, *we get*

$$\left|\hat{L}_{n,\lambda}^{(\alpha)}(f;x)-f(x)\right|\leq M\left(\frac{\mu_{n,\lambda}^{(\alpha)}(x)}{x}\right)^{\frac{\beta}{2}}$$

where

$$\mu_{n,\lambda}^{(\alpha)}(x)=\hat{L}_{n,\lambda}^{(\alpha)}((t-x)^2;x)$$

Proof Assuming $\beta=1$,
Then for $f\in Lip_M^*(1)$, we have

$$\left|\hat{L}_{n,\lambda}^{(\alpha)}(f;x)-f(x)\right|\leq\left|\sum_k w_{n,k}^{(\alpha)}(u_n(x))f\left(\frac{k}{n}\right)-f(x)\right|$$
$$\leq\sum_{k=0}^{\infty}w_{n,k}^{(\alpha)}(u_n(x))\left|f\left(\frac{k}{n}\right)-f(x)\right|$$

$$\leq M\sum_{k=0}^{\infty} w_{n,k}^{(\alpha)}(u_n(x))\frac{\left|\frac{k}{n}-x\right|}{\left(\frac{k}{n}+x\right)^{1/2}}$$

Using Cauchy–Schwarz inequality and $\frac{1}{\left(\frac{k}{n}+x\right)^{1/2}}\leq\frac{1}{x^{1/2}}$ and linearity of $\hat{L}_{n,\lambda}^{(\alpha)}(f;x)$, we have

$$\begin{aligned}\left|\hat{L}_{n,\lambda}^{(\alpha)}(f;x)-f(x)\right| &\leq \frac{M}{\sqrt{x}}\sum_{k=0}^{\infty} w_{n,k}^{(\alpha)}(u_n(x))\left\{\left(\frac{k}{n}-x\right)^2\right\}^{1/2}\\ &\leq \frac{M}{\sqrt{x}}\left\{\sum_{k=0}^{\infty} w_{n,k}^{(\alpha)}(u_n(x))\right\}^{1/2}\left\{\sum_{k=0}^{\infty} w_{n,k}^{(\alpha)}\left(\frac{k}{n}-x\right)^2\right\}^{1/2}\\ &= M\left(\frac{\mu_{n,\lambda}^{(\alpha)}(x)}{x}\right)^{\frac{1}{2}}.\end{aligned}$$

Therefore, the result is true for $\beta=1$.

Now, to prove result for $0<\beta<1$, consider $f\in Lip_M^*(\beta)$.

$$\begin{aligned}\left|\hat{L}_{n,\lambda}^{(\alpha)}(f;x)-f(x)\right| &\leq \sum_{k=0}^{\infty} w_{n,k}^{(\alpha)}(u_n(x))\left|f\left(\frac{k}{n}\right)-f(x)\right|\\ &\leq M\sum_{k=0}^{\infty} w_{n,k}^{(\alpha)}(u_n(x))\frac{\left|\frac{k}{n}-x\right|^{\beta}}{\left(\frac{k}{n}+x\right)^{\beta/2}}\end{aligned}$$

Using Hölders inequality for $p=\frac{2}{\beta}$, $q=\frac{2}{2-\beta}$ and inequality $\frac{1}{\sqrt{\frac{k}{n}+x}}\leq\frac{1}{\sqrt{x}}$, we have

$$\begin{aligned}\left|\hat{L}_{n,\lambda}^{(\alpha)}(f;x)-f(x)\right| &\leq \frac{M}{x^{\beta/2}}\sum_{k=0}^{\infty} w_{n,k}^{(\alpha)}(u_n(x))\left\{\left(\frac{k}{n}-x\right)^2\right\}^{\beta/2}\\ &\leq \frac{M}{x^{\beta/2}}\left\{\sum_{k=0}^{\infty} w_{n,k}^{(\alpha)}(u_n(x))\left(\frac{k}{n}-x\right)^2\right\}^{\beta/2}\left\{\sum_{k=0}^{\infty} w_{n,k}^{(\alpha)}(u_n(x))\right\}^{2-\beta/2}\\ &\leq M\left\{\frac{\hat{L}_{n,\lambda}^{(\alpha)}\left((t-x)^2;x\right)}{x}\right\}^{\beta/2}\\ &= M\left\{\frac{\mu_{n,\lambda}^{(\alpha)}(x)}{x}\right\}^{\beta/2}\end{aligned}$$

□

References

1. F. Altomare, M. Campiti, *Korovkin-Type Approximation Theory and its Applications* (Walter de Gruyter, Berlin, 1994)
2. G.A. Anastassiou, S.G. Gal, *Approximation Theory* (Birkhäuser, Boston, 2000)
3. M.M. Birou, Bernstein type operators with a better approximation for some functions. Appl. Math. Comput. **219**(17), 9493–9499 (2013)
4. Q.B. Cai, B.Y. Lian, G. Zhou, Approximation properties of λ-Bernstein operators. J. Inequal. Appl. **12**, 1–11 (2018)
5. D. Cárdenas-Morales, P. Garrancho, I. Raŝa, Bernstein-type operators which preserve polynomials. Comput. Math. Appl. **62**(1), 158–163 (2011)
6. N. Deo, N. Bhardwaj, Some approximation results for Durrmeyer operators. Appl. Math. Comput. **217**, 5531–5536 (2011)
7. N. Deo, M. Dhamija, Generalized positive linear operators. Iran. J. Sci. Technol. Trans. Sci. **26**(175–179), 123–134 (2017)
8. N. Deo, M.A. Noor, M.A. Siddiqui, On approximation by a class of new Bernstein type operators. Appl. Math. Comput. **201**, 604–612 (2008)
9. N. Deo, M. Dhamija, D. Miclaus, Stancu–Kantorovich operators based on inverse Pólya-Eggenberger distribution. Appl. Math. Comput. **273**, 281–289 (2016)
10. R.A. DeVore, G.G. Lorentz, *Constructive Approximation* (Springer, Berlin, 1993)
11. O. Doğru, M. Örkcü, King type modification of Meyer-König and Zeller operators based on the q-integers. Math. Comput. Model. **50**, 1245–1251 (2009)
12. J.L. Durrmeyer, *Une formule d'inversion de la transformée de Laplace-applications à la théorie des moments*, Thése de 3e cycle, Faculté des Sciences de l' Université de Paris, (1967)
13. V. Gupta, T.M. Rassias, P.N. Agrawal, A.M. Acu, *Recent Advances in Constructive Approximation Theory* (Springer, Cham, 2018)
14. J.P. King, Positive linear operators which preserve x^2. Acta Math. Hung. **99**, 203–208 (2003)
15. Y.C. Kwun, A.M. Acu, A. Rafiq, V.A. Radu, F. Ali, S.M. Kang, Bernstein–Stancu type operators which preserve polynomials. J. Comput. Anal. Appl. **23**(1), 758–770 (2017)
16. M.A. Özarslan, O. Duman, C. Kaanoğlu, Rate of convergence of certain King-type operators for functions with derivative of bounded variation. Math. Comput. Model. **52**, 334–345 (2010)
17. V.A. Radu, Quantitative estimates for some modified Bernstein–Stancu operators. Miskolc Math. Notes **19**(1), 517–525 (2018)
18. L. Rempulska, K. Tomczak, Approximation by certain linear operators preserving x^2. Turkish J. Math. **33**, 273–281 (2009)
19. D.D. Stancu, Approximation of functions by a new class of linear polynomial operators. Rev. Roum. Math. Pure. Appl. **13**, 1173–1194 (1968)
20. D.D. Stancu, Two classes of positive linear operators. Anal. Univ. Timisoara ser stin Matem. **8**, 213–220 (1970)

Q-Analogue of Generalized Bernstein–Kantorovich Operators

Ram Pratap and Naokant Deo

Abstract In the present article, we consider the q-analogue of generalized Bernstein–Kantorovich operators. For the proposed operators, we studied some convergence properties by using first- and second-order modulus of continuity.

Keywords Bernstein operators · Kantorovich operators · Modulus of continuity

2010 Mathematics Subject Classification 41A25 · 41A36

1 Introduction

In the year 1912, Bernstein [5] introduced the Bernstein operators and provided the constructive proof of Weierstrass theorem. Later, several researchers have generalized Bernstein operators using different parameters and studied various convergence properties. For more (see [6, 7, 16]).

Recently, Chen et al. [7] defined a family of Bernstein operators, for the functions $f \in [0, 1]$, α is fixed and $n \in \mathbb{N}$ are as follows:

$$B_n^{(\alpha)}(f; x) = \sum_{k=0}^{n} f_k p_{n,k}^{(\alpha)}(x), \tag{1.1}$$

where $f_k = f\left(\frac{k}{n}\right)$. For $n > 2$ the α-Bernstein polynomial $p_{n,k}^{(\alpha)}(x)$ of degree n is defined by

$$p_{1,0}^{(\alpha)}(x) = 1 - x, \; p_{1,1}^{(\alpha)}(x) = x,$$

R. Pratap (✉) · N. Deo
Department of Applied Mathematics, Delhi Technological University (Delhi College of Engineering), Bawana Road, Delhi 110042, India
e-mail: rampratapiitr@gmail.com

N. Deo
e-mail: naokantdeo@dce.ac.in

N. Deo et al. (eds.), *Mathematical Analysis I: Approximation Theory*, Springer Proceedings in Mathematics & Statistics 306,
https://doi.org/10.1007/978-981-15-1153-0_6

and

$$p_{n,k}^{(\alpha)}(x) = \left[\binom{n-2}{k}(1-\alpha)x + \binom{n-2}{k-2}(1-\alpha)(1-x) + \binom{n}{k}\alpha x(1-x)\right] x^{k-1}(1-x)^{n-k-1}, \qquad x \in [0,1].$$

For the first time in 1987, Bernstein operators based on q-integers were introduced by Lupas [12] and they are rational functions. Again in 1997, Phillips [14] introduced the q-Bernstein polynomials known as Phillips q-Bernstein operators. In past decade, linear positive operators based on q-integers is an active area of research. For more (see [4, 8, 11]).

Chai et al. [8] have considered the q-analouge of (1.1) is as follows:

$$B_{n,q}^{(\alpha)}(f;x) = \sum_{k=0}^{n} f_k p_{n,q,k}^{(\alpha)}(x), \tag{1.2}$$

where

$$p_{n,q,k}^{(\alpha)}(x) = \left(\begin{bmatrix} n-2 \\ k \end{bmatrix}_q (1-\alpha)x + \begin{bmatrix} n-2 \\ k-2 \end{bmatrix}_q (1-\alpha)q^{n-k-2}\left(1-q^{n-k-1}x\right) + \begin{bmatrix} n \\ k \end{bmatrix}_q \alpha x\left(1-q^{n-k-1}x\right)\right) x^{k-1}(1-x)_q^{n-k-1},$$

$q \in (0,1]$ and $f_k = f\left(\frac{[k]_q}{[n]_q}\right)$. For detailed explanation (see [3]).

Dhamija et al. [10] proposed the Kantorovich form of modified Szász–Mirakyan operators. Several researchers have also studied Kantorovich form of different linear positive operators and established local and global approximation results. More details (see [1, 2, 13, 15]).

Mohiuddine et al. [13] proposed the Kantorovich form of the operators (1.1), which is given as

$$K_n^{(\alpha)}(f;x) = (n+1)\sum_{k=0}^{n} p_{n,k}^{(\alpha)}(x) \int_{k/(n+1)}^{(k+1)/(n+1)} f(t)dt, \tag{1.3}$$

where $p_{n,k}^{(\alpha)}(x)$ is defined in (1.1).

For $\alpha = 1$ and $q = 1$ the operators (1.4) reduces to Bernstein–Kantorovich operators.

Motivated from the above stated work, we consider the q-analogue of the operators (1.3) as follows:

$$K_{n,q}^{(\alpha)}(f;x) = [n+1]_q \sum_{k=0}^{n} p_{n,q,k}^{(\alpha)}(x) \int\limits_{\frac{q[k]_q}{[n+1]_q}}^{\frac{[k+1]_q}{[n+1]_q}} f(t)d_qt, \tag{1.4}$$

and $p_{n,q,k}^{(\alpha)}(x)$ is given in (1.2).

In this paper, we estimated the moments of the proposed operators and discuss the rate of convergence using modulus of continuity.

2 Basic Results

In this section, we prove some auxiliary result to prove our main results.

Lemma 2.1 *From* [8], *we have* $B_{n,q}^{(\alpha)}(1;x) = 1$, $B_{n,q}^{(\alpha)}(t;x) = x$ *and*

$$B_{n,q}^{(\alpha)}(t^2;x) = x^2 + \frac{x(1-x)}{[n]_q} + \frac{(1-\alpha)q^{n-1}[2]_q x(1-x)}{[n]_q^2}.$$

Lemma 2.2 *(i)* $K_{n,q}^{(\alpha)}(1;x) = 1$;
(ii) $K_{n,q}^{(\alpha)}(t;x) = \frac{2q[n]_q}{[2]_q[n+1]_q}x + \frac{1}{[2]_q[n+1]_q}$;
(iii) $K_{n,q}^{(\alpha)}(t^2;x) = \frac{3q^2[n]_q^2}{[3]_q[n+1]_q^2}x^2 + \frac{3q^2}{[3]_q[n+1]_q^2}\left([n]_q + (1-\alpha)q^{n-1}[2]_q\right)x(1-x)$
$+ \frac{3q[n]_q x}{[3]_q[n+1]_q^2} + \frac{1}{[3]_q[n+1]_q^2}$.

Proof From [15], $\int\limits_{\frac{q[k]_q}{[n+1]_q}}^{\frac{[k+1]_q}{[n+1]_q}} 1d_qt = \frac{1}{[n+1]_q}$, $\int\limits_{\frac{q[k]_q}{[n+1]_q}}^{\frac{[k+1]_q}{[n+1]_q}} td_qt = \frac{2q[k]_q}{[2]_q[n+1]_q^2} + \frac{1}{[2]_q[n+1]_q^2}$ and

$$\int\limits_{\frac{q[k]_q}{[n+1]_q}}^{\frac{[k+1]_q}{[n+1]_q}} t^2d_qt = \frac{3q^2\,[k]_q^2}{[3]_q\,[n+1]_q^3} + \frac{3q[k]_q}{[3]_q\,[n+1]_q^3} + \frac{1}{[3]_q\,[n+1]_q^3}.$$

It is easy to say that $K_{n,q}^{(\alpha)}(1;x) = 1$.

For $f(t) = t$ and using Lemma 2.1, we have

$$
\begin{aligned}
K_{n,q}^{(\alpha)}(t;x) &= [n+1]_q \sum_{k=0}^{n} p_{n,q,k}^{(\alpha)}(x) \int\limits_{\frac{q[k]_q}{[n+1]_q}}^{\frac{[k+1]_q}{[n+1]_q}} t d_q t \\
&= [n+1]_q \sum_{k=0}^{n} p_{n,q,k}^{(\alpha)}(x) \left(\frac{2q[k]_q}{[2]_q[n+1]_q^2} + \frac{1}{[2]_q[n+1]_q^2} \right) \\
&= \frac{[n]_q}{[n+1]_q} \left(\frac{2q}{[2]_q} \sum_{k=0}^{n} p_{n,q,k}^{(\alpha)}(x) \frac{[k]_q}{[n]_q} + \frac{1}{[2]_q[n]_q} \sum_{k=0}^{n} p_{n,q,k}^{(\alpha)}(x) \right) \\
&= \frac{2q[n]_q x + 1}{[2]_q[n+1]_q}.
\end{aligned}
$$

Similarly, for $f(t) = t^2$, we can estimate. So here we skip. □

Lemma 2.3 *The central moments for the operators (1.4) are as follows:*

(i) $K_{n,q}^{(\alpha)}(t-x;x) = \frac{2q[n]_q}{[2]_q[n+1]_q} x + \frac{1}{[2]_q[n+1]_q}$;

(ii) $K_{n,q}^{(\alpha)}((t-x)^2;x) = \left(\frac{3q^2[n]_q^2}{[3]_q[n+1]_q^2} - \frac{4q[n]_q}{[2]_q[n+1]_q} + 1 \right) x^2$
$+ \frac{3q^2}{[3]_q[n+1]_q} \left([n]_q + [2]_q(1-\alpha)q^{n-1} \right) x(1-x) + \left(\frac{3q[n]_q}{[3]_q[n+1]_q^2} - \frac{2}{[3]_q[n+1]_q} \right) x$
$+ \frac{1}{[3]_q[n+1]_q^2}$.

Proof Using linearity property of the operators (1.4) and Lemma 2.2, we get the required results. □

Lemma 2.4 *Let* $0 < q < 1$ *and* $c \in [0, qd]$, $d > 0$. *Then the inequality*

$$
\int\limits_c^d |t-x| \, d_q t \le \left(\int\limits_c^d (t-x)^2 d_q t \right)^{\frac{1}{2}} \left(\int\limits_c^d d_q t \right)^{\frac{1}{2}}.
$$

Proof For the proof of the Lemma (see [15]). □

3 Main Results

Let $C[0,1]$ be the space of all continuous functions on $[0,1]$ with sup-norm $\|f\| := \sup_{x \in [0,1]} |f(x)|$. Let $f \in C[0,1]$ and $\delta > 0$. Then the modulus of continuity $\omega(f, \delta)$ is given as:

$$
\omega(f,\delta) = \sup_{\substack{|v-w| \le \delta \\ v, w \in [0,1]}} |f(v) - f(w)|.
$$

It is well-known $\lim_{\delta \to 0} \omega(f; \delta) = 0$. For $f \in C[0, 1]$ and $x, t \in [0, 1]$, we have

$$|f(t) - f(x)| \leq \omega(f; \delta)\left(1 + \frac{|t - x|}{\delta}\right) \tag{3.1}$$

For $f \in C[0, 1]$ the Peetre K-functional is given by

$$K_2(f; \delta) = \inf_{g \in W^2} \left\{|f - g| + \delta \left\|g''\right\|\right\},$$

where $\delta > 0$ and $W^2 = \left\{g \in C[0, 1] : g', g'' \in C[0, 1]\right\}$. In [9], there exists an absolute constant $\lambda > 0$, such that

$$K_2(f; \delta) \leq \lambda \omega_2(f; \sqrt{\delta}). \tag{3.2}$$

and the second-order modulus of continuity $\omega_2(.; \delta)$ for $f \in C[0, 1]$ as follows:

$$\omega_2(f; \delta) = \sup_{0<h\leq\delta} \sup_{x,x+h,x+2h\in[0,1]} |f(x + 2h) - 2f(x + h) + f(x)|.$$

Theorem 3.1 *For $0 < q \leq 1$, $q = \{q_n\}$ be a sequence converging to 1 as $n \to \infty$. Then, for all $f \in C[0, 1]$ and $\alpha \in [0, 1]$, it implies $K_{n,q}^{(\alpha)}(f; x)$ converges to $f(x)$ uniformly on $[0, 1]$ for sufficiently large n.*

Proof From Lemma 2.2, $\lim_{n\to\infty} q_n = 1$, we have $\lim_{n\to\infty} K_{n,q}^{(\alpha)}(1; x) = 1$, $\lim_{n\to\infty} K_{n,q}^{(\alpha)}(t; x) = x$ and $\lim_{n\to\infty} K_{n,q}^{(\alpha)}(t^2; x) = x^2$. Then by Bohaman–Korovkin theorem $\lim_{n\to\infty} K_{n,q}^{(\alpha)}(f(t); x) = f(x)$ converges uniformly on $[0, 1]$. □

Theorem 3.2 *For $f \in C[0, 1]$, $q \in (0, 1)$ and $\alpha \in [0, 1]$, we have*

$$\left|K_{n,q}^{(\alpha)}(f; x) - f(x)\right| \leq \lambda\omega_2\left(f; \sqrt{\mu_{n,2}^q(x) + {\mu_{n,1}^q}^2(x)}\right) + \omega\left(f; \omega_{n,1}^q(x)\right),$$

where $\mu_{n,2}^q(x)$ and $\mu_{n,1}^q(x)$ are second- and first-central moments of the operators (1.4).

Proof We define an auxiliary operators

$$\hat{K}_{n,q}^{(\alpha)}(f; x) = K_{n,q}^{(\alpha)}(f; x) - f\left(\frac{2q[n + 1]_q x + 1}{[2]_q[n + 1]_q}\right) + f(x). \tag{3.3}$$

For the operators $\hat{K}_{n,q}^{(\alpha)}(.; x)$, we get

$$\hat{K}_{n,q}^{(\alpha)}(t - x; x) = 0. \tag{3.4}$$

Suppose, $g \in W^2$, $x, t \in [0, 1]$. Then by Tylor's expansion, we have

$$g(t) = g(x) + (t - x)g'(x) + \int_x^t (t - u)g''(u)du.$$

Applying $\hat{K}_{n,q}^{(\alpha)}(.; x)$ in above equation, we have

$$\hat{K}_{n,q}^{(\alpha)}(g; x) = g(x) + \hat{K}_{n,q}^{(\alpha)}\left(\int_x^t (t - u)g^{''}(u)du; x\right).$$

Therefore,

$$\begin{aligned}\left|\hat{K}_{n,q}^{(\alpha)}(g; x) - g(x)\right| &\leq \left|K_{n,q}^{(\alpha)}\left(\int_x^t (t - u)g''(u)du; x\right)\right| \\ &+ \left|\left(\int_x^{\frac{2q[n+1]_q x+1}{[2]_q[n+1]_q}} \left(\frac{2q[n+1]_q x + 1}{[2]_q[n+1]_q} - x\right) g^{''}(u)du; x\right)\right| \\ &\leq K_{n,q}^{(\alpha)}\left(\int_x^t |t - x|\, g^{''}(u)du; x\right) \\ &+ \left|\left(\int_x^{\frac{2q[n+1]_q x+1}{[2]_q[n+1]_q}} \left|\frac{2q[n+1]_q x + 1}{[2]_q[n+1]_q} - u\right| \left|g''(x)\right| du; x\right)\right| \\ &\leq \left[K_{n,q}^{(\alpha)}((t - x)^2; x) + \left(\frac{2q[n+1]_q x + 1}{[2]_q[n+1]_q} - x\right)^2\right] \left\|g''\right\|.\end{aligned} \tag{3.5}$$

From (3.3), we have

$$\left|K_{n,q}^{(\alpha)}(f; x) \leq \|f\|\right| K_{n,q}^{(\alpha)}(1; x) + 2\|f\| = 3\|f\|. \tag{3.6}$$

From (3.3), (3.5) and (3.6), we have

$$\begin{aligned}\left|K_{n,q}^{(\alpha)}(f;x)-f(x)\right| &\leq \left|K_{n,q}^{(\alpha)}(f-g;x)\right|+|f-g| \\ &\quad+\left|f\left(\frac{2q[n+1]_q x+1}{[2]_q[n+1]_q}\right)-f(x)\right| \\ &\leq 4\|f-g\|+\left(\mu_{n,2}^q(x)+{\mu_{n,1}^q}^2(x)\right) \\ &\quad+\left|f\left(\frac{2q[n+1]_q x+1}{[2]_q[n+1]_q}\right)-f(x)\right|\end{aligned}$$

Now taking infimum on the right-hand side of the above inequality over $g \in W^2$, we get

$$\leq 4K_2\left(f;\mu_{n,2}^q(x)+{\mu_{n,1}^q}^2(x)\right)+\omega\left(f;\mu_{n,1}^q(x)\right)$$

From (3.2), we get

$$\left|K_{n,q}^{(\alpha)}(f;x)-f(x)\right| \leq \lambda\omega_2\left(f;\sqrt{\mu_{n,2}^q(x)+{\mu_{n,1}^q}^2(x)}\right)+\omega\left(f;\omega_{n,1}^q(x)\right).$$

Hence, this is our required result. □

Theorem 3.3 *Let $q_n \in (0,1)$ be a sequence converging to* 1 *and α is fixed. Then for $f \in C[0,1]$, we have*

$$\left|K_{n,q}^{(\alpha)}(f;x)-f(x)\right| \leq 2\omega(f;\delta_n(x)),$$

where $\delta_n(x)=\left(K_{n,q}^{(\alpha)}((t-x)^2;x)\right)^{\frac{1}{2}}$.

Proof For nondecreasing function $f \in C[0,1]$. Using linearity and monotonicity of $K_{n,q}^{(\alpha)}$, we have

$$\begin{aligned}\left|K_{n,q}^{(\alpha)}(f;x)-f(x)\right| &\leq K_{n,q}^{(\alpha)}\left(|f(t)-f(x)|;x\right) \\ &\leq \omega(f;\delta)\left(1+\frac{1}{\delta}K_{n,q}^{(\alpha)}\left(|t-x|;x\right)\right)\end{aligned}$$

Applying Lemma 2.4 with $c=\frac{q[k]_q}{[n+1]_q}$ and $d=\frac{[k+1]_q}{[n+1]_q}$, we get

$$\left|K_{n,q}^{(\alpha)}(f;x)-f(x)\right| \leq \omega(f;x)\left\{1+\frac{[n+1]_q}{\delta}\sum_{k=0}^{n} p_{n,q,k}^{(\alpha)}(x)\left(\int_{\frac{q[k]_q}{[n+1]_q}}^{\frac{[k+1]_q}{[n+1]_q}}(t-x)^2 d_q t\right)^{\frac{1}{2}}\right.$$

$$\times \left(\int_{\frac{q[k]_q}{[n+1]_q}}^{\frac{[k+1]_q}{[n+1]_q}} d_q t \right)^{\frac{1}{2}} \Bigg\}$$

Using Hölder's inequality for sums, we have

$$= \omega(f;x) \left\{ 1 + \frac{1}{\delta} \left([n+1]_q \sum_{k=0}^{n} p_{n,q,k}^{(\alpha)}(x) \int_{\frac{q[k]_q}{[n+1]_q}}^{\frac{[k+1]_q}{[n+1]_q}} (t-x)^2 d_q t \right)^{\frac{1}{2}} \right.$$

$$\left. \times \left([n+1]_q \sum_{k=0}^{n} p_{n,q,k}^{(\alpha)}(x) \int_{\frac{q[k]_q}{[n+1]_q}}^{\frac{[k+1]_q}{[n+1]_q}} d_q t \right)^{\frac{1}{2}} \right\}$$

$$= \omega(f;x) \left\{ 1 + \frac{1}{\delta} \left(K_{n,q}^{(\alpha)}((t-x)^2;x) \right)^{\frac{1}{2}} \right\}.$$

By choosing $\delta = \delta_n(x)$, we get the required result. □

References

1. O. Agratini, Kantorovich-type operators preserving affine functions. Hacet. J. Math. Stat. **45**(6), 1657–1663 (2016)
2. T. Acar, A.M. Acu, N. Manav, Approximation of functions by genuine Bernstein-Durrmeyer type operators. J. Math. Inequal. **12**(4), 975–987 (2018)
3. A. Aral, V. Gupta, R.P. Agrawal, *Applications of q-Calculus in Operator Theory* (Springer, New York, 1993)
4. P.N. Agrawal, V. Gupta, A.S. Kumar, On q-analogue of Bernstein-Schurer-Stancu operators. Appl. Math. Comput. **219**(14), 7754–7764 (2013)
5. S.N. Bernstein, Demonstration of the Weierstrass theorem based on the calculation of probabilities. Common Soc. Math. Charkow Ser. 2t **13**, 1–2 (1912)
6. Q.B. Cai, B.Y. Lian, G. Zho, Approximation properties of Lambda-Bernstein operators. J. Inequal. Appl. **61** (2018). https://doi.org/10.1186/s13660-018-1653-7
7. X. Chen, J. Tan, Z. Liu, J. Xie, Approximation of functions by a new family of generalized Bernstein operators. J. Math. Anal. Appl. **450**, 244–261 (2017)
8. Q.B. Cai, X.W. Xu, Shape-preserving properties of a new family of generalized Bernstein operators. J. Inequal. Appl. (2018)
9. R.A. Devore, G.G. Lorentz, *Constructive Approximation* (Springer, Berlin, 1993)
10. M. Dhamija, R. Pratap, N. Deo, Approximation by Kantorovich form of modified Szász-Mirakyan operators. Appl. Math. Comput. **317**, 109–120 (2018)
11. V. Gupta, T.M. Rassias, H. Sharma, q-Durrmeyer operators based on Pólya distribution. J. Nonlinear Sci. Appl. **9**, 1497–1504 (2016)

12. A. Lupas, A q-analogue of the Bernstein operator, *Seminar on Numerical and Statistical Calculus*, vol. 9 (University of Cluj-Napoca, 1987)
13. S.A. Mohiuddine, T. Acar, A. Alotaibi, Construction of a new family of Bernstein-Kantorovich operators. Math. Methods Appl. Sci. 1–11 (2017). https://doi.org/10.1002/mma.4559
14. G.M. Phillips, Bernstein polynomials based on the q-integers. Ann. Numer. Math. **4**, 511–518 (1997)
15. M. Örkcü, O. Doğru, q-Szasz-Mirakyan-Kantorovich operators preserving test functions. Appl. Math. Lett. **24**, 1588–1593 (2011)
16. X. Zeng, F. Cheng, On the rate of Bernstein type operators. J. Approx. Theory **109**(2), 242–256 (2001)

Approximation by Certain Operators Linking the α-Bernstein and the Genuine α-Bernstein–Durrmeyer Operators

Ana Maria Acu and Voichiţa Adriana Radu

Abstract This paper presents a new family of operators which constitute the link between α-Bernstein operators and genuine α-Bernstein–Durrmeyer operators. Some approximation results, which include local approximation and error estimation in terms of the modulus of continuity are given. Finally, a quantitative Voronovskaya type theorem is established and some Grüss type inequalities are obtained.

Keywords α-Bernstein operators · U_n^ρ operators · Modulus of smoothness · Rate of convergence · Voronovskaya type theorem

2010 MSC 41A10 · 41A25 · 41A36

1 Introduction

In 1912, Bernstein [8] defined the Bernstein polynomials in order to prove Weierstrass's fundamental theorem. These operators are the foundation of approximation theory by positive linear operators. There is a rich literature connecting with these remarkable operators, given for any $n \in \mathbb{N}$ and $f \in C[0, 1]$ by

$$B_n(f; x) = \sum_{k=0}^{n} f\left(\frac{k}{n}\right) p_{n,k}(x), \tag{1}$$

where

A. M. Acu (✉)
Department of Mathematics and Informatics, Lucian Blaga University of Sibiu,
Str. Dr. I. Ratiu, No.5-7, 550012 Sibiu, Romania
e-mail: anamaria.acu@ulbsibiu.ro

V. A. Radu
Department of Statistics-Forecasts-Mathematics, FSEGA, Babes-Bolyai University,
Cluj-Napoca, Romania
e-mail: voichita.radu@econ.ubbcluj.ro; voichita.radu@gmail.com

N. Deo et al. (eds.), *Mathematical Analysis I: Approximation Theory*,
Springer Proceedings in Mathematics & Statistics 306,
https://doi.org/10.1007/978-981-15-1153-0_7

$$p_{n,k}(x)=\binom{n}{k}x^k(1-x)^{n-k},\ \ x\in[0,1].$$

These operators appear as a powerful tool in solving differential equations, as well in the domain of numerical analysis and computer aided geometric design. They are the prototype of positive linear operators of all kinds used in the theory of approximation. The generalizations of Bernstein type operators in order to have better approximation properties were established in the research papers of the present authors [2, 4, 5, 7, 16, 18, 21, 22].

Genuine Bernstein–Durrmeyer operators were introduced independently, by Chen [9] and Goodman and Sharma [15] and were intensively studied by numerous authors (see for example [3, 5, 12, 13, 19, 20]).

The genuine Bernstein–Durrmeyer operators are given by

$$U_n(f;x)=(n-1)\sum_{k=1}^{n-1}\left(\int_0^1 f(t)p_{n-2,k-1}(t)dt\right)p_{n,k}(x)+(1-x)^nf(0)+x^nf(1),\ \ f\in C[0,1].$$

These operators are limits of the Bernstein–Durrmeyer operators with Jacobi weights, namely,

$$U_nf=\lim_{\alpha\to-1,\beta\to-1}M_n^{<\alpha,\beta>}f,\text{ where}$$

$$M_n^{<\alpha,\beta>}:C[0,1]\to\Pi_n,\ \ M_n^{<\alpha,\beta>}(f;x)=\sum_{k=0}^n p_{n,k}(x)\frac{\int_0^1 w^{(\alpha,\beta)}(t)p_{n,k}(t)f(t)dt}{\int_0^1 w^{(\alpha,\beta)}(t)p_{n,k}(t)dt},$$

$$w^{(\alpha,\beta)}(t)=x^\beta(1-x)^\alpha,\ x\in(0,1),\ \ \alpha,\beta>-1.$$

On the other hand, the genuine Bernstein–Durrmeyer operators can be written as a composition of Bernstein operators and Beta operators:

$$U_n=B_n\circ\overline{\mathbb{B}}_n,$$

where the Beta-type operators $\overline{\mathbb{B}}_n$ were introduced by Lupaş [17], as follows

$$\overline{\mathbb{B}}_n(f;x):=\begin{cases}f(0),\ \ x=0,\\[2mm] \dfrac{1}{B(nx,n-nx)}\displaystyle\int_0^1 t^{nx-1}(1-t)^{n-1-nx}f(t)dt,\ \ 0<x<1,\\[2mm] f(1),\ \ x=1,\end{cases}$$

with $n=1,2,3,\ldots$, $f\in C[0,1]$ and $B(\cdot,\cdot)$ is the Euler's Beta function.

Further, let $\rho>0$, $n\geq 1$ be fixed and the functionals $F_{n,k}^{\rho}:C[0,1]\to\mathbb{R}$, $k=\overline{0,n}$ defined by

$$F^{\rho}_{n,0}(f) = f(0),$$
$$F^{\rho}_{n,n}(f) = f(1),$$
$$F^{\rho}_{n,k}(f) = \int_0^1 \frac{t^{k\rho-1}(1-t)^{(n-k)\rho-1}}{B(k\rho,(n-k)\rho)} f(t)dt, \quad k = \overline{1, n-1}.$$

The operator $U_n^{\rho} : C[0,1] \to \Pi_n$, where Π_n is the linear space of all real polynomials of degree at most n, $n \in \mathbb{N}_0$ is given by

$$U_n^{\rho}(f;x) = \sum_{k=0}^{n} F^{\rho}_{n,k}(f) p_{n,k}(x), \quad f \in C[0,1].$$

These operators were introduced by Păltănea [20] and represent a link between Bernstein operators and the genuine Bernstein–Durrmeyer operators. For $\rho = 1$ and $f \in C[0,1]$ we obtain $U_n(f;x)$ which is the genuine Bernstein–Durrmeyer operators, while for $\rho \to \infty$, for each $f \in C[0,1]$ the sequence U_n^{ρ} converges uniformly to $B_n(f;x)$.

A new family of generalized Bernstein operators depending on α which is a nonnegative real parameter was introduced by Chen et al. in [10]. The form of this operator is

$$T_{n,\alpha}(f;x) = \sum_{i=0}^{n} f\left(\frac{i}{n}\right) p^{(\alpha)}_{n,i}(x), \quad f \in C[0,1]. \tag{2}$$

Here, for $i = \overline{0,n}$ the α-Bernstein polynomial $p^{(\alpha)}_{n,i}(x)$ of degree n is defined by

$$\begin{cases} p^{(\alpha)}_{1,0}(x) = 1 - x, \\ p^{(\alpha)}_{1,1}(x) = x, \\ p^{(\alpha)}_{n,i}(x) = \left[\binom{n-2}{i}(1-\alpha)x + \binom{n-2}{i-2}(1-\alpha)(1-x) + \binom{n}{i}\alpha x(1-x)\right] x^{i-1}(1-x)^{n-i-1}, n \geq 2, \end{cases} \tag{3}$$

where $x \in [0,1]$ and the binomial coefficients are given by

$$\binom{n}{i} = \begin{cases} \dfrac{n!}{(n-i)!\,i!}, & \text{if} \quad 0 \leq i \leq n, \\ 0, & \text{else}. \end{cases}$$

We can observe that for $\alpha = 1$, the α-Bernstein operator reduces to the classical Bernstein polynomial given in (1). Also, for $\alpha \in [0,1]$ the operators $T_{n,\alpha}$ are positive linear operators. In the following we will consider $\alpha \in [0,1]$.

Very recently, Acar et al. [1] introduce genuine α-Bernstein–Durrmeyer operators as follows

$$U_{n,\alpha} := T_{n,\alpha} \circ \overline{\mathbb{B}}_n.$$

Our aim in this paper is to introduce a new class of operators which represent a link between α-Bernstein operators and genuine α-Bernstein–Durrmeyer operators. We define the operators

$$U_{n,\alpha}^{\rho}(f;x) := \sum_{k=0}^{n} F_{n,k}^{\rho}(f) p_{n,i}^{(\alpha)}(x),$$

and in explicit form

$$U_{n,\alpha}^{\rho}(f;x) = f(0)\cdot p_{n,0}^{(\alpha)}(x) + f(1)\cdot p_{n,n}^{(\alpha)}(x) + \sum_{k=1}^{n-1}\left(\int_0^1 \frac{t^{k\rho-1}(1-t)^{(n-k)\rho-1}}{B(k\rho,(n-k)\rho)} f(t)dt\right)\cdot p_{n,k}^{(\alpha)}(x), \tag{4}$$

$f \in C[0,1]$ and $x \in [0,1]$.

For $\rho = 1$ we obtain $U_{n,\alpha}$ and for $\rho \to \infty$ the sequences $U_{n,\alpha}^{\rho}$ converges to $T_{n,\alpha}$.

2 Approximation Properties of $U_{n,\alpha}^{\rho}$ Operators

The approximation properties of $U_{n,\alpha}^{\rho}$ operators are investigated in the present section. Also, the rate of convergence is estimated using classical moduli of smoothness.

Throughout this paper, we will use a positive constant C, not necessarily the same at each occurrence.

Lemma 2.1 *The $U_{n,\alpha}^{\rho}$ operators have the end point interpolation properties*

$$U_{n,\alpha}^{\rho}(f;0) = f(0) \quad \text{and} \quad U_{n,\alpha}^{\rho}(f;1) = f(1).$$

Lemma 2.2 *The $U_{n,\alpha}^{\rho}$ operators verify*

(i) $U_{n,\alpha}^{\rho}(e_0;x) = 1;$
(ii) $U_{n,\alpha}^{\rho}(e_1;x) = x;$
(iii) $U_{n,\alpha}^{\rho}(e_2;x) = x^2 + \frac{x(1-x)}{n\rho+1}\cdot\left(\rho+1+\frac{2(1-\alpha)\rho}{n}\right);$
(iv) $U_{n,\alpha}^{\rho}(e_3;x) = x^3 + \frac{x(1-x)}{(n\rho+2)(n\rho+1)n}\{3\rho x(\rho+1)n^2 + (4\rho^2 x + \rho^2 + 3\rho + 2x + 2 - 6\alpha\rho^2 x)n + 6\rho(1-\alpha)(1+\rho-2\rho x)\};$
(v) $U_{n,\alpha}^{\rho}(e_4;x) = x^4 + \frac{x(1-x)}{(n\rho+3)(n\rho+2)(n\rho+1)n}\{6\rho^2x^2(\rho+1)n^3 - \rho x(12\alpha\rho^2 x - \rho^2 x - 7\rho^2 - 18\rho - 11x - 11)n^2 + (60\alpha\rho^3x^2 - 36\alpha\rho^3 x - 54\rho^3x^2 - 36\alpha\rho^2 x + 30\rho^3 x + \rho^3 + 24\rho^2 x + 6\rho^2 + 6x^2 + 11\rho + 6x + 6)n + 2\rho(1-\alpha)(36\rho^2x^2 - 36\rho^2 x + 7\rho^2 - 36\rho x + 18\rho + 11)\}.$

Lemma 2.3 *The central moments are the following*

(i) $U^{\rho}_{n,\alpha}(t-x;x)=0;$

(ii) $U^{\rho}_{n,\alpha}((t-x)^2;x)=\dfrac{x(1-x)}{n\rho+1}\cdot\left(\rho+1+\dfrac{2(1-\alpha)\rho}{n}\right).$

Denote by

$$\phi(x):=\sqrt{x(1-x)}\quad\text{and}\quad \vartheta_{n,\alpha}(x):=\frac{\phi^2(x)}{n\rho+1}\cdot\left(\rho+1+\frac{2(1-\alpha)\rho}{n}\right).$$

Lemma 2.4 *The $U^{\rho}_{n,\alpha}$ operators verify*

(i) $\lim\limits_{n\to\infty} nU^{\rho}_{n,\alpha}(t-x;x)=0;$

(ii) $\lim\limits_{n\to\infty} nU^{\rho}_{n,\alpha}\left((t-x)^2;x\right)=\dfrac{\rho+1}{\rho}\cdot\phi^2(x);$

(iii) $\lim\limits_{n\to\infty} n^2U^{\rho}_{n,\alpha}\left((t-x)^4;x\right)=\dfrac{3(\rho+1)^2}{\rho^2}\cdot\phi^4(x).$

Lemma 2.5 *Let $f\in C[0,1]$, $x\in[0,1]$ and $n\in\mathbb{N}$. Then*

$$||U^{\rho}_{n,\alpha}(f;\cdot)||\le||f||,$$

where $||\cdot||$ is the uniform norm on [0, 1].

Proof From Lemma 2.2 we have $U^{\rho}_{n,\alpha}(e_0;x)=1$ so,

$$|U^{\rho}_{n,\alpha}(f;x)|\le U^{\rho}_{n,\alpha}(e_0;x)||f||=||f||.$$

□

Theorem 2.1 *If $f\in C[0,1]$, then $\lim\limits_{n\to\infty} U^{\rho}_{n,\alpha}(f;x)=f(x)$ uniformly on [0, 1].*

Proof From Lemma 2.2 it is easy to observe that $\lim\limits_{n\to\infty} U^{\rho}_{n,\alpha}(e_k;x)=e_k(x)$ uniformly on $[0,1]$, for $k\in\{0,1,2\}$ and applying the Bohmann–Korovkin theorem, we get the result. □

Theorem 2.2 *If $f\in C[0,1]$, then*

$$\left|U^{\rho}_{n,\alpha}(f;x)-f(x)\right|\le 2\omega\left(f;\vartheta^{\frac{1}{2}}_{n,\alpha}(x)\right),$$

where ω is the usual modulus of continuity.

Proof The inequality is trivially true if $x \in \{0, 1\}$. Otherwise, using the well known property of modulus of continuity

$$|f(t) - f(x)| \leq \omega(f; \delta)\left(\frac{(t-x)^2}{\delta^2} + 1\right),$$

we obtain

$$|U^{\rho}_{n,\alpha}(f; x) - f(x)| \leq U^{\rho}_{n,\alpha}(|f(t) - f(x)|; x) \leq \omega(f; \delta)\left(1 + \frac{1}{\delta^2}U^{\rho}_{n,\alpha}((t-x)^2; x)\right).$$

We have the desired result, by choosing $\delta = \vartheta^{\frac{1}{2}}_{n,\alpha}(x),\ x \in (0, 1)$. □

Theorem 2.3 *If* $f \in C^1[0, 1]$, *then*

$$\left|U^{\rho}_{n,\alpha}(f; x) - f(x)\right| \leq 2\vartheta^{\frac{1}{2}}_{n,\alpha}(x)\omega\left(f', \vartheta^{\frac{1}{2}}_{n,\alpha}(x)\right). \tag{5}$$

Proof For $f \in C^1[0, 1]$ and any $x, t \in [0, 1]$, we have

$$f(t) - f(x) = f'(x)(t - x) + \int_x^t \left(f'(y) - f'(x)\right) dy,$$

and it follows

$$U^{\rho}_{n,\alpha}\left(f(t) - f(x); x\right) = f'(x)U^{\rho}_{n,\alpha}(t - x; x) + U^{\rho}_{n,\alpha}\left(\int_x^t (f'(y) - f'(x))dy; x\right).$$

Using the property of modulus of continuity that

$$|f(y) - f(x)| \leq \omega(f; \delta)\left(\frac{|y-x|}{\delta} + 1\right), \quad \delta > 0,$$

we have

$$\left|\int_x^t |f'(y) - f'(x)|dy\right| \leq \omega(f'; \delta)\left[\frac{(t-x)^2}{\delta} + |t - x|\right].$$

Therefore,

$$\begin{aligned}|U^{\rho}_{n,\alpha}(f; x) - f(x)| &\leq |f'(x)| \cdot |U^{\rho}_{n,\alpha}(t - x; x)| \\ &\quad + \omega(f'; \delta)\left\{\frac{1}{\delta}U^{\rho}_{n,\alpha}\left((t-x)^2; x\right) + U^{\rho}_{n,\alpha}(|t - x|; x)\right\}.\end{aligned}$$

Using Cauchy–Schwartz inequality we obtain

$$\begin{aligned}|U^{\rho}_{n,\alpha}(f;x)-f(x)| &\leq |f'(x)||U^{\rho}_{n,\alpha}(t-x;x)| \\ &\quad +\omega(f',\delta)\left\{\frac{1}{\delta}\sqrt{U^{\rho}_{n,\alpha}\left((t-x)^2;x\right)}+1\right\}\sqrt{U^{\rho}_{n,\alpha}\left((t-x)^2;x\right)} \\ &\leq \omega(f',\delta)\cdot\left\{\frac{1}{\delta}\vartheta^{\frac{1}{2}}_{n,\alpha}(x)+1\right\}\vartheta^{\frac{1}{2}}_{n,\alpha}(x). \qquad (6)\end{aligned}$$

For $x \in \{0, 1\}$, the inequality (5) is true. Otherwise, for $\delta = \vartheta^{\frac{1}{2}}_{n,\alpha}(x)$, $x \in (0, 1)$ in the relation (6) we get to the desired result. □

We recall the definition of K-functional, in order to give the next result

$$K_2(f,\delta) := \inf\left\{\|f-g\|+\delta\left\|g''\right\| : g \in W^2[0,1]\right\},$$

where $W^2[0,1] = \left\{g \in C[0,1] : g'' \in C[0,1]\right\}$, $\delta \geq 0$ and $\|\cdot\|$ is the uniform norm on $C[0, 1]$.

The second-order modulus of continuity is defined as follows

$$\omega_2\left(f,\sqrt{\delta}\right) = \sup_{0<h\leq\sqrt{\delta}}\ \sup_{x,x+2h\in[0,1]}\left\{|f(x+2h)-2f(x+h)+f(x)|\right\}.$$

It is well known that K-functional $K_2(f, \delta)$ and the second-order modulus of continuity $\omega_2\left(f, \sqrt{\delta}\right)$ are equivalent:

$$K_2(f,\delta) \leq C\omega_2\left(f,\sqrt{\delta}\right), \quad \delta \geq 0,\ C > 0. \qquad (7)$$

Theorem 2.4 *If $f \in C[0, 1]$, then*

$$\left|U^{\rho}_{n,\alpha}(f;x)-f(x)\right| \leq C\omega_2\left(f,\frac{1}{\sqrt{2}}\vartheta^{\frac{1}{2}}_{n,\alpha}(x)\right),$$

where C is a positive constant.

Proof Using Lemma 2.2 and applying $U^{\rho}_{n,\alpha}$ to Taylor's formula, we get

$$U^{\rho}_{n,\alpha}(g;x) = g(x) + U^{\rho}_{n,\alpha}\left(\int_x^t (t-y)g''(y)dy;x\right).$$

This implies that

$$\begin{aligned}|U^{\rho}_{n,\alpha}(g;x)-g(x)| &\leq \left|U^{\rho}_{n,\alpha}\left(\int_x^t (t-y)g''(y)dy;x\right)\right| \\ &\leq U^{\rho}_{n,\alpha}((t-x)^2;x)\left\|g''\right\| \leq \vartheta_{n,\alpha}(x)\|g''\|.\end{aligned}$$

In view of Lemma 2.5 we have

$$|U_{n,\alpha}^{\rho}(f;x)| \le \|f\|. \tag{8}$$

For $f \in C[0,1]$ and $g \in W^2[0,1]$ and using (8) we get

$$|U_{n,\alpha}^{\rho}(f;x) - f(x)| \le \left|U_{n,\alpha}^{\rho}(f-g;x)\right| + \left|U_{n,\alpha}^{\rho}(g;x) - g(x)\right| + |g(x) - f(x)|$$
$$\le 2\|f-g\| + \vartheta_{n,\alpha}(x)\left\|g''\right\|.$$

Taking the infimum on right side over all $g \in W^2[0,1]$, we have

$$|U_{n,\alpha}^{\rho}(f;x) - f(x)| \le 2K_2\left(f, \frac{1}{2}\vartheta_{n,\alpha}(x)\right).$$

Finally, using the equivalence between K-functional and the second-order modulus of continuity, given by relation (7), the proof is completed. □

3 Voronovskaja Type Theorem

In this section, we prove a Voronovskaja type asymptotic formula for the operator $U_{n,\alpha}^{\rho}$. In order to give the main result we recall the definition of the Ditzian–Totik first-order modulus of smoothness:

$$\omega_1^{\phi}(f;t) = \sup_{0<h\le t}\left\{\left|f\left(x + \frac{h\phi(x)}{2}\right) - f\left(x - \frac{h\phi(x)}{2}\right)\right|, x \pm \frac{h\phi(x)}{2} \in [0,1]\right\}, \tag{9}$$

where $\phi(x) = \sqrt{x(1-x)}$ and $f \in C[0,1]$.

The corresponding K-functional of the Ditzian–Totik first-order modulus of smoothness is given by

$$K_{\phi}(f;t) = \inf_{g \in W_{\phi}[0,1]}\{||f-g|| + t||\phi g'||\} \ (t > 0), \tag{10}$$

where $W_{\phi}[0,1] = \{g : g \in AC_{loc}[0,1], \|\phi g'\| < \infty\}$ and $AC_{loc}[0,1]$ is the class of absolutely continuous functions on every interval $[a,b] \subset [0,1]$.

Between K-functional and the Ditzian–Totik first-order modulus of smoothness there is the following relation

$$K_{\phi}(f;t) \le C\omega_1^{\phi}(f;t), \tag{11}$$

where $C > 0$ is a constant.

Theorem 3.1 *For any $f \in C^2[0,1]$ and sufficiently large n the following inequality holds*

$$\left|U^{\rho}_{n,\alpha}(f;x)-f(x)-\frac{1}{2}\vartheta_{n,\alpha}(x)f''(x)\right|\leq\frac{1}{n}C\phi^2(x)\omega_1^{\phi}\left(f'',\sqrt{\frac{\rho+1}{n\rho}}\right),\tag{12}$$

where C is a positive constant.

Proof For $f\in C^2[0,1],\ t,x\in[0,1]$, by Taylor's expansion, we have

$$f(t)-f(x)=(t-x)f'(x)+\int_x^t(t-y)f''(y)dy.$$

Hence

$$\begin{aligned}f(t)-f(x)-(t-x)f'(x)-\frac{1}{2}(t-x)^2f''(x)&=\int_x^t(t-y)f''(y)dy-\int_x^t(t-y)f''(x)dy\\&=\int_x^t(t-y)[f''(y)-f''(x)]dy.\end{aligned}$$

Applying $U^{\rho}_{n,\alpha}(\cdot;x)$ to both sides of the above relation, we obtain

$$\left|U^{\rho}_{n,\alpha}(f;x)-f(x)-\frac{1}{2}\vartheta_{n,\alpha}(x)f''(x)\right|\leq U^{\rho}_{n,\alpha}\left(\left|\int_x^t|t-y||f''(y)-f''(x)|dy\right|;x\right).\tag{13}$$

In [11, p. 337], the quantity $\left|\int_x^t\left|f''(y)-f''(x)\right||t-y|dy\right|$ was estimated as

$$\left|\int_x^t|f''(y)-f''(x)||t-y|dy\right|\leq 2\|f''-g\|(t-x)^2+2\|\phi g'\|\phi^{-1}(x)|t-x|^3,\ x\in(0,1),\ g\in W_{\phi}[0,1].\tag{14}$$

Note that for $x\in\{0,1\}$ the inequality (12) is verified.

There exists a constant $C>0$ such that for n sufficiently large and using Lemma 2.4 we obtain

$$U^{\rho}_{n,\alpha}\left((t-x)^2;x\right)\leq\frac{C(\rho+1)}{n\rho}\phi^2(x)\text{ and }U^{\rho}_{n,\alpha}\left((t-x)^4;x\right)\leq\frac{C(\rho+1)^2}{n^2\rho^2}\phi^4(x).\tag{15}$$

Applying the Cauchy–Schwarz inequality and from relations (13–15), we get

$$\begin{aligned}&\left|U^{\rho}_{n,\alpha}(f;x)-f(x)-\frac{1}{2}\vartheta_{n,\alpha}(x)f''(x)\right|\\&\leq 2\|f''-g\|U^{\rho}_{n,\alpha}\left((t-x)^2;x\right)+2\|\phi g'\|\phi^{-1}(x)U^{\rho}_{n,\alpha}\left(|t-x|^3;x\right)\\&\leq\frac{C(\rho+1)}{n\rho}\phi^2(x)\|f''-g\|+2\|\phi g'\|\phi^{-1}(x)\left\{U^{\rho}_{n,\alpha}(t-x)^2;x\right\}^{1/2}\left\{U^{\rho}_{n,\alpha}\left((t-x)^4;x\right)\right\}^{1/2}\\&\leq\frac{C(\rho+1)}{n\rho}\phi^2(x)\|f''-g\|+\phi^2(x)\frac{C(\rho+1)\sqrt{\rho+1}}{n\rho\sqrt{n\rho}}\|\phi g'\|\\&\leq\frac{C(\rho+1)}{n\rho}\phi^2(x)\left\{\|f''-g\|+\sqrt{\frac{\rho+1}{n\rho}}\|\phi g'\|\right\}.\end{aligned}$$

Taking the infimum on right hand side of the above relations over $g \in W_\phi[0,1]$, we have the proof complete. □

Corollary 3.1 *If $f \in C^2[0,1]$, then*

$$\lim_{n\to\infty} n\left\{U_{n,\alpha}^{\rho}(f;x) - f(x) - \frac{1}{2}\vartheta_{n,\alpha}(x)f''(x)\right\} = 0.$$

Motivated by the Grüss type inequalities for certain positive linear operators studied in [6, 14], in the following we prove a Grüss–Voronovskaya type theorem of $U_{n,\alpha}^{\rho}$ operators.

Denote by $\Omega(f,g;x) := U_{n,\alpha}^{\rho}(fg;x) - U_{n,\alpha}^{\rho}(f;x)U_{n,\alpha}^{\rho}(g;x)$.

Theorem 3.2 *Let $f, g \in C^2[0,1]$ and $\rho > 0$. Then, for each $x \in [0,1]$,*

$$\lim_{n\to\infty} n\cdot\Omega(f,g;x) = \frac{\rho+1}{\rho}x(1-x)f'(x)g'(x).$$

Proof Since

$$(fg)(x) = f(x)g(x), \quad (fg)'(x) = f'(x)g(x) + f(x)g'(x)$$

and

$$(fg)''(x) = f''(x)g(x) + 2f'(x)g'(x) + f(x)g''(x),$$

we can write

$$\begin{aligned}
\Omega(f,g;x) &= U_{n,\alpha}^{\rho}(fg;x) - U_{n,\alpha}^{\rho}(f;x)U_{n,\alpha}^{\rho}(g;x)\\
&= \left\{U_{n,\alpha}^{\rho}(fg;x) - f(x)g(x) - (fg)'(x)U_{n,\alpha}^{\rho}(t-x;x) - \frac{(fg)''(x)}{2!}U_{n,\alpha}^{\rho}\left((t-x)^2;x\right)\right\}\\
&- g(x)\left\{U_{n,\alpha}^{\rho}(f;x) - f(x) - f'(x)U_{n,\alpha}^{\rho}(t-x;x) - \frac{f''(x)}{2!}U_{n,\alpha}^{\rho}\left((t-x)^2;x\right)\right\}\\
&- U_{n,\alpha}^{\rho}(f;x)\left\{U_{n,\alpha}^{\rho}(g;x) - g(x) - g'(x)U_{n,\alpha}^{\rho}(t-x;x) - \frac{g''(x)}{2!}U_{n,\alpha}^{\rho}\left((t-x)^2;x\right)\right\}\\
&+ \frac{1}{2!}U_{n,\alpha}^{\rho}\left((t-x)^2;x\right)\{f(x)g''(x) + 2f'(x)g'(x) - g''(x)U_{n,\alpha}^{\rho}(f;x)\}\\
&+ U_{n,\alpha}^{\rho}(t-x;x)\left\{f(x)g'(x) - g'(x)U_{n,\alpha}^{\rho}(f;x)\right\}.
\end{aligned}$$

Consequently,

$$\begin{aligned}
\lim_{n\to\infty} n\cdot\Omega(f,g;x) &= \lim_{n\to\infty} n\left\{U_{n,\alpha}^{\rho}(fg;x) - U_{n,\alpha}^{\rho}(f;x)U_{n,\alpha}^{\rho}(g;x)\right\}\\
&= \lim_{n\to\infty} nf'(x)g'(x)\,U_{n,\alpha}^{\rho}\left((t-x)^2;x\right) + \lim_{n\to\infty} n\frac{g''(x)}{2!}\left\{f(x) - U_{n,\alpha}^{\rho}(f;x)\right\}U_{n,\alpha}^{\rho}\left((t-x)^2;x\right).
\end{aligned}$$

From Theorem 2.1 it follows that for each $x \in [0,1]$, $U_{n,\alpha}^{\rho}(f;x)$ converges to the function f, as $n \to \infty$ and in view of Lemma 2.4, $\lim\limits_{n\to\infty} n\,U_{n,\alpha}^{\rho}\left((t-x)^2;x\right)$ is finite. Hence, the second term in the right hand side of the above relation is zero and

$$\lim_{n\to\infty} n\cdot\Omega(f,g;x)=\frac{\rho+1}{\rho}x(1-x)f'(x)g'(x),$$

which completes the proof. □

Theorem 3.3 *Let $f, g\in C[0,1]$, $x\in[0,1]$ and $\rho>0$. Then*

$$|\Omega(f,g;x)|\le\frac{3}{2}\cdot\sqrt{\xi_1(f,x)\cdot\xi_1(g,x)}$$

where $\xi_1(f,x):=\omega_2\left(f^2;\sqrt{\vartheta_{n,\alpha}(x)}\right)+2||f||\cdot\omega_2\left(f;\sqrt{\vartheta_{n,\alpha}(x)}\right)$ and $\xi_1(g,x)$ is analogously defined.

Proof In [14, Theorem 1] we consider $H=U_{n,\alpha}^{\rho}$. □

Let $\kappa:=\frac{1}{\phi(x)}\cdot\sqrt{\frac{\vartheta_{n,\alpha}(x)}{2}}\ge 0$, $x\in(0,1)$, $\phi(x)=\sqrt{x(1-x)}$. The following estimate for $\Omega(f,g;x)$ can be obtained.

Theorem 3.4 *Let $f, g\in C[0,1]$, $x\in[0,1]$ and $\rho>0$. Then*

$$|\Omega(f,g;x)|\le\frac{9}{2}\cdot\sqrt{\xi_2(f,x)\cdot\xi_2(g,x)}$$

where $\xi_2(f,x):=\omega_2^{\phi}(f^2;\kappa)+2||f||\cdot\omega_2^{\phi}(f;\kappa)$ and

$$\omega_2^{\phi}(f;\delta)=\sup_{0<h\le\delta}\left\{\left|f\left(x+\frac{h\phi(x)}{2}\right)-2f(x)+f\left(x-\frac{h\phi(x)}{2}\right)\right|;x\pm\frac{h\phi(x)}{2}\in[0,1]\right\},\ \delta>0.$$

Proof In [14, Theorem 2] we consider $H=U_{n,\alpha}^{\rho}$. □

Acknowledgements The work of the first author was financed from Lucian Blaga University of Sibiu research grant LBUS-IRG-2018-04.

References

1. T. Acar, A.M. Acu, N. Manav, Approximation of functions by genuine Bernstein–Durrmeyer type operators. J. Math. Inequal. **12**(4), 975–987 (2018)
2. A.M. Acu, V.A. Radu, About the iterates of some operators depending on a parameter and preserving the affine functions. Iran. J. Sci. Technol. Trans. A Sci. **43**(1), 265–271 (2019)
3. A.M. Acu, I. Rasa, New estimates for the differences of positive linear operators. Numer. Algorithms **73**(3), 775–789 (2016)
4. A.M. Acu, T. Acar, C.V. Muraru, V.A. Radu, Some approximation properties by a class of bivariate operators. Math. Methods Appl. Sci. **42**, 1–15 (2019). https://doi.org/10.1002/mma.5515
5. A.M. Acu, T. Acar, V.A. Radu, Approximation by modified U_n^{ρ} operators. RACSAM (2019). https://doi.org/10.1007/s13398-019-00655-y

6. A.M. Acu, H.H. Gonska, I. Rasa, Grüss-type and Ostrowski-type inequalities in approximation theory. Ukr. Math. J. **63**(6), 843–864 (2011)
7. A.M. Acu, C.V. Muraru, D.F. Sofonea, V.A. Radu, Some approximation properties of a Durrmeyer variant of q-Bernstein-Schurer operators. Math. Methods Appl. Sci. **39**(18), 5636–5650 (2016)
8. S.N. Bernstein, Démonstration du théorème de Weierstrass fondée sur le calcul des probabilités. Communications de la Société Mathematique de Kharkov **13**, 1–2 (1913)
9. W. Chen, On the modified Bernstein-Durrmeyer operator, in *Report of the Fifth Chinese Conference on Approximation Theory* (ZhengZhou, China, 1987)
10. X. Chen, J. Tan, Z. Liu, J. Xie, Approximation of functions by a new family of generalized Bernstein operators. J. Math. Anal. Appl. **450**, 244–261 (2017)
11. Z. Finta, Remark on Voronovskaja theorem for q -Bernstein operators. Stud. Univ. Babeş-Bolyai Math. **56**, 335–339 (2011)
12. H.H. Gonska, R. Păltănea, Quantitative convergence theorems for a class of Bernstein–Durrmeyer operators preserving linear functions. Ukr. Math. J. **62**, 913–922 (2010)
13. H.H. Gonska, R. Păltănea, Simultaneous approximation by a class of Bernstein–Durrmeyer operators preserving linear functions. Czechoslov. Math. J. **60**(135), 783–799 (2010)
14. H.H. Gonska, G. Tachev, Grüss-type inequalities for positive linear operators with second order moduli. Mat. Vesn. **63**(4), 247–252 (2011)
15. T.N.T. Goodman, A. Sharma, A modified Bernstein–Schoenberg operator, in *Proceedings of the Conference on Constructive Theory of Functions, Varna 1987,* eds. by B.l. Sendov et al. (Publishing House of Bulgarian Academy of Science, Sofia, 1988), pp. 166–173
16. Y.C. Kwun, A.M. Acu, A. Rafiq, V.A. Radu, F. Ali, S.M. Kang, Bernstein-Stancu type operators which preserve polynomials. J. Comput. Anal. Appl. **23**(1), 758–770 (2017)
17. A. Lupaş, Die Folge der Betaoperatoren, Ph.D. Dissertation, Universität Stuttgart, 1972
18. C.V. Muraru, A.M. Acu, V.A. Radu, On the monotonicity of q-Schurer-Stancu type polynomials. Miskolc Math. Notes **19**(1), 19–28 (2018)
19. T. Neer, A.M. Acu, P.N. Agrawal, Bezier variant of genuine-Durrmeyer type operators based on Polya distribution. Carpathian J. Math. **33**(1), 73–86 (2017)
20. R. Păltănea, A class of Durrmeyer type operators preserving linear functions. Ann. Tiberiu Popoviciu Semin. Funct. Equ. Approx. Convex. (Cluj-Napoca) **5**, 109–117 (2007)
21. V.A. Radu, Note on the iterates of q and (p;q) Bernstein operators. Sci. Stud. Res. Ser. Math. Inform. **26**(2), 83–94 (2016)
22. V.A. Radu, Quantitative estimates for some modified Bernstein-Stancu operators. Miskolc Math. Notes **19**(1), 517–525 (2018)

Note on a Proof for the Representation of the kth Order Kantorovich Modification of Linking Baskakov Type Operators

Margareta Heilmann and Ioan Raşa

Abstract The topic of this note is a simplification of proofs for the representation of kth order Kantorovich modifications of linking Baskakov type operators given in our previous papers (Heilmann and Raşa in Mathematics and Computing. Springer, Berlin, pp. 312–320, 2017, [1], Heilmann and Raşa in Results Math. 74:9, 2019, [2]).

Keywords Linking Baskakov type operators · Kantorovich type modifications

MSC 2010: 41A36 · 41A10 · 41A30 · 41A28

1 Introduction

During the last years the investigation of so-called linking operators came into the focus of research in approximation theory. Starting with the consideration of a non-trivial link between genuine Bernstein Durrmeyer operators and classical Bernstein operators (see [3]) the study was generalized to genuine Baskakov Durrmeyer type operators and their classical counterparts as well as to kth order Kantorovich modifications. For a survey of the available literature we refer to [1, 2, 4] and the references given there.

In [1, 2] the authors of this paper used different methods for the proof of useful representations for the kth order Kantorovich modifications when the linking parameter ρ is assumed to be a natural number. By observing that the linking basis

M. Heilmann (✉)
School of Mathematics and Natural Sciences, University of Wuppertal, Gaußstraße 20, 42119 Wuppertal, Germany
e-mail: heilmann@math.uni-wuppertal.de

I. Raşa
Department of Mathematics, Technical University, Str. Memorandumului 28, 400114 Cluj-Napoca, Romania
e-mail: Ioan.Rasa@math.utcluj.ro

N. Deo et al. (eds.), *Mathematical Analysis I: Approximation Theory*, Springer Proceedings in Mathematics & Statistics 306,
https://doi.org/10.1007/978-981-15-1153-0_8

functions $\mu_{n,j,\rho}$ can be expressed as basis functions of the classical operators we are now able to simplify the proof significantly.

Let $c \in \mathbb{R}$, $n \in \mathbb{R}$, $n > c$ for $c \geq 0$ and $-n/c \in \mathbb{N}$ for $c < 0$. Furthermore let $\rho \in \mathbb{R}^+$, $j \in \mathbb{N}_0$, $x \in I_c$ with $I_c = [0, \infty)$ for $c \geq 0$ and $I_c = [0, -1/c]$ for $c < 0$. By $n^{c,\overline{j}}$ we denote $n^{c,\overline{j}} = \prod_{l=0}^{j-1}(n + cl)$, $j \in \mathbb{N}$, $n^{c,\overline{0}} = 1$. Then the basis functions are given by

$$p_{n,j}(x) = \begin{cases} \frac{n^j}{j!} x^j e^{-nx} & , c = 0, \\ \frac{n^{c,\overline{j}}}{j!} x^j (1 + cx)^{-(\frac{n}{c}+j)} & , c \neq 0. \end{cases} \tag{1}$$

$$= \begin{cases} \frac{1}{n-c} \frac{(-c)^{j+1}}{B(j+1, -\frac{n}{c}-j+1)} x^j (1 + cx)^{-(\frac{n}{c}+j)} & , c < 0, \\ \frac{n^j}{\Gamma(j+1)} x^j e^{-nx} & , c = 0, \\ \frac{1}{n-c} \frac{c^{j+1}}{B(j+1, \frac{n}{c}-1)} x^j (1 + cx)^{-(\frac{n}{c}+j)} & , c > 0. \end{cases} \tag{2}$$

We remark that (2) is well defined also for $j \in \mathbb{R}$, $j \geq 0$ which will be used in (5).

In the following definitions of the operators we omit the parameter c in the notations in order to reduce the necessary sub and superscripts.

We assume that $f : I_c \longrightarrow \mathbb{R}$ is given in such a way that the corresponding integrals and series are convergent.

Definition 1 The operators of Baskakov-type are defined by

$$(B_n f)(x) = \sum_{j=0}^{\infty} p_{n,j}(x) f\left(\frac{j}{n}\right),$$

and the genuine Baskakov–Durrmeyer type operators are denoted by

$$(B_{n,1} f)(x) = f(0) p_{n,0}(x) + f\left(-\frac{1}{c}\right) p_{n,-\frac{n}{c}}(x) + \sum_{j=1}^{-\frac{n}{c}-1} p_{n,j}(x)(n + c) \int_0^{-\frac{1}{c}} p_{n+2c,j-1}(t) f(t) dt \tag{3}$$

for $c < 0$ and by

$$(B_{n,1} f)(x) = f(0) p_{n,0}(x) + \sum_{j=1}^{\infty} p_{n,j}(x)(n + c) \int_0^{\infty} p_{n+2c,j-1}(t) f(t) dt$$

for $c \geq 0$.

Depending on a parameter $\rho \in \mathbb{R}^+$ the linking operators are given by

$$(B_{n,\rho} f)(x) = \sum_{j=0}^{\infty} F_{n,j}^{\rho}(f) p_{n,j}(x) \tag{4}$$

where

$$F_{n,j}^{\rho}(f)=\begin{cases} f(0) & , j=0,\ c\in\mathbb{R},\\ f\left(-\dfrac{1}{c}\right) & , j=-\frac{n}{c},\ c<0,\\ \displaystyle\int_{I_c}\mu_{n,j,\rho}(t)f(t)dt & , \textit{otherwise}, \end{cases}$$

with

$$\mu_{n,j,\rho}(t)=\begin{cases} \dfrac{(-c)^{j\rho}}{B\left(j\rho,-\left(\frac{n}{c}+j\right)\rho\right)}t^{j\rho-1}(1+ct)^{-\left(\frac{n}{c}+j\right)\rho-1} & , c<0,\\ \dfrac{(n\rho)^{j\rho}}{\Gamma(j\rho)}t^{j\rho-1}e^{-n\rho t} & , c=0,\\ \dfrac{c^{j\rho}}{B\left(j\rho,\frac{n}{c}\rho+1\right)}t^{j\rho-1}(1+ct)^{-\left(\frac{n}{c}+j\right)\rho-1} & , c>0. \end{cases}$$

The *k*th order Kantorovich modifications of the operators $B_{n,\rho}$ are given by

$$B_{n,\rho}^{(k)}:=D^k\circ B_{n,\rho}\circ I_k$$

where D^k denotes the *k*th order ordinary differential operator and I_k the corresponding antiderivative, i.e.,

$$I_kf=f,\ \text{if}\ k=0,\ \text{and}\ (I_kf)(x)=\int_0^x\frac{(x-t)^{k-1}}{(k-1)!}f(t)dt,\ \text{if}\ k\in\mathbb{N}.$$

For $k=0$ we omit the superscript (k) as indicated by the definition above.

By studying again the papers which were published so far, we noticed that it is not necessary to use a different notation for the functions related to the linking parameter ρ. By using (2) they can be written also in terms of the basis functions, i.e.,

$$\mu_{n,j,\rho}(t)=(n\rho+c)p_{n\rho+2c,j\rho-1}(t). \tag{5}$$

Therefore,

$$(B_{n,\rho}f)(x)=f(0)p_{n,0}(x)+f\left(-\frac{1}{c}\right)p_{n,-\frac{n}{c}}(x) \tag{6}$$
$$+\sum_{j=1}^{-\frac{n}{c}-1}p_{n,j}(x)(n\rho+c)\int_0^{-\frac{1}{c}}p_{n\rho+2c,j\rho-1}(t)f(t)dt$$

for $c<0$ and

$$(B_{n,\rho}f)(x) = f(0)p_{n,0}(x) + \sum_{j=1}^{\infty} p_{n,j}(x)(n\rho + c) \int_0^{\infty} p_{n\rho+2c,j\rho-1}(t)f(t)dt$$

for $c \geq 0$.

In [1, 2] we proved the following representations for $B_{n,\rho}(f; x)$ in case of $\rho \in \mathbb{N}$.

Let $c = -1$, $n, k \in \mathbb{N}$, $n - k \geq 1$, $\rho \in \mathbb{N}$ and $f \in L_1[0, 1]$. Then

$$B_{n,\rho}^{(k)}(f; x) = \frac{n!(n\rho - 1)!}{(n-k)!(n\rho + k - 2)!} \sum_{j=0}^{n-k} p_{n-k,j}(x)$$
$$\times \int_0^1 \sum_{i_1=0}^{\rho-1} \cdots \sum_{i_k=0}^{\rho-1} p_{n\rho+k-2,j\rho+i_1+\cdots+i_k+k-1}(t)f(t)dt.$$

Let $c \geq 0$, $n, k \in \mathbb{N}$, $n - k \geq 1$, $\rho \in \mathbb{N}$ and $f \in W_n^{\rho}$. Here W_n^{ρ} denotes the space of functions $f \in L_{1,loc}[0, \infty)$ satisfying certain growth conditions, i.e., there exist constants $M > 0, 0 \leq q < n\rho + c$, such that $|f(t)| \leq Me^{qt}$ for $c = 0$, $|f(t)| \leq Mt^{\frac{q}{c}}$ for $c > 0$ a. e. on $[0, \infty)$. Then

$$B_{n,\rho}^{(k)}(f; x) = \frac{n^{c,\overline{k}}}{(n\rho)^{c,\underline{k-1}}} \sum_{j=0}^{\infty} p_{n+kc,j}(x)$$
$$\times \int_0^{\infty} \sum_{i_1=0}^{\rho-1} \cdots \sum_{i_k=0}^{\rho-1} p_{n\rho-c(k-2),j\rho+i_1+\cdots+i_k+k-1}(t)f(t)dt.$$

For $k = 1$ the proofs given in [1, 2] can be simplified by using the well-known formula

$$p'_{m,l}(x) = m\left[p_{m+c,l-1}(x) - p_{m+c,l}(x)\right] \tag{7}$$

with $l \in \mathbb{N}_0$ and setting $p_{m+c,l-1}(x) \equiv 0$ for $l = 0$.

To be more precise, we have for $\rho \in \mathbb{N}$, $c \geq 0$,

$$B_{n,\rho}^{(1)}(f; x) = \sum_{j=1}^{\infty} p'_{n,j}(x)(n\rho + c) \int_0^{\infty} p_{n\rho+2c,j\rho-1}(t)I_1(f; t)dt$$
$$= np_{n+c,0}(x)(n\rho + c) \int_0^{\infty} p_{n\rho+2c,j\rho-1}I_1(f; t)dt$$
$$+n \sum_{j=1}^{\infty} p_{n+c,j}(x)(n\rho + c) \int_0^{\infty} [p_{n\rho+2c,(j+1)\rho-1}(t) - p_{n\rho+2c,j\rho-1}(t)]I_1(f; t)dt$$
$$= -\sum_{j=1}^{\infty} p_{n+c,j}(x)n \int_0^{\infty} \sum_{i=0}^{\rho-1} p'_{n\rho+c,j\rho+i}(t)I_1(f; t)dt$$

as, with $p_{m,l} = 0$ if $l < 0$, for each $j \in \mathbb{N}_0$

$$\begin{aligned}
&(n\rho + c)[p_{n\rho+2c,(j+1)\rho-1}(t) - p_{n\rho+2c,j\rho-1}(t)] \\
&= -\sum_{i=0}^{\rho-1}(n\rho + c)[p_{n\rho+2c,j\rho-1+i}(t) - p_{n\rho+2c,j\rho+i}(t)] \\
&= -\sum_{i=0}^{\rho-1} p'_{n\rho+c,j\rho+i}(t).
\end{aligned}$$

Note that the representation of $p_{n\rho+2c,(j+1)\rho-1}(t) - p_{n\rho+2c,j\rho-1}(t)$ as a telescoping sum is only possible for $\rho \in \mathbb{N}$.

Thus, as $I_1(f;0) = 0$ and $\lim_{t\to\infty} p_{n\rho+c,j\rho+i}(t) = 0$, $j\rho + i \in \mathbb{N}_0$, integration by parts leads to

$$B^{(1)}_{n,\rho}(f;x) = \sum_{j=0}^{\infty} p_{n+c,j}(x) n \int_0^{\infty} \sum_{i=0}^{\rho-1} p_{n\rho+c,j\rho+i}(t) f(t) dt.$$

The case $\rho \in \mathbb{N}$, $c < 0$, can be treated analogously.

References

1. M. Heilmann, I. Raşa, A nice representation for a link between Bernstein-Durrmeyer and Kantorovich operators, in *Proceedings of ICMC 2017, Haldia, India, January 2017*, ed. by D. Giri et.al. Springer Proceedings in Mathematics and Computing (Springer, 2017), pp. 312–320
2. M. Heilmann, I. Raşa, A nice representation for a link between Baskakov- and Szász-Mirakjan-Durrmeyer operators and their Kantorovich variants. Results Math. **74**, 9 (2019). https://doi.org/10.1007/s00025-018-0932-4
3. R. Păltănea, A class of Durrmeyer type operators preserving linear functions, Ann. Tiberiu Popoviciu Sem. Funct. Eq. Approx. Conv. (Cluj-Napoca) **5**, 109–117 (2007)
4. E. Stanila, On Bernstein-Euler-Jacobi operators. PhD thesis, Universität Duisburg-Essen, 2014

Degree of Approximation by Generalized Boolean Sum of λ-Bernstein Operators

Ruchi Chauhan and P. N. Agrawal

Abstract The purpose of the present paper is to investigate the degree of approximation of the λ-Bernstein operators introduced by Cai et al. (J Inequal Appl 61:1–11, 2018 [9]) by means of the Steklov mean, the Ditizian–Totik modulus of smoothness and the approximation of functions with derivatives of bounded variation. We introduce the bivariate case of the above operators and investigate the rate of convergence with the aid of the total and partial modulus of continuity and the Peetre's K-functional. Furthermore, we define the associated GBS (Generalized Boolean Sum) operator of the bivariate operators and establish the degree of approximation in terms of the mixed modulus of smoothness for Bögel continuous and Bögel differentiable functions.

Keywords λ-Bernstein operators · Partial moduli of continuity · Total modulus of continuity · Bögel continuous · Bögel differentiable · GBS operators · Modulus of smoothness

Mathematics Subject Classification (2010) 26A15 · 41A25 · 41A36

1 Introduction

In, 1912, for a bounded real valued function on I, $I = [0, 1]$, Bernstein [1] introduced a sequence of polynomials as

$$B_n(g; y) = \sum_{k=0}^{n} b_{n,k}(y) g\left(\frac{k}{n}\right), \quad \lambda \in [-1, 1] \text{ and } y \geq 0,$$

R. Chauhan (✉) · P. N. Agrawal
Department of Mathematics, Indian Institute of Technology Roorkee, Roorkee 247667, India
e-mail: ruchichauhan753@gmail.com

P. N. Agrawal
e-mail: pnappfma@gmail.com

N. Deo et al. (eds.), *Mathematical Analysis I: Approximation Theory*, Springer Proceedings in Mathematics & Statistics 306,
https://doi.org/10.1007/978-981-15-1153-0_9

where $b_{n,k}(y) = \binom{n}{k} y^k (1-y)^{n-k}$, $y \in I$.

Recently, Cai et al. [9] defined a sequence of λ-Bernstein operators as

$$B_{n,\lambda}(g; y) = \sum_{k=0}^{n} \tilde{b}_{n,k}(\lambda; y) g\left(\frac{k}{n}\right), \tag{1.1}$$

where Bezier basis $\tilde{b}_{n,k}(\lambda; y)$, [13] for $\lambda \in [-1, 1]$ is defined as

$$\tilde{b}_{n,0}(\lambda; y) = b_{n,0}(y) - \frac{\lambda b_{n+1,1}(y)}{n+1},$$
$$\tilde{b}_{n,i}(\lambda, y) = b_{n,i}(y) + \lambda\left(\frac{(n-2i+1)b_{n+1,i}(y)}{n^2-1} - \frac{(n-2i-1)b_{n+1,i+1}(y)}{n^2-1}\right), \quad (1 \leq i \leq n-1)$$
$$\tilde{b}_{n,n}(\lambda; y) = b_{n,n}(y) - \frac{\lambda b_{n+1,n}(y)}{n+1}.$$

In the same paper, the authors considered a Kantorovich variant of these operators and studied some of their approximation properties. Further, Cai et al. [10] defined the Bezier variant of the Kantorovich type operators and studied the rate of approximation with the aid of the Ditzian–Totik modulus of smoothness and also determined the degree of approximation for functions of bounded variation. In this paper, we investigate the rate of convergence in terms of the modulus of continuity and for functions having derivatives of bounded variation. Also, we define the bivariate generalization of the operators (1.1) and study the convergence properties. Lastly, the associated GBS operator is introduced and the rate of approximation of these operators is discussed by means of the mixed modulus of smoothness.

2 Preliminaries

Lemma 1 ([9]) *For λ-Bernstein operators given by (1.1), the following equalities hold:*

(i) $B_{n,\lambda}(1; y) = 1$;

(ii) $B_{n,\lambda}(z; y) = y + \frac{1-2y+y^{n+1}-(1-y)^{n+1}}{n(n-1)}\lambda$;

(iii) $B_{n;\lambda}(z^2; y) = y^2 + \frac{y(1-y)}{n} + \lambda\left[\frac{2y-4y^2+2y^{n+1}}{n(n-1)} + \frac{y^{n+1}+(1-y)^{n+1}-1}{n^2(n-1)}\right]$;

(iv) $B_{n;\lambda}(z^3; y) = y^3 + \frac{3y^2(1-y)}{n} + \frac{2y^3-3y^2+y}{n^2} + \lambda\left[\frac{-6y^3+6y^{n+1}}{n^2} + \frac{3y^2-3y^{n+1}}{n(n-1)}\right.$
$\left. + \frac{-9y^2+9y^{n+1}}{n^2(n-1)} + \frac{-4y+4y^{n+1}}{n^3(n-1)} + \frac{(1-y^{n+1}-(1-y)^{n+1}(n+3)}{n^3(n^2-1)}\right]$;

(v) $B_{n;\lambda}(z^4; y) = y^4 + \frac{6(y^3-y^4)}{n} + \frac{7y^2-18y^3+11y^4}{n^2} + \frac{y-7y^2+12y^3-6y^4}{n^3}$
$+\lambda\Big[\frac{6y^2-2y^3-8y^4+4y^{n+1}}{n^2} + \frac{-y^2-32y^3+16y^4+17y^{n+1}}{n^3} + \frac{y-y^{n+1}}{n^4}$
$+\frac{7y^2-7y^{n+1}}{n^2(n-1)} + \frac{y-23y^2+22y^{n+1}}{n^3(n-1)} + \frac{(1-y)^{n+1}+y-1}{n^4(n-1)}\Big].$

Lemma 2 ([9]) *For $y \in I$ and $\lambda \in [-1, 1]$, there hold the following:*

(i) $\lim_{n\to\infty} nB_{n,\lambda}(z - y; y) = 0$;
(ii) $\lim_{n\to\infty} nB_{n,\lambda}((z - y)^2; y) = y(1 - y)$;
(iii) $\lim_{n\to\infty} n^2 B_{n,\lambda}((z - y)^4; y) = 3y^2 - 6y^3 + 3y^4 + 6(y^2 - y^3)\lambda$.

Proof From Lemma 1, the proof of this lemma can be given by simple computation, hence, we skip the details.

Consequently, we have

$$B_{n,\lambda}((z - y)^2; y) \le \frac{C}{n}\left[y(1 - y) + \frac{1}{n}\right] = \eta^2(y). \tag{2.1}$$

3 Main Results

In what follows, let $B_{n,\lambda}(g; y) - g(y) = \mathfrak{K}(g)$, $\Psi_n = \sup_{y\in[0,1]} B_{n,\lambda}((z - y); y)$ and $\gamma_n^2 = \sup_{y\in[0,1]} B_{n,\lambda}((z - y)^2; y)$. In the following theorem, we estimate the rate of convergence of the operators defined by (1.1) in terms of the first-order modulus of continuity.

For any $\delta > 0$, the first-order modulus of continuity of $f \in C(I)$, is given by

$$\omega(f; \delta) = \sup_{0<|h|<\delta} \sup_{x,x+h\in[0,1]} |f(x + h) - f(x)|.$$

Theorem 1 *For $g \in C[0, 1]$, there holds the following inequality:*

$$||\mathfrak{K}(g)|| \le 4\omega(g; \gamma_n).$$

Proof Considering the properties of the first-order modulus of continuity, and using Cauchy–Schwarz inequality, Lemma 1,

$$||\mathfrak{K}(g)|| \le 2\omega(g; \gamma)\left(\frac{\gamma_n}{\gamma} + 1\right).$$

Now, choosing $\gamma = \gamma_n$, we obtain the desired result. □

Next, let $C^r(I)$, $r = 1, 2, \ldots$ denote the space of r-times continuously differentiable functions on I. In the next result we determine the degree of approximation of the operators $B_{n,\lambda}$ for the continuously differentiable functions in I.

Theorem 2 *Let $g \in C^1(I)$, then we have*

$$||\mathfrak{K}(g)|| \leq |\Psi_n| \parallel g' \parallel + 2\gamma_n \omega(g'; \gamma_n).$$

Proof The Taylor's formula for $g \in C^1[0, 1]$ yields

$$g(z) - g(y) = g'(y)(z - y) + \int_y^z (g'(u) - g'(y))du.$$

Now applying the operator $B_{n,\lambda}$ on both sides of the above equality and the fact that

$$|g'(u) - g'(y)| \leq \omega(g'; \gamma)\left(1 + \frac{|z - y|}{\gamma}\right),$$

we are led to

$$|\mathfrak{K}(g)| \leq |g'(y)||\Psi_n| + \omega(g'; \gamma)\left\{\frac{1}{\gamma} B_{n,\lambda}((z - y)^2; y) + B_{n,\lambda}(|z - y|; y)\right\}.$$

Now, using Cauchy–Schwarz inequality and choosing $\gamma = \gamma_n$, we reach the required result. □

In the following theorem, we show that by employing a smoothing process, e.g., Steklov means we achieve a better estimate of the error in the approximation of a function in $C(I)$ by the operators (1.1) in terms of the second-order modulus of continuity than the estimate is given by Theorem 1.

Theorem 3 *Let $g \in C(I)$, there holds the following inequality:*

$$||\mathfrak{K}(g)|| \leq 5\omega(g; \gamma_n) + \frac{13}{2}\omega_2(g; \gamma_n).$$

Proof The Steklov mean for $g \in C(I)$ is defined as

$$g_h(y) = \frac{4}{h^2} \int_0^{\frac{h}{2}} \int_0^{\frac{h}{2}} (2g(y + u + v) - g(y + 2u + 2v))dudv, h > 0,$$

which implies that

$$g(y) - g_h(y) = \frac{4}{h^2} \int_0^{\frac{h}{2}} \int_0^{\frac{h}{2}} \Delta^2_{u+v} g(y)dudv, \ \ h > 0,$$

hence,

$$||g_h - g|| \leq \omega_2(g; h). \tag{3.1}$$

Also, it follows that

$$||g_h^{'}|| \leq \frac{5}{h}\omega(g; h); \ \ ||g_h^{''}|| \leq \frac{9}{h^2}\omega_2(g; h). \tag{3.2}$$

Now, we may write

$$|\mathfrak{K}(g)| \leq |B_{n,\lambda}((g - g_h)(z); y)| + |g_h(y) - g(y)| + |B_{n,\lambda}(g_h(z) - g_h(y); y)|. \tag{3.3}$$

By using relation (3.1), we are led to

$$B_{n,\lambda}(|g - g_h|; y) \leq ||g - g_h|| \leq \omega_2(g; h). \tag{3.4}$$

By the Taylor's expansion and the Cauchy–Schwarz inequality

$$\begin{aligned} |B_{n,\lambda}(g_h(z) - g_h(y); y)| &\leq |B_{n,\lambda}(g_h'(y)(z - y); y)| + \left| B_{n,\lambda}\left(\int_y^z (z - u)g_h''(u)du; y \right) \right| \\ &= ||g_h'||\sqrt{B_{n;\lambda}((z - y)^2; y)} + \frac{1}{2}||g_h''||B_{n,\lambda}((z - y)^2; y). \end{aligned} \tag{3.5}$$

Now, assuming $h = \gamma_n$, collecting (3.3)–(3.5) and using the inequalities (3.1), (3.2), we get the desired result. □

Our last result of this section is to obtain the rate of approximation of the operators $B_{n,\lambda}$ for functions with derivatives of bounded variation.

The operators $B_{n,\lambda}(g; y)$ can be expressed in an integral form as follows:

$$B_{n,\lambda}(g; y) = \int_0^1 \frac{\partial}{\partial z} K_{n,\lambda}(y, z)g(z)dz, \tag{3.6}$$

where the kernel $K_{n,\lambda}$ is given by $K_{n,\lambda}(y, u) = \begin{cases} \sum\limits_{k \leq nu} \tilde{b}_{n,k}(\lambda; u), & 0 < u \leq 1 \\ 0, & u = 0 \end{cases}$.

Lemma 3 *For a fixed $y \in (0, 1)$ and sufficiently large n, we have*

(i) $\xi_{n,\lambda}(y, t) = \int_0^t \frac{\partial}{\partial u} K_{n,\lambda}(y, u)du \leq \frac{\eta^2(y)}{(t-y)^2}, \ 0 \leq t < y,$
(ii) $1 - \xi_{n,\lambda}(y, s) = \int_s^1 \frac{\partial}{\partial u} K_{n,\lambda}(y, u)du \leq \frac{\eta^2(y)}{(y-s)^2}, \ y < s \leq 1,$

where $\eta^2(y)$ is defined in (2.1).

Proof Using Lemma 1 and (2.1), the proof of the lemma is straightforward. □

Let $DBV[0, 1]$ be the space of all functions which have a derivative of bounded variation on $[0, 1]$. Such a function $f \in DBV[0, 1]$ can be represented as

$$f(x)=\int_0^x g(t)dt+f(0),$$

where g denotes a function of bounded variation on every finite subinterval of $[0,\infty)$.

Theorem 4 *Let $g\in DBV([0,1])$ then for any $y\in(0,1)$, we have*

$$|B_{n,\lambda}(g;y)-g(y)|\le\frac{1}{2}|g'(y+)+g'(y-)|\frac{C_1}{n}+\frac{1}{2}\eta y|g'(y+)-g'(y-)|$$
$$+\frac{\eta^2(y)}{y}\sum_{k=1}^{\sqrt{n}}\bigvee_{y-\frac{y}{k}}^{y}(g'_y)+\frac{\eta^2(y)}{(1-y)}\sum_{k=1}^{\sqrt{n}}\bigvee_{y}^{y+\frac{(1-y)}{k}}g'_y+\frac{(1-y)}{\sqrt{n}}\bigvee_{y}^{y+\frac{(1-y)}{\sqrt{n}}}g'_y,$$

where $\eta^2(y)$ is defined in (2.1).

Proof Using Lemma 3 and proceeding in a manner similar to the proof of (Theorem 3.10, [8]), the theorem is established. Hence, the details are omitted. □

4 Bivariate Extension of the Operator

For $g\in C(I^2)$, $I^2=I\times I$ endowed with the norm $||g||=\sup_{(y_1,y_2)\in I^2}|g(y_1,y_2)|$, the bivariate case of the operators defined in (1.1) is given by

$$B_{m,n,\lambda_1,\lambda_2}(g;y_1,y_2)=\sum_{j=0}^{m}\sum_{k=0}^{n}\tilde{b}_{m,n,j,k}(\lambda_1,\lambda_2,y_1,y_2)g\left(\frac{j}{m},\frac{k}{n}\right), \tag{4.1}$$

where $\tilde{b}_{m,n,j,k}(\lambda_1,\lambda_2,y_1,y_2)=\tilde{b}_{m,j}(\lambda_1;y_1)\tilde{b}_{n,k}(\lambda_2;y_2)$, $\lambda_1,\lambda_2\in[-1,1]$.

Lemma 4 *For the operator (4.1), the following equalities hold:*

(i) $B_{m,n,\lambda_1,\lambda_2}(1;y_1,y_2)=1$;
(ii) $B_{m,n,\lambda_1,\lambda_2}(z;y_1,y_2)=y_1+\frac{1-2y_1+y_1^{m+1}-(1-y_1)^{m+1}}{m(m-1)}\lambda_1$;
(iii) $B_{m,n,\lambda_1,\lambda_2}(w;y_1,y_2)=y_2+\frac{1-2y_2+y_2^{n+1}-(1-y_2)^{n+1}}{n(n-1)}\lambda_2$;
(iv) $B_{m,n,\lambda_1,\lambda_2}(z^2;y_1,y_2)=y_1^2+\frac{y_1(1-y_1)}{m}+\lambda_1\left[\frac{2y_1-4y_1^2+2y_1^{m+1}}{m(m-1)}+\frac{y_1^{m+1}+(1-y_1)^{m+1}-1}{m^2(m-1)}\right]$;
(v) $B_{m,n,\lambda_1,\lambda_2}(w^2;y_1,y_2)=y_2^2+\frac{y_2(1-y_2)}{n}+\lambda_2\left[\frac{2y_2-4y_2^2+2y_2^{n+1}}{n(n-1)}+\frac{y_2^{n+1}+(1-y_2)^{n+1}-1}{n^2(n-1)}\right]$.

Proof By using Lemma 1, the proof of the lemma is straightforward. Hence, we omit the details. □

Lemma 5 ([12]) *Let J_1 and J_2 be compact intervals of the real line and $e_{ij} = y_1^i y_2^j$. Let $L_{m,n} : C(J_1 \times J_2) \to C(J_1 \times J_2)$ be linear positive operators. If $\lim_{m,n\to\infty} L_{m,n}(e_{ij}; y_1, y_2) = y_1^i y_2^j$, $(i, j) \in \{(0, 0), (1, 0), (0, 1)\}$ and*

$$\lim_{m,n\to\infty} L_{m,n}(e_{20} + e_{02}; y_1, y_2) = y_1^2 + y_2^2,$$

uniformly in $J_1 \times J_2$ then the sequence $L_{m,n}(g)$ converges to g uniformly on $J_1 \times J_2$ for any $g \in C(J_1 \times J_2)$.

For $g \in C(I^2))$, $I^2 = I \times I$ the total modulus of continuity for the bivariate case is defined as

$$\bar{\omega}(g; \gamma_1, \gamma_2) = \sup\Big\{|g(z, w) - g(y_1, y_2) : |z - y_1| < \gamma_1, |w - y_2| < \gamma_2\},$$

where $\gamma_1, \gamma_2 > 0$. The properties of $\bar{\omega}(g; \gamma_1, \gamma_2)$ are given below:

(a) $\bar{\omega}(g; \gamma_1, \gamma_2) \to 0$ if $\gamma_1 \to 0$ and $\gamma_2 \to 0$,

(b) $|g(z, w) - g(y_1, y_2)| \leq \bar{\omega}(g; \gamma_1, \gamma_2)\Big(1 + \frac{|z-y_1|}{\gamma_1}\Big)\Big(1 + \frac{|w-y_2|}{\gamma_2}\Big)$.

In what follows, we assume $B_{n,m,\lambda_1,\lambda_2}(g(z, w); y_1, y_2) - g(y_1, y_2) = \mathfrak{J}(f)$.

Further, let $\eta_m = \sqrt{\sup_{y_1\in I} B_{m,\lambda_1}((z-y_1)^2; y_1)}$, $\zeta_n = \sqrt{\sup_{y_2\in I} B_{n,\lambda_2}((w-y_2)^2; y_2)}$, $\Psi_m = \sup_{y_1\in I} |B_{m,\lambda_1}((z - y_1); y_1)|$ and $\phi_n = \sup_{y_2\in I} |B_{n,\lambda_2}((w - y_2); y_2)|$.

Theorem 5 *Let $g \in C(I^2)$, then*

$$||\mathfrak{J}(g)|| \leq 4\, \bar{\omega}(g; \eta_m, \zeta_n).$$

Proof By using the linearity and positivity of the operator (4.1) and by the property (b) of the total modulus of continuity

$$|\mathfrak{J}(g)| \leq \bar{\omega}(g; \gamma_1, \gamma_2)\Bigg(1 + \frac{1}{\gamma_1}(B_{m,n,\lambda_1,\lambda_2}(|z - y_1|; y_1) + \frac{1}{\gamma_2} B_{m,n,\lambda_1,\lambda_2}(|w - y_2|; y_2) + \frac{1}{\gamma_1\gamma_2} B_{m,n,\lambda_1,\lambda_2}(|z - y_1||w - y_2|; y_1, y_2)\Bigg)$$

Now, applying Cauchy–Schwarz inequality and choosing $\gamma_1 = \eta_m$ and $\gamma_2 = \zeta_n$, we are led to desired result. □

5 Degree of Approximation

Now, we give an upper bound of the degree of approximation by the operator (1.1) for the Lipschitz class functions of two variables.

For $0 < \beta_1 \leq 1$ and $0 < \beta_2 \leq 1$, we define the $Lip_M(\beta_1, \beta_2)$ for the operator (1.1) as follows:

$$|f(z, w) - f(y_1, y_2)| \leq M|z - y_1|^{\beta_1}|w - y_2|^{\beta_2},$$

where $(t, s), (x, y) \in I^2$ are arbitrary.

Theorem 6 *Let $g \in Lip_M(\beta_1, \beta_2)$, then we have*

$$||\mathfrak{I}(g)|| \leq M(\eta_m)^{\beta_1}(\zeta_n)^{\beta_2}.$$

Proof By our hypothesis, and the Hölder's inequality with $p_1 = \frac{2}{\beta_1}, q_1 = \frac{2}{2-\beta_1}, p_2 = \frac{2}{\beta_2}$ and $q_2 = \frac{2}{2-\beta_2}$, the desired result is easily obtained. Hence we omit the details. □

Let $C^1(I^2)$ denote the space of all functions in $C(I^2)$ whose first-order partial derivative is continuous in I^2.

Theorem 7 *For $g \in C^1(I^2)$, the following inequality holds:*

$$||\mathfrak{I}(g)|| \leq ||g'_{y_1}||\eta_m + ||g'_{y_2}||\zeta_n.$$

Proof In view of equality

$$g(z, w) - g(y_1, y_2) = \int_{y_1}^{z} g'_u(u, w)du + \int_{y_2}^{w} f'_v(y_1, v)dv,$$

and taking into account the inequalities,

$$|\int_{y_1}^{z} g'_u(u, w)du| \leq ||g'_{y_1}|||z - y_1| \text{ and } |\int_{y_2}^{w} g'_v(y_1, v)dv| \leq ||g'_{y_2}|||w - y_2|,$$

on an application of Cauchy–Schwarz inequality, we obtain the required result. □

For $g \in C(I^2)$ and $\gamma > 0$, the partial moduli of continuity for bivariate case are defined as

$$\omega_1(g; \gamma) = \sup\left\{g(y_{1'}, y_2) - g(y_{1''}, y_2)| : y_2 \in I \text{ and } |y_{1'} - y_{1''}| \leq \gamma\right\}$$

$$\omega_2(g; \gamma) = \sup\left\{g(y_1, y_{2'}) - g(y_1, y_{2''}) : y_1 \in I \text{ and } |y_{2'} - y_{2''}| \leq \gamma\right\}.$$

Let $C^2(I^2)$ denote the space of all functions in $C(I^2)$ whose second-order partial derivative is continuous in I^2. The norm on the space $C^2(I^2)$ is defined as

$$\|f\|_{C^2(I^2)} = \|f\| + \sum_{i=1}^{2}\left(\left\|\frac{\partial^i f}{\partial x^i}\right\| + \left\|\frac{\partial^i f}{\partial y^i}\right\|\right).$$

The Peetre's K-functional of the function $g \in C(I^2)$ is defined as

$$\mathcal{K}(g;\delta) = \inf_{f\in C^2(I^2)} \{||g-f|| + \delta||f||_{C^2(I^2)}\}, \delta > 0.$$

Also by [7], it follows that

$$\mathcal{K}(g;\delta) \le M\left\{\tilde{\omega_2}(g;\sqrt{\delta}) + min(1,\delta)||g||\right\}, \tag{5.1}$$

holds for all $\delta > 0$.

Theorem 8 *Let $g \in C(I^2)$, then*

$$||\Im(g)|| \le 2\{\omega_1(g;\eta_m) + \omega_2(g;\zeta_n)\}.$$

Proof By the definition of partial moduli of continuity and the Cauchy–Schwarz inequality, we get

$$\begin{aligned}||\Im(g)|| &\le \omega_1(g;\gamma_1)\left(1 + \frac{1}{\gamma_1}\sqrt{B_{m,n,\lambda_1,\lambda_2}((z-y_1)^2; y_1, y_2)}\right)\\ &+ \omega_2(g;\gamma_2)\left(1 + \frac{1}{\gamma_2}\sqrt{B_{m,n,\lambda_1,\lambda_2}((w-y_2)^2; y_1, y_2)}\right)\\ &= 2\{\omega_1(g;\gamma_1) + \omega_2(g;\gamma_2)\}.\end{aligned}$$

Now, choosing $\gamma_1 = \eta_m$ and $\gamma_2 = \eta_n$, the proof easily follows. □

Theorem 9 *For the function $g \in C(I^2)$, the following inequality holds:*

$$\begin{aligned}||\Im(g)|| \le M\Bigg\{&\tilde{\omega_2}(g;\Theta_{m,n}) + \\ &min\{1,\Theta_{m,n}\}||g||\Bigg\}\\ &+\omega\left(g;\sqrt{\Psi_m^2 + \phi_n^2}\right),\end{aligned}$$

where $\Theta_{m,n} = \eta_m^2 + \zeta_n^2 + \Psi_m^2 + \phi_n^2$ and the constant $M(>0)$, is independent of g.

Proof First, we define the auxiliary operators as

$$B^*_{m,n,\lambda_1,\lambda_2}(g;x,y) = B_{m,n,\lambda_1,\lambda_2}(g;x,y) - g(B_{m,\lambda_1}(z;y_1), B_{n,\lambda_2}(w;y_2)) + g(y_1,y_2). \quad (5.2)$$

From Lemma 4, $B_{m,n,\lambda_1,\lambda_2}(1;y_1,y_2) = 1$,
$B^*_{m,n,\lambda_1,\lambda_2}(z-y_1);y_1,y_2) = 0$ and $B^*_{m,n,\lambda_1,\lambda_2}((w-y_2);y_1,y_2) = 0$.
For any $f \in C^2(I^2)$, by Taylor's theorem

$$\begin{aligned} f(z,w) - f(y_1,y_2) &= f(z,y_2) - f(y_1,y_2) + f(z,w) - f(z,y_2) \\ &= \frac{\partial f(y_1,y_2)}{\partial y_1}(z-y_1) + \int_{y_1}^{z}(z-u)\frac{\partial^2 f(u,y_2)}{\partial u^2}du \\ &\quad + \frac{\partial f(y_1,y_2)}{\partial y_2}(w-y_2) + \int_{y_2}^{w}(w-v)\frac{\partial^2 f(y_1,v)}{\partial v^2}dv. \end{aligned}$$

Now applying the auxiliary operator on the above equation and using (5.2)

$$\begin{aligned} B^*_{m,n,\lambda_1,\lambda_2}(f;y_1,y_2) - f(y_1,y_2) &= B_{m,n,\lambda_1,\lambda_2}\left(\int_{y_1}^{z}(z-u)\frac{\partial^2 f(u,y_2)}{\partial u^2}du, y_2\right) \\ &\quad - \int_{y_1}^{B_{m,\lambda_1}(z;y_1)}(B_{m,\lambda_1}(z;y_1)-u)\frac{\partial^2 f(u,y_2)}{\partial u^2}du \\ &\quad + B_{n,\lambda_2}\left(\int_{y_2}^{w}(w-v)\frac{\partial^2 f(y_1,v)}{\partial v^2}dv; y_1, y_2\right) \\ &\quad - \int_{y_2}^{B_{n,\lambda_2}(w;y_2)}(B_{n,\lambda_2}(w;y_2)-v)\frac{\partial^2 f(y_1,v)}{\partial v^2}dv. \end{aligned}$$

Hence,
$|B^*_{m,n,\lambda_1,\lambda_2}(f;y_1,y_2) - f(y_1,y_2)|$

$$\begin{aligned} &\leq \left\{ B_{m,n,\lambda_1,\lambda_2}((z-y_1)^2;y_1,y_2) + \left(B_{m,\lambda_1}(z;y_1) - y_1\right)^2 \right. \\ &\quad \left. + B_{m,n,\lambda_1,\lambda_2}((w-y_2)^2;y_1,y_2) + \left(B_{n,\lambda_2}(w;y_2) - y_2\right)^2 \right\} ||f||_{C^2(I^2)} \\ &= (\eta_m^2 + \zeta_n^2 + \Psi_m^2 + \phi_n^2)||f||_{C^2(I^2)} \\ &= \Theta_{m,n}||f||_{C^2(I^2)}. \end{aligned} \quad (5.3)$$

Now, using (5.2)
$|B^*_{m,n,\lambda_1,\lambda_2}(g;y_1,y_2)|$

$$\begin{aligned} &\leq |B_{m,n,\lambda_1,\lambda_2}(g;y_1,y_2)| + |g(B_{m,\lambda_1}(z;y_1), B_{n,\lambda_2})(w;y_2))| + |g(y_1,y_2)| \\ &\leq 3||g||. \end{aligned} \quad (5.4)$$

Hence, taking into account (5.2)–(5.4), for $g \in C(I^2)$ and any $f \in C^2(I^2)$
$|B_{m,n,\lambda_1,\lambda_2}(g; y_1, y_2) - g(y_1, y_2)|$

$$\leq 4||g - f|| + |B^*_{m,n,\lambda_1,\lambda_2}(f; y_1, y_2) - f(y_1, y_2)|$$
$$+ \left| g(B_{m,n,\lambda_1,\lambda_2}(z; y_1, y_2), B_{m,n,\lambda_1,\lambda_2}(w; y_1, y_2)) - g(y_1, y_2) \right|$$
$$\leq \left(4||g - f|| + (\eta_m^2 + \zeta_n^2 + \Psi_m^2 + \phi_n^2)||f||_{C^2(I^2)} \right)$$
$$+ \omega\left(g; \sqrt{\Psi_m^2 + \phi_n^2} \right).$$

Now, taking the infimum on the right side of the above inequality over all $f \in C^2(I^2)$, and using the equivalence (5.1) between K-functional and $\tilde{\omega_2}$, we obtain the desired result. □

6 GBS of the Operator $B_{m,n,\lambda_1,\lambda_2}$

Bögel ([5, 6]) pioneered the study of B-continuous and B-differentiable functions. Dobrescu and Matei [11] showed that the GBS operator associated to the bivariate Bernstein polynomial converges uniformly to the B-continuous function. Badea and Cottin [2] established Korovkin-type theorem for GBS operators. Subsequently, Badea et al. [3] proved the very famous "Test function theorem" to approximate these kind of functions. A quantitative variant of the Korovkin-type theorem for these functions was established by Badea and Badea in [4].

A real valued function f defined on I^2 is called B-continuous at (y_1, y_2) if

$$\lim_{(z,w)\to(y_1,y_2)} \Delta_{(z,w)} g(y_1, y_2) = 0,$$

where $\Delta_{(z,w)} g(y_1, y_2) = g(z, w) - g(z, y_1) - g(y_1, w) + g(y_1, y_2)$.

Let $C_b(I^2) := \{g : g\, is\, \mathrm{B}{-}continuous\ on\ I^2\}$ and $B_b(I^2)$ be the set of all B-bounded functions on I^2, equipped with the norm $||g||_B = \sup_{(z,w)(y_1,y_2)\in I^2} |\Delta_{(z,w)} g(y_1, y_2)|$. Let $B(I^2)$ denote the space of all bounded functions on I^2 endowed with the norm $||f||_\infty = \sup_{(y_1,y_2)\in I^2} |f(y_1, y_2)|$.

A function g is said to be B-differentiable at (y_1, y_2) if

$$\lim_{(z,w)\to(y_1,y_2)} \frac{\Delta_{(z,w)} g(y_1, y_2)}{(z - y_1)(w - y_2)} = D_B g(y_1, y_2) < \infty.$$

Here, $D_B g$ is called B-derivative of g and the space of all B-differentiable functions is denoted by $D_b(I^2)$. For any $g \in C_b(I^2)$, the GBS operator associated with $B_{m,n,\lambda_1,\lambda_2}$ is given by

$$C_{m,n,\lambda_1,\lambda_2}(g; y_1, y_2) = B_{m,n,\lambda_1,\lambda_2}(g(y, w) + g(z, y_2) - g(z, w); y_1, y_2)$$
$$= \sum_{j=0}^{m}\sum_{k=0}^{n} \tilde{b}_{m,n,j,k}(\lambda_1, \lambda_2; y_1, y_2)\left(g\left(y_1, \frac{k}{n}\right) + g\left(\frac{j}{m}, y_2\right) - g\left(\frac{j}{m}, \frac{k}{n}\right)\right). \quad (6.1)$$

Hence, the operator (6.1) is a linear operator and is well defined from the space $C_b(I^2)$ into $C(I^2)$. The mixed modulus of smoothness of $g \in C_b(I^2)$ is defined as

$$\omega_B(g; \delta_1, \delta_2) := \sup\left\{\left|\Delta_{(z,w)}g(y_1, y_2)\right| : |z - y_1| < \delta_1, |w - y_2| < \delta_2\right\},$$

for all $(y_1, y_2), (z, w) \in I^2$ and for any $(\delta_1, \delta_2) \in (0, \infty) \times (0, \infty)$ with $\omega_B : [0, \infty) \times [0, \infty) \to \mathbb{R}$.

Theorem 10 *Let $\{D_{m,n}\}$, $D_{m,n} : C_b(J^2) \to B(J^2)$, $m, n \in N$ be a sequence of bivariate linear positive operators, $H_{m,n}$ be the GBS-operators associated with $D_{m,n}$ and the following conditions are satisfied:*

(1) $H_{m,n}(1, y_1, y_2) = 1;$
(2) $H_{m,n}(z, y_1, y_2) = y_1 + u_{m,n}(y_1, y_2);$
(3) $H_{m,n}(w, y_1, y_2) = y_1 + v_{m,n}(y_1, y_2);$
(4) $H_{m,n}(z^2 + w^2, y_1, y_2) = y_1^2 + y_2^2 + w_{m,n}(y_1, y_2);$

for all $y_1, y_2 \in J^2$. If all the sequences $u_{m,n}(y_1, y_2)$, $v_{m,n,\lambda_1,\lambda_2}(y_1, y_2)$ and $w_{m,n}(y_1, y_2)$ converge to zero uniformly in J^2, then the sequence $\{H_{m,n}g\}$ converges to g uniformly on J^2 for all $g \in C_b(J^2)$.

As a consequence of the above theorem and applying Lemma 5, we have

Theorem 11 *For $g \in C_b(I^2)$, the operator $C_{m,n,\lambda_1,\lambda_2}(g; y_1, y_2)$ converges to g uniformly in I^2.*

Now, we determine the degree of approximation in terms of the mixed modulus of smoothness for the GBS operators $C_{m,n,\lambda_1,\lambda_2}$.

Theorem 12 *For $g \in C_b(I^2)$, the following inequality holds,*

$$||C_{m,n,\lambda_1,\lambda_2}(g) - g|| \leq 4\omega_B(g; \eta_m, \zeta_n).$$

Proof By using the definition of $\omega_B(g, \gamma_1, \gamma_2)$ and the inequality

$$\omega_B(g; \mu_1\gamma_1, \mu_2\gamma_2) \leq (1 + \mu_1)(1 + \mu_2)\ \omega_B(g, \gamma_1, \gamma_2); \ \ \mu_1, \mu_2 > 0,$$

we are led to

$$|\Delta_{(z,w)}g(y_1, y_2)| \leq \omega_B(g; |z - y_1|, |w - y_2|)$$
$$\leq \left(1 + \frac{|z - y_1|}{\gamma_1}\right)\left(1 + \frac{|w - y_2|}{\gamma_2}\right)\omega_B(g; \gamma_1, \gamma_2), \quad (6.2)$$

From the definition of $\Delta_{(z,w)}g(y_1, y_2)$, we get

$$g(y_1, w) + g(z, y_2) - g(z, w) = g(y_1, y_2) - \Delta_{(z,w)}g(y_1, y_2).$$

On applying the linear positive operator (4.1) to this equality and using the definition of operator (6.1), we can write

$$C_{m,n,\lambda_1,\lambda_2}(g; y_1, y_2) = g(y_1, y_2)\, B_{m,n,\lambda_1,\lambda_2}(e_{00}; y_1, y_2) - B_{m,n,\lambda_1,\lambda_2}\left(\Delta_{(z,w)}g(y_1, y_2); y_1, y_2\right).$$

Since $B_{m,n,\lambda_1,\lambda_2}(1; y_1, y_2) = 1$, considering the inequality (6.2), we obtain,
$|C_{m,n,\lambda_1,\lambda_2}(g; y_1, y_2) - g(y_1, y_2)|$

$$\begin{aligned}
&\leq \bigg(B_{m,n,\lambda_1,\lambda_2}(e_{00}; y_1, y_2) + \frac{1}{\gamma_1} B_{m,n,\lambda_1,\lambda_2}(|z - y_1|; y_1, y_2) \\
&+ \frac{1}{\gamma_2} B_{m,n,\lambda_1,\lambda_2}(|w - y_2|; y_1, y_2) + \frac{1}{\gamma_1\gamma_2} B_{m,n,\lambda_1,\lambda_2}(|z - y_1|; y_1, y_2) \\
&\times B_{m,n,\lambda_1,\lambda_2}(|w - y_2|; y_1, y_2) \bigg) \omega_B(g; \gamma_1, \gamma_2).
\end{aligned}$$

Now, applying the Cauchy–Schwarz inequality and choosing $\gamma_1 = \eta_m$, $\gamma_2 = \zeta_n$, we reach the required result. □

Now, let us define the Lipschitz class for B-continuous functions. For $g \in C_b(I^2)$, the Lipschitz class $Lip_M(\xi, \eta)$ with $\xi, \eta \in (0, 1]$ is given by

$$Lip_M(\xi, \eta) \doteqdot \left\{ g \in C_b(I^2) : \left|\Delta_{(z,w)}g(y_1, y_2)\right| \leq M\,|z - y_1|^{\xi}\,|w - y_2|^{\eta},\ \text{for}\,(z, w), (y_1, y_2) \in I^2 \right\}.$$

Our next theorem gives the degree of approximation for the operators (6.1) by means of the Lipschitz class of Bögel continuous functions.

Theorem 13 *For $g \in Lip_M(\xi, \eta)$, we have*

$$||C_{m,n,\lambda_1,\lambda_2}(g) - g|| \leq M(\eta_m)^{\xi}(\zeta_n)^{\eta},$$

for $M > 0$, $\xi, \eta \in (0, 1]$.

Proof By the definition of the operator (6.1) and the linearity of the operator $B_{m,n,\lambda_!,\lambda_2}$, we can write

$$\begin{aligned}
C_{m,n,\lambda_1,\lambda_2}(g; y_1, y_2) &= B_{m,n,\lambda_1,\lambda_2}(g(y_1, w) + g(z, y_2) - g(z, w); y_1, y_2) \\
&= B_{m,n,\lambda_1,\lambda_2}(g(y_1, y_2) - \Delta_{(z,w)}g(y_1, y_2); y_1, y_2) \\
&= g(y_1, y_2) B_{m,n,\lambda_1,\lambda_2}(1; y_1, y_2) - B_{m,n,\lambda_1,\lambda_2}(\Delta_{(z,w)}g(y_1, y_2); y_1, y_2).
\end{aligned}$$

Hence, by our hypothesis

$$\begin{aligned}\left|C_{m,n,\lambda_1,\lambda_2}(g; y_1, y_2) - g(y_1, y_2)\right| &\le B_{m,n,\lambda_1,\lambda_2}\left(\left|\Delta_{(z,w)}g(y_1, y_2)\right|; y_1, y_2\right)\\ &\le M B_{m,n,\lambda_1,\lambda_2}\left(|z - y_1|^{\xi}\,|w - y_2|^{\eta}; y_1, y_2\right)\\ &= M B_{m,n,\lambda_1,\lambda_2}\left(|z - y_1|^{\xi}; y_1, y_2\right) B_{m,n,\lambda_1,\lambda_2}\left(|w - y_2|^{\eta}; y_1, y_2\right).\end{aligned}$$

Now, applying Hölder's inequality with $p_1 = 2/\xi, q_1 = 2/(2-\xi)$ and $p_2 = 2/\eta, q_2 = 2/(2-\eta)$, the desired result is obtained. □

In the following theorem, we investigate the rate of approximation for Bögel differentiable functions with a bounded Bögel derivative by the operators $C_{m,n,\lambda_1,\lambda_2}$.

Theorem 14 *Let the function $g \in D_b(I^2)$ with $D_B g \in B(I^2)$. Then,*

$$||C_{m,n,\lambda_1,\lambda_2}(g) - g|| \le \frac{M}{m^{\frac{1}{2}} n^{\frac{1}{2}}}\left(||D_B g||_\infty + \omega_B(D_B g; m^{\frac{-1}{2}}, n^{\frac{-1}{2}})\right),$$

for some constant $M > 0$.

Proof Since $g \in D_b(I^2)$, we have

$$\Delta_{(z,w)}g(y_1, y_2) = (z - y_1)(w - y_2)D_B g(\xi, \eta), \quad with\ y_1 < \xi < z;\ \ y_2 < \eta < w.$$

Also, we note that

$$D_B g(\xi, \eta) = \Delta D_B g(\xi, \eta) + D_B g(\xi, y_2) + D_B g(y_1, \eta) - D_B g(y_1, y_2).$$

Since $D_B g \in B(I^2)$, we can write

$$\begin{aligned}|B_{m,n,\lambda_1,\lambda_2}(\Delta_{(z,w)}g(y_1, y_2); y_1, y_2)| &= |B_{m,n,\lambda_1,\lambda_2}((z - y_1)(w - y_2)D_B g(\xi, \eta); y_1, y_2)|\\ &\le B_{m,n,\lambda_1,\lambda_2}(|z - y_1||w - y_2|\omega_B(D_B g; |\xi - y_1|, |\eta - y_2|); y_1, y_2)\\ &\quad +3\,||D_B g||_\infty\, B_{m,n,\lambda_1,\lambda_2}(|z - y_1||w - y_2|; y_1, y_2).\end{aligned}$$

Now, by using the inequality (6.2) and applying the Cauchy–Schwarz inequality we obtain

$$\begin{aligned}|C_{m,n,\lambda_1,\lambda_2}(g; y_1, y_2) - g(y_1, y_2)| &= |B_{m,n,\lambda_1,\lambda_2}\Delta_{(z,w)}(y_1, y_2); y_1, y_2|\\ &\le 3||D_B g||_\infty \sqrt{B_{m,n,\lambda_1,\lambda_2}((z - y_1)^2(w - y_2)^2; y_1, y_2)}\\ &\quad +\Bigg(\sqrt{B_{m,n,\lambda_1,\lambda_2}((z - y_1)^2(w - y_2)^2; y_1, y_2)}\\ &\quad +\gamma_1^{-1}\sqrt{B_{m,n,\lambda_1,\lambda_2}((z - y_1)^4(w - y_2)^2; y_1, y_2)}\\ &\quad +\gamma_2^{-1}\sqrt{B_{m,n,\lambda_1,\lambda_2}((z - y_1)^2(w - y_2)^4; y_1, y_2)}\\ &\quad +\gamma_1^{-1}\gamma_2^{-1} B_{m,n,\lambda_1,\lambda_2}((z - y_1)^2(w - y_2)^2; y_1, y_2)\Bigg)\omega_B(D_B g; \gamma_1, \gamma_2).\end{aligned} \tag{6.3}$$

Considering Lemma 2, for $(z, w), (y_1, y_2) \in I^2$ and $i, j \in \{1, 2\}$,

$$B_{m,n,\lambda_1,\lambda_2}((z-y_1)^{2i}(w-y_2)^{2j};y_1,y_2) = B_{m,\lambda_1}((z-y_1)^{2i};y_1)B_{n,\lambda_2}((w-y_2)^{2j};y_2),$$
$$\leq \frac{M_1}{m^i}\frac{M_2}{n^j}, \tag{6.4}$$

for some constants $M_1, M_2 > 0$.

Let $\gamma_1 = m^{\frac{-1}{2}}$, and $\gamma_2 = n^{\frac{-1}{2}}$.

Then, combining (6.3), (6.4)

$$|C_{m,n,\lambda_1,\lambda_2}(g;y_1,y_2) - g(y_1,y_2)| = 3||D_B g||_\infty O\left(m^{\frac{-1}{2}}\right)O\left(n^{\frac{-1}{2}}\right)$$
$$+O\left(\frac{1}{m^{\frac{1}{2}}}\right)O\left(\frac{1}{n^{\frac{1}{2}}}\right)\omega_B(D_B g; m^{\frac{-1}{2}}, n^{\frac{-1}{2}}),$$

uniformly in $(y_1, y_2) \in I^2$. This completes the proof. □

Acknowledgements The first author is thankful to the "Ministry of Human Resource and Development", New Delhi, India for financial support to carry out the above work.

References

1. S.N. Bernstein, Démonstration du théorème de Weierstrass fondée sur la calcul des probabilités. Comm. Soc. Math. Charkow Sér. 2 t. **13**, 1–2 (1912)
2. C. Badea, C. Cottin, Korovkin-type theorems for generalised boolean sum operators, colloquia mathematica societatis jános bolyai, in *Approximation Theory*, vol. 58 (Kecskemét, Hungary, 1990), pp. 51–67
3. C. Badea, I. Badea, H.H. Gonska, A test function theorem and approximation by pseudopolynomials. Bull. Austral. Math. Soc. **34**, 95–108 (1986)
4. C. Badea, I. Badea, H.H. Gonska, Notes on the degree of approximation of B-continuous and B-differentiable functions. J. Approx. Theory Appl. **4**, 95–108 (1988)
5. K. Bögel, Mehrdimensionale differentiation von funtionen mehrerer Veränderlicher. J. Reine Angew. Math. **170**, 197–217 (1934)
6. K. Bögel, Über die mehrdimensionale differentiation, integration und beschränkte variation. J. Reine Angew. Math. **173**, 5–29 (1935)
7. P.L. Butzer, H. Berens, *Semi-Groups of Operators and Approximation.* (Springer, New York, 1967)Xit 318 p
8. A. Kajla, P.N. Agrawal, Szász-Kantorovich type operators based on Charlier polynomials. Kyungpook Math. J **56**(3), 877–897 (2016)
9. Q.B. Cai, B.Y. Lian, G. Zhou, Approximation properties of λ-Bernstein operators. J. Inequal. Appl. **61**, 1–11 (2018)
10. Q.B. Cai, The Bézier variant of Kantorovich type λ-Bernstein operators. J. Inequal. Appl. **90**, 1–10 (2018)
11. E. Dobrescu, I. Matei, The approximation by Bernstein type polynomials of bidimensionally continuous functions (Romannian). An. Univ. Timişoara Ser. Sti. Mat.-Fiz. **4**, 85–90 (1966)
12. V.I. Volkov, On the convergence of sequences of linear positive operators in the space of continuous functions of two variables. Dokl. Akad. Nauk. **115**, 17–19 (1957)
13. Z. Ye, X. Long, X.M. Zeng, Adjustment algorithms for Bézier curve and surface, *International Conference on Computer Science and Education* (2010), pp. 1712–1716

Durrmeyer Modification of Lupaş Type Baskakov Operators Based on IPED

Minakshi Dhamija

Abstract The purpose of this paper is to consider Durrmeyer variant of Lupaş type Baskakov operators having inverse Pólya–Eggenberger distribution basis function. We derive some direct results which include uniform convergence, pointwise approximation via modulus of continuity and asymptotic formula.

Keywords Stancu operators · Baskakov operators · Durrmeyer operators · Inverse Pólya–Eggenberger distribution

2010 Mathematics Subject Classication 41A25 · 41A36

1 Introduction

In 1968, a new sequence of positive linear operators $P_n^{[\alpha]} : C[0, 1] \to C[0, 1]$ introduced by Stancu [17]. This sequence was based on Pólya–Eggenberger distribution:

$$P_n^{[\alpha]}(f; x) = \sum_{k=0}^{n} p_{n.k}^{[\alpha]}(x) f\left(\frac{k}{n}\right), \tag{1.1}$$

where

$$p_{n.k}^{[\alpha]}(x) = \binom{n}{k} \frac{x^{[k,-\alpha]}(1-x)^{[n-k,-\alpha]}}{1^{[n,-\alpha]}},$$

$\alpha \geq 0$ such that it may depend on natural number $n \in \mathbb{N}$ and $t^{[n,h]} = t(t-h)(t-2h) \cdot \dots \cdot (t-\overline{n-1}h)$, $t^{[0,h]} = 1$ denotes the factorial power of t with increment h.

For special case $\alpha = 0$, operators (1.1) reduce to the classical Bernstein operators [4] and when $\alpha = \frac{1}{n}$ we get another particular case

M. Dhamija (✉)
Department of Mathematics, Shaheed Rajguru College of Applied Sciences for Women, Vasundhara Enclave, East Delhi, New Delhi 110096, Delhi, India
e-mail: minakshidhamija11@gmail.com

N. Deo et al. (eds.), *Mathematical Analysis I: Approximation Theory*, Springer Proceedings in Mathematics & Statistics 306,
https://doi.org/10.1007/978-981-15-1153-0_10

$$P_n^{[\frac{1}{n}]}(f;x)=\sum_{k=0}^{n}\binom{n}{k}\frac{x^{[k,-\frac{1}{n}]}(1-x)^{[n-k,-\frac{1}{n}]}}{1^{[n,-\frac{1}{n}]}}f\left(\frac{k}{n}\right), \tag{1.2}$$

considered by Lupaş and Lupaş [16].

Recently, Gupta and Rassias [13] introduced the Durrmeyer type integral modification of operators (1.2) and established local and global approximation results.

Stancu [18] also considered inverse Pólya–Eggenberger distribution and gave a generalization of the Baskakov operators:

$$V_n^{[\alpha]}(f;x)=\sum_{k=0}^{\infty}v_{n,k}^{[\alpha]}(x)f\left(\frac{k}{n}\right), \tag{1.3}$$

where

$$v_{n,k}^{[\alpha]}(x)=\binom{n+k-1}{k}\frac{1^{[n,-\alpha]}x^{[k,-\alpha]}}{(1+x)^{[n+k,-\alpha]}}$$

and $f\in C_B[0,\infty)$. The operators (1.3) also have special cases:

(1) For $\alpha=0$, we get classical Baskakov operators [3] as:

$$V_n(f,x)=\sum_{k=0}^{\infty}v_{n,k}(x)f\left(\frac{k}{n}\right), \tag{1.4}$$

with

$$v_{n,k}(x)=\binom{n+k-1}{k}\frac{x^k}{(1+x)^{n+k}}.$$

(2) For $\alpha=\frac{1}{n}$, the operators (1.3) reduces to following Lupaş type Baskakov operators considered by Dhamija et al. [9]:

$$V_n^{[\frac{1}{n}]}(f;x)=\sum_{k=0}^{\infty}v_{n,k}^{[\frac{1}{n}]}(x)f\left(\frac{k}{n}\right) \tag{1.5}$$

where

$$v_{n,k}^{[\frac{1}{n}]}(x)=\binom{n+k-1}{k}\frac{1^{[n,-\frac{1}{n}]}x^{[k,-\frac{1}{n}]}}{(1+x)^{[n+k,-\frac{1}{n}]}}.$$

Inspired by Gupta and Rassias [13], we now consider Durrmeyer type modification of Generalized Baskakov operators (1.5):

$$D_n^{[\frac{1}{n}]}(f;x)=(n-1)\sum_{k=0}^{\infty}v_{n,k}^{[\frac{1}{n}]}(x)\int_0^{\infty}v_{n,k}(t)f(t)\,dt \tag{1.6}$$

where

$$v_{n,k}(t) = \binom{n+k-1}{k} \frac{t^k}{(1+t)^{n+k}}$$

In recent years, many researchers are working on Durrmeyer type operators and connecting these operators in the direction of q-calculus (see [1]). The aim of this paper is to obtain moments of operators (1.6) and then establish direct results including pointwise approximation by using classical modulus of continuity, second order modulus of smoothness and asymptotic formula. These approximation properties have also been discussed in [5, 7, 11, 12, 14].

For details of Pólya and Inverse Pólya–Eggenberger distribution, see [6, 10].

2 Auxiliary Results

Consider the set of monomials $e_i = t^i, i = 0, 1, 2$ known as test functions. Now we find the moments of these test functions for our operators (1.6) which is necessarily in order to provide our main results.

Lemma 2.1 *For Durremeyer operators* (1.6) *hold*

$$D_n^{[\frac{1}{n}]}(e_0; x) = 1,$$

$$D_n^{[\frac{1}{n}]}(e_1; x) = \frac{n^2}{(n-1)(n-2)}x + \frac{1}{n-2},$$

$$D_n^{[\frac{1}{n}]}(e_2; x) = \frac{1}{(n-2)(n-3)}\left[\frac{n^3(n+1)}{(n-1)(n-2)}x^2 + \frac{n^2(5n-7)}{(n-1)(n-2)}x + 2\right]$$

Proof To obtain these moments, first we prove the following result:

$$\int_0^\infty v_{n,k}(t)\, t^r dt = \frac{(k+r)!\,(n-r-2)!}{k!\,(n-1)!}$$

By using the above result, we can get desired moments. □

Lemma 2.2 *For the operators* (1.6),

$$D_n^{[\frac{1}{n}]}(t-x; x) = \frac{3n-2}{(n-1)(n-2)}x + \frac{1}{n-2}$$

$$\begin{aligned} D_n^{[\frac{1}{n}]}\left((t-x)^2; x\right) &= \tfrac{1}{(n-1)(n-2)^2(n-3)}[(3n^3+11n^2-28n+12)x^2 \\ &\quad +(3n^3+5n^2-22n+12)x + 2(n^2-3n+2)] \end{aligned}$$

Proof Taking into account Lemma 2.1, we can have above central moments. □

Lemma 2.3 *Let f be a bounded function defined on $[0,\infty)$, with $\|f\| = \sup_{x\in[0,\infty)} |f(x)|$, then*

$$\left|D_n^{\left[\frac{1}{n}\right]}(f;x)\right| \leq \|f\|.$$

Proof Using the definition of operators (1.6) and Lemma 2.1, it follows

$$\left|D_n^{\left[\frac{1}{n}\right]}(f;x)\right| = \left|(n-1)\sum_{k=0}^{\infty} v_{n,k}^{\left[\frac{1}{n}\right]}(x)\int_0^{\infty} v_{n,k}(t)\, f(t)\, dt\right| \leq \|f\|\, D_n^{\left[\frac{1}{n}\right]}(e_0;x)$$
$$= \|f\|. \quad □$$

3 Direct Results

Using the well-known Bohman–Korovkin–Popoviciu theorem (see [15]) we get the uniform convergence of operators (1.6).

Theorem 3.1 *Let $f \in C[0,\infty) \cap E$, then we have*

$$\lim_{n\to\infty} D_n^{\left[\frac{1}{n}\right]}(f;x) = f(x)$$

uniformly on each compact subset of $[0,\infty)$, where $C[0,\infty)$ is the space of all real-valued continuous functions on $[0,\infty)$ and

$$E := \left\{ f : x \in [0,\infty),\ \frac{f(x)}{1+x^2} \text{ is convergent as } x \to \infty \right\}$$

Proof Taking Lemma 2.1 into the account it is clear that

$$\lim_{n\to\infty} D_n^{\left[\frac{1}{n}\right]}(e_i;x) = x^i,\quad i = 0, 1, 2$$

uniformly on each compact subset of $[0,\infty)$. Hence, applying the well-known Korovkin-type theorem [2] regarding the convergence of a sequence of positive linear operators, we get the desired result. □

Modulus of continuity is the main tool to measure the degree of approximation of linear positive operators towards the identity operators.

Definition 3.1 Let $f \in C_B[0,\infty)$ be given. The modulus of continuity of the function f is defined by

$$\omega(f,\delta) := \sup\{|f(x) - f(y)| : x, y \in [0,+\infty),\ |x-y| \leq \delta\}, \tag{3.1}$$

where $\delta > 0$ and $C_B[0, \infty)$ is the space of all real-valued functions continuous and bounded on $[0, \infty)$.

Moreover, we have a following useful property:

$$|f(x) - f(y)| \leq \omega(f, |x - y|) \leq \left(1 + \tfrac{1}{\delta}|x - y|\right) \cdot \omega(f, \delta). \tag{3.2}$$

Definition 3.2 For $f \in C[0, \infty)$ and $\delta \geq 0$ and

$$\omega_2(f, \delta) := \sup\{|f(x + h) - 2f(x) + f(x - h)| : x, x \pm h \in [0, \infty), 0 \leq h \leq \delta\} \tag{3.3}$$

are the moduli of smoothness of second order.

Definition 3.3 Let f be a function from the space $C_B[0, \infty)$ endowed with the norm $\|f\| = \sup_{x \in [0,\infty)} |f(x)|$ then Peetre's K-functional is defined as

$$K_2(f, \delta) = \inf_{g \in W_\infty^2} \left\{\|f - g\| + \delta \left\|g''\right\|\right\}, \tag{3.4}$$

where $\delta > 0$ and $W_\infty^2 = \left\{g \in C_B[0, \infty) : g', g'' \in C_B[0, \infty)\right\}$. Also, from ([8], p. 177, Theorem 2.4), we can write

$$K_2(f, \delta) \leq M\omega_2\left(f, \sqrt{\delta}\right), \tag{3.5}$$

$M > 0$ is an absolute constant.

We start our direct results in terms of moduli of continuity.

Theorem 3.2 *If $f \in C_B[0, \infty)$, then for any $x \in [0, \infty)$ and $\delta > 0$, it follows*

$$\left|D_n^{\left[\frac{1}{n}\right]}(f; x) - f(x)\right| \leq 2 \cdot \omega(f, \delta), \quad \textit{with } \delta = \left(D_n^{\left[\frac{1}{n}\right]}\left((e_1 - x)^2; x\right)\right)^{\frac{1}{2}}.$$

Proof By using the property (3.2) and Lemma 2.1, we get

$$\begin{aligned}
\left|D_n^{\left[\frac{1}{n}\right]}(f; x) - f(x)\right| &\leq (n-1) \sum_{k=0}^{\infty} v_{n,k}^{\left[\frac{1}{n}\right]}(x) \int_{\frac{k}{n}}^{\frac{k+1}{n}} v_{n,k}(t)|f(t) - f(x)|\,dt \\
&\leq \left(1 + \tfrac{1}{\delta}(n-1) \sum_{k=0}^{\infty} v_{n,k}^{\left[\frac{1}{n}\right]}(x) \int_{\frac{k}{n}}^{\frac{k+1}{n}} v_{n,k}(t)|t - x|\,dt\right) \omega(f, \delta).
\end{aligned}$$

Applying Cauchy–Schwarz inequality for integral, we get

$$\left|D_n^{\left[\frac{1}{n}\right]}(f;x)-f(x)\right|$$

$$\leq \left[1+\frac{1}{\delta}(n-1)\sum_{k=0}^{\infty}v_{n,k}^{\left[\frac{1}{n}\right]}(x)\left(\int_{\frac{k}{n}}^{\frac{k+1}{n}}v_{n,k}(t)dt\right)^{1/2}\left(\int_{\frac{k}{n}}^{\frac{k+1}{n}}v_{n,k}(t)(t-x)^2dt\right)^{1/2}\right]\omega(f,\delta).$$

Again, applying Cauchy–Schwarz inequality for sum, it follows

$$\left|D_n^{\left[\frac{1}{n}\right]}(f;x)-f(x)\right| \leq \left[1+\frac{1}{\delta}\left((n-1)\sum_{k=0}^{\infty}v_{n,k}^{\left[\frac{1}{n}\right]}(x)\int_{\frac{k}{n}}^{\frac{k+1}{n}}v_{n,k}(t)dt\right)^{1/2}\right.$$

$$\left.\times\left((n-1)\sum_{k=0}^{\infty}v_{n,k}^{\left[\frac{1}{n}\right]}(x)\int_{\frac{k}{n}}^{\frac{k+1}{n}}v_{n,k}(t)(t-x)^2dt\right)^{1/2}\right]\omega(f,\delta)$$

$$=\left[1+\frac{1}{\delta}\left(D_n^{\left[\frac{1}{n}\right]}(e_0;x)\right)^{1/2}\left(D_n^{\left[\frac{1}{n}\right]}((e_1-x)^2;x)\right)^{1/2}\right]\omega(f,\delta)=2\cdot\omega(f,\delta),$$

with $\delta := \left(D_n^{\left[\frac{1}{n}\right]}((e_1-x)^2;x)\right)^{1/2}$. □

Next we estimate the degree of approximation by using Peetre's K-functional.

Theorem 3.3 *Let be $f \in C[0,\infty)$, then for any $x \in [0,\infty)$ it follows*

$$\left|D_n^{\left[\frac{1}{n}\right]}(f;x)-f(x)\right| \leq M\omega_2\left(f,\tfrac{1}{2}\delta_n(x)\right)+\omega(f,\delta_\omega),$$

where M is an absolute constant and

$$\delta_n(x)=\left(D_n^{\left[\frac{1}{n}\right]}((e_1-x)^2;x)+\left(D_n^{\left[\frac{1}{n}\right]}(e_1-x;x)\right)^2\right)^{\frac{1}{2}},\quad \delta_\omega=\left|D_n^{\left[\frac{1}{n}\right]}(e_1-x;x)\right|.$$

Proof For $x \in [0,\infty)$, consider the operators

$$\hat{D}_n^{\left[\frac{1}{n}\right]}(f;x)=D_n^{\left[\frac{1}{n}\right]}(f;x)-f\left(\frac{n^2x}{(n-1)(n-2)}+\frac{1}{n-2}\right)+f(x). \tag{3.6}$$

We note that $\hat{D}_n^{\left[\frac{1}{n}\right]}(e_0;x)=1$ and $\hat{D}_n^{\left[\frac{1}{n}\right]}(e_1;x)=x$, i.e., the operators $\hat{D}_n^{\left[\frac{1}{n}\right]}$ preserve constants and linear functions. Therefore

$$\hat{D}_n^{\left[\frac{1}{n}\right]}(e_1-x;x)=0. \tag{3.7}$$

Let $g \in W_\infty^2$ and $x,t \in [0,\infty)$. By Taylor's expansion, we have

$$g(t)=g(x)+(t-x)g'(x)+\int_x^t (t-u)g''(u)\,du$$

Operating $\hat{D}_n^{\left[\frac{1}{n}\right]}$ on both sides of the above equation, we get

$$\hat{D}_n^{\left[\frac{1}{n}\right]}(g;x)-g(x)=g'(x)\cdot\hat{D}_n^{\left[\frac{1}{n}\right]}(e_1-x;x)+\hat{D}_n^{\left[\frac{1}{n}\right]}\left(\int_x^t(t-u)\,g''(u)\,du;x\right)$$

$$=D_n^{\left[\frac{1}{n}\right]}\left(\int_x^t(t-u)\,g''(u)\,du;x\right)-\int_x^{\frac{n^2x}{(n-1)(n-2)}+\frac{1}{n-2}}\left(\frac{n^2x}{(n-1)(n-2)}+\frac{1}{n-2}-u\right)g''(u)du.$$

On the other hand

$$\left|\int_x^t(t-u)g''(u)\right|\le(t-x)^2\cdot\|g''\|,$$

then

$$\left|\hat{D}_n^{\left[\frac{1}{n}\right]}(g;x)-g(x)\right|\le\left(D_n^{\left[\frac{1}{n}\right]}((e_1-x)^2;x)+\left(D_n^{\left[\frac{1}{n}\right]}(e_1-x;x)\right)^2\right)\cdot\|g''\|.$$

Making use of definition (3.6) of the operators $\hat{D}_n^{\left[\frac{1}{n}\right]}$ and Lemma 2.3, we have

$$\begin{aligned}\left|D_n^{\left[\frac{1}{n}\right]}(f;x)-f(x)\right|&\le\left|\hat{D}_n^{\left[\frac{1}{n}\right]}(f-g;x)\right|+\left|\hat{D}_n^{\left[\frac{1}{n}\right]}(g;x)-g(x)\right|\\&\quad+|g(x)-f(x)|+\left|f\left(\frac{n^2x}{(n-1)(n-2)}+\frac{1}{n-2}\right)-f(x)\right|\\&\le4\|f-g\|+\delta_n^2(x)\|g''\|+\omega(f,\delta_\omega),\end{aligned}$$

with $\delta_n^2(x)=D_n^{\left[\frac{1}{n}\right]}\left((e_1-x)^2;x\right)+\left(D_n^{\left[\frac{1}{n}\right]}(e_1-x;x)\right)^2$ and $\delta_\omega=\left|D_n^{\left[\frac{1}{n}\right]}(e_1-x;x)\right|$. Now, taking infimum on the right-hand side over all $g\in W_\infty^2$ and using the relation (3.5), we get

$$\begin{aligned}\left|D_n^{\left[\frac{1}{n}\right]}(f;x)-f(x)\right|&\le4K_2\left(f,\frac{\delta_n{}^2(x)}{4}\right)+\omega(f,\delta_\omega)\\&\le M\omega_2\left(f,\tfrac{1}{2}\delta_n(x)\right)+\omega(f,\delta_\omega).\end{aligned}$$

Further, we present asymptotic formula for the operators (1.6). □

Theorem 3.4 *Let f be a bounded and integrable function on $[0,\infty)$ such that there exists first and second derivative of the function f at a fixed point $x\in[0,\infty)$, then*

$$\lim_{n\to\infty}n\left(D_n^{\left[\frac{1}{n}\right]}(f;x)-f(x)\right)=(3x+1)f'(x)+\frac{3}{2}x(x+1)f''(x).$$

Proof Taylor's expansion of function f gives:

$$f(t) = f(x) + (t - x)f'(x) + \frac{1}{2!}(t - x)^2 f''(x) + \varepsilon(t, x)(t - x)^2,$$

where $\varepsilon(t, x) \to 0$ as $t \to x$.

Applying operators $D_n^{[\frac{1}{n}]}$ on both sides of above equation and in view of Lemma 2.2, we get

$$D_n^{[\frac{1}{n}]}(f; x) - f(x) = D_n^{[\frac{1}{n}]}(t - x; x)\, f'(x) + \frac{1}{2} D_n^{[\frac{1}{n}]}\left((t - x)^2; x\right) f''(x) + \\ + D_n^{[\frac{1}{n}]}\left(\varepsilon(t, x) \cdot (t - x)^2; x\right).$$

Therefore,

$$\lim_{n\to\infty} n\left(D_n^{[\frac{1}{n}]}(f; x) - f(x)\right) = (3x + 1)f'(x) + \frac{3}{2}x(1 + x)f''(x), + \\ + \lim_{n\to\infty} n\left(D_n^{[\frac{1}{n}]}\left(\varepsilon(t, x) \cdot (t - x)^2; x\right)\right), \tag{3.8}$$

To obtain the desired result, it is sufficient to prove that

$$\lim_{n\to\infty} n\left(D_n^{[\frac{1}{n}]}\left(\varepsilon(t, x) \cdot (t - x)^2; x\right)\right) = 0$$

Application of Cauchy–Schwarz inequality gives

$$D_n^{[\frac{1}{n}]}\left(\varepsilon(t, x)(t - x)^2; x\right) \le \sqrt{D_n^{[\frac{1}{n}]}\left(\varepsilon^2(t, x); x\right)}\sqrt{D_n^{[\frac{1}{n}]}\left((t - x)^4; x\right)}. \tag{3.9}$$

Since $\varepsilon^2(x, x) = 0$ and $\varepsilon^2(\cdot, x) \in C[0, \infty) \cap E$, thus we can use convergence of operators $D_n^{[\frac{1}{n}]}$ from Theorem 3.1,

$$\lim_{n\to\infty} D_n^{[\frac{1}{n}]}\left(\varepsilon^2(t, x); x\right) = \varepsilon^2(x, x) = 0. \tag{3.10}$$

Therefore, from (3.9) and (3.10) yields

$$\lim_{n\to\infty} n\left(D_n^{[\frac{1}{n}]}\left(\varepsilon(t, x) \cdot (e_1 - x)^2; x\right)\right) = 0$$

and using (3.8) we obtain required result. □

References

1. A.M. Acu, C.V. Muraru, D.F. Sofonea, V.A. Radu, Some approximation properties of a Durrmeyer variant of q-Bernstein-Schurer operators. Math. Methods Appl. Sci. **39**(18), 5636–5650 (2016)
2. F. Altomare, M. Campiti, *Korovkin-Type Approximation Theory and its Application*, de Gruyter Studies in Mathematics, vol. 17 (Walter de Gruyter & Co., Berlin, 1994)
3. V.A. Baskakov, A sequence of linear positive operators in the space of continuous functions. Dokl. Acad. Nauk. SSSR **113**, 249–251 (1957)
4. S.N. Bernstein, Démonstration du théoréme de Weierstrass fondée sur le calcul de probabilités. Commun. Sco. Math. Charkov **13**(2), 1–2 (1912)
5. N. Deo, M. Dhamija, Generalized positive linear operators based on PED and IPED. Iran. J. Sci. Technol. Trans. Sci. **43** (2018), https://doi.org/10.1007/s40995-017-0477-5
6. N. Deo, M. Dhamija, D. Miclăuş, Stancu-Kantorovich operators based on inverse Pólya-Eggenberger distribution. Appl. Math. Comput. **273**, 281–289 (2015)
7. N. Deo, M. Dhamija, Jain–Durrmeyer operators associated with the inverse Pólya-Eggenberger distribution. Appl. Math. Comput. **286**, 15–22 (2016)
8. R.A. DeVore, G.G. Lorentz, *Constructive Approximation* (Springer, Berlin, 1993)
9. M. Dhamija, N. Deo, D. Miclăuş, Modified Baskakov operators based on inverse Pólya-Eggenberger Distribution. (Communicated)
10. F. Eggenberger, G. Pólya, Über die statistik verkerter vorgänge. Z. Angew. Math. Mech. **1**, 279–289 (1923)
11. Z. Finta, On approximation properties of Stancu'soperators. Stud. Univ. Bolyai Math. **XLVII**(4), 47–55 (2002)
12. V. Gupta, R.P. Agarwal, *Convergence Estimates in Approximation Theory* (Springer, New York, 2014)
13. V. Gupta, T.M. Rassias, Lupaş-Durrmeyer operators based on Polya distribution. Banach J. Math. Anal. **8**(2), 145–155 (2014)
14. H.S. Jung, N. Deo, M. Dhamija, Pointwise approximation by Bernstein type operators in mobile interval. Appl. Math. Comput. **214**(1), 683–694 (2014)
15. P.P. Korovkin, Convergence of linear positive operators in the spaces of continuous functions(Russian). Doklady Akad. Nauk. SSSR(N.N.) **90**, 961–964 (1953)
16. L. Lupaş, A. Lupaş, Polynomials of binomial type and approximation operators, Stud. Univ. Babeş-Bolyai Math. **32**(4), 61–69 (1987)
17. D.D. Stancu, Approximation of functions by a new class of linear polynomial operators. Rev. Roum. Math. Pures Appl. **13**, 1173–1194 (1968)
18. D.D. Stancu, Two classes of positive linear operators. Anal. Univ. Timişoara Ser. Mat. **8**, 213–220 (1970)

On Bernstein–Chlodowsky Type Operators Preserving Exponential Functions

Firat Ozsarac, Ali Aral and Harun Karsli

Abstract In this paper, we introduce, analyze, and obtain some features of a new type of Bernstein–Chlodowsky operators using a different technique that is utilized as the classical Chlodowsky operators. These operators preserve the functions exp (μt) and exp $(2\mu t)$, $\mu > 0$. As a first result, the rate of convergence of the operator using an appropriately weighted modulus of continuity is obtained. Later, Quantitative-Voronovskaya type and Grüss–Voronovskaya type theorems for the new operators are presented. Then, we prove that the first derivative of the Bernstein–Chlodowsky operators applied to a function converges to the function itself. Finally, the variation detracting property of the operators is presented. It is proved that the variation seminorm property is preserved. Also, it is shown that the operators converge to $f/\exp_{\mu}$ in variation seminorm is valid if and only if the function is absolutely continuous.

Keywords Bernstein operators · Bernstein–Chlodowsky operators · Voronovskaya theorem · Generalized convexity

2000 Mathematics Subject Classification Primary 41A36 · Secondary 41A25

F. Ozsarac (✉) · A. Aral
Department of Mathematics, Faculty of Science and Arts, Kirikkale University, 71450 Yahsihan, Kirikkale, Turkey
e-mail: firat_ozsarac@hotmail.com

A. Aral
e-mail: aliaral73@yahoo.com

H. Karsli
Department of Mathematics, Faculty of Science and Arts, Bolu Abant Izzet Baysal University, 14030 Merkez, Bolu, Turkey
e-mail: karsli_h@ibu.edu.tr

N. Deo et al. (eds.), *Mathematical Analysis I: Approximation Theory*, Springer Proceedings in Mathematics & Statistics 306,
https://doi.org/10.1007/978-981-15-1153-0_11

1 Introduction

Recall that the classical Bernstein–Chlodowsky operator C_n defined from $C[0,\infty) \to C[0,\infty)$ is given by

$$C_n f(x) = \sum_{k=0}^{n} f\left(\frac{kb_n}{n}\right)\binom{n}{k}\left(\frac{x}{b_n}\right)^k\left(1-\frac{x}{b_n}\right)^{n-k}, \quad x \in [0, b_n] \tag{1.1}$$

where f is a function defined on $[0,\infty)$ and bounded on every finite interval $[0, b_n] \subset [0,\infty)$ with a certain rate, and b_n is a monotone increasing, positive and real sequence such that $\lim_{n\to\infty} b_n = \infty$ and $\lim_{n\to\infty} \frac{b_n}{n} = 0$.

The classical Bernstein–Chlodowsky polynomials were introduced by I. Chlodovsky in 1937 as a generalization of the Bernstein polynomials. Note that the case $b_n = 1, n \in \mathbb{N}$, in Eq. 1.1, defines an approximation to the function f on the interval $[0, 1]$ (or, suitably modified on any fixed finite interval $[-b, b]$).

For $b > 0$, let $M(b; f) := \sup_{0\le t\le b} |f(t)|$. It is shown by Chlodowsky that when $f \in C[0,\infty)$ and $\lim_{n\to\infty} M(b; f)\exp\left(-\frac{\sigma n}{b_n}\right) = 0$ for each $\sigma > 0$, then the classical Bernstein–Chlodowsky operator converges to $f(x)$ at each point where f is continuous. Chlodovsky also showed that the simultaneous convergence of the derivative $(C_n f)'(x)$ to $f'(x)$ at points x, where the derivative of $f(x)$ exists, a result taken up by Butzer [4, 5]. Due to these two former results, the classical Bernstein–Chlodowsky operators and their generalizations have been an increasing interest in the field of approximation theory.

During the paper, $\mu > 0$ is a fixed real parameter and $\exp_\mu$ represents the exponential function defined by $\exp_\mu(t) = e^{\mu t}$.

Herein, we consider a generalization of Bernstein–Chlodowsky operators of the form

$$\mathcal{C}_n f(x) = \sum_{k=0}^{n} \alpha_{n,k}(x) f\left(\frac{kb_n}{n}\right) p_{n,k}(a_n(x)), \quad x \in [0, b_n] \tag{1.2}$$

$$\alpha_{n,k}(x) = e^{\mu x} e^{-\frac{\mu k b_n}{n}} \text{ and } p_{n,k}(x) = \binom{n}{k}\left(\frac{x}{b_n}\right)^k\left(1-\frac{x}{b_n}\right)^{n-k}$$

with the property that

$$\mathcal{C}_n(\exp_\mu; x) = e^{\mu x}, \quad \mathcal{C}_n(\exp_\mu^2; x) = e^{2\mu x}. \tag{1.3}$$

Then, the operator $\mathcal{C}_n$ is more explicitly given by

$$\mathcal{C}_n f(x) = e^{\mu x}\left(e^{\frac{\mu b_n}{n}} - 1\right)^{-n} \sum_{k=0}^{n} f\left(\frac{kb_n}{n}\right)\binom{n}{k} e^{-\frac{\mu k b_n}{n}}\left(e^{\frac{\mu x}{n}} - 1\right)^k\left(e^{\frac{\mu b_n}{n}} - e^{\frac{\mu x}{n}}\right)^{n-k}, \tag{1.4}$$

with

$$a_n(x) = b_n \frac{e^{\frac{\mu x}{n}} - 1}{e^{\frac{\mu b_n}{n}} - 1}.$$

Note that the connection of this operator with the classical Bernstein–Chlodowsky operator can be expressed as

$$\mathcal{C}_n f(x) = f_0(x) C_n(f/f_0)(a_n(x)), \qquad f_0(x) = e^{\mu x}. \tag{1.5}$$

Namely,

$$\begin{aligned}
f_0(x) C_n\left(\frac{f}{f_0}\right)(a_n(x)) &= e^{\mu x} \sum_{k=0}^{n} \left(\frac{f}{f_0}\right)\left(\frac{kb_n}{n}\right)\binom{n}{k}\left(\frac{a_n(x)}{b_n}\right)^k \left(1 - \frac{a_n(x)}{b_n}\right)^{n-k} \\
&= e^{\mu x} \sum_{k=0}^{n} \frac{f\left(\frac{kb_n}{n}\right)}{f_0\left(\frac{kb_n}{n}\right)} p_{n,k}(a_n(x)) \\
&= e^{\mu x} \sum_{k=0}^{n} \frac{f\left(\frac{kb_n}{n}\right)}{e^{\mu \frac{kb_n}{n}}} p_{n,k}(a_n(x)) \\
&= e^{\mu x} \sum_{k=0}^{n} e^{-\mu \frac{kb_n}{n}} f\left(\frac{kb_n}{n}\right) p_{n,k}(a_n(x)) \\
&= \sum_{k=0}^{n} e^{\mu x} e^{-\mu \frac{kb_n}{n}} f\left(\frac{kb_n}{n}\right) p_{n,k}(a_n(x)) \\
&= \sum_{k=0}^{n} \alpha_{n,k}(x) f\left(\frac{kb_n}{n}\right) p_{n,k}(a_n(x)) \\
&= \mathcal{C}_n f(x).
\end{aligned}$$

Also note that the Bernstein–Chlodovsky operators $\mathcal{C}_n$, based on functions defined on $[0, \infty)$, are bounded on every $[0, b_n] \subset [0, \infty)$ with a certain rate. Thus, they are a very natural polynomial process in approximating unbounded functions on the unbounded infinite interval $[0, \infty)$; but this approximation process is not so easy to handle.

We know that the classical Bernstein–Chlodowsky operators have the degree of exactness one, that is, they preserve the monomials 1 and x. On the other side, the operator (1.4) does not preserve 1 and x, but it satisfies the exponential moments (1.3) that play an important role in our calculations.

The aim of the present paper is to investigate the operators $\mathcal{C}_n$, $n \in \mathbb{N}$ in deeper to reveal, in addition to elementary properties, their advanced properties. Moreover, the development of the some theoretical results of the generalized operator is within the aim of the paper. After Voronovskaya type theorems for the generalized operator is stated , it is compared to the classical Bernstein–Chlodovsky operators in terms of effectiveness. For this purpose, the convergence of the derivative $(\mathcal{C}_n f)'(x)$ to $f'(x)$

is also considered. Finally, in the last section, the variation detracting property of the operators and variation seminorm property is stated. Moreover, it is proved that the operators converge to $f/\exp_\mu$ in variation seminorm is valid if and only if the function is absolutely continuous.

2 Preliminary Results

For the operator $\mathcal{C}_n, n \in \mathbb{N}$, we give here some of their properties and results. At first, we calculate all the moments of operator (1.4).

Lemma 1 *For each $n \in \mathbb{N}$ and $x \in [0, b_n]$, the following identities hold:*

$$\mathcal{C}_n e_0(x) = e^{\mu x - \mu b_n} \left(e^{\frac{\mu b_n}{n}} + 1 - e^{\frac{\mu x}{n}} \right)^n,$$

$$\mathcal{C}_n(\exp_\mu^3; x) = e^{\mu x} \left(e^{\frac{\mu x}{n}} \left(e^{\frac{\mu b_n}{n}} + 1 \right) - e^{\frac{\mu b_n}{n}} \right)^n,$$

$$\mathcal{C}_n(\exp_\mu^4, x) = e^{\mu x} \left(e^{\frac{\mu b_n}{n}} \left(e^{\frac{\mu x}{n}} - 1 \right) \left(e^{\frac{\mu b_n}{n}} + 1 \right) + e^{\frac{\mu x}{n}} \right)^n.$$

Using *Mathematica*, we give two limits, which play an important role in both the uniform approximation of operator to functions and Voronoskaya type result.

For each $x \in (0, \infty)$, we shall consider the function $\exp_{\mu,x}$, defined for $t \in (0, \infty)$ by

$$\exp_{\mu,x}(t) = e^{\mu t} - e^{\mu x}.$$

Using Lemma 1 and (1.3) , one easily finds that

$$\begin{aligned} \mathcal{C}_n(\exp_{\mu,x}; x) &= \mathcal{C}_n(\exp_\mu; x) - e^{\mu x} \mathcal{C}_n e_0(x) \\ &= e^{\mu x} (1 - \mathcal{C}_n e_0(x)) \end{aligned} \tag{2.1}$$

and

$$\begin{aligned} \mathcal{C}_n(\exp_{\mu,x}^2; x) &= \mathcal{C}_n(\exp_\mu^2; x) - 2e^{\mu x} \mathcal{C}_n(\exp_\mu; x) + e^{2\mu x} \mathcal{C}_n e_0(x) \\ &= e^{2\mu x} (\mathcal{C}_n e_0(x) - 1). \end{aligned} \tag{2.2}$$

Lemma 2 *For each $x \in [0, \infty)$, the following identities hold:*

$$\lim_{n\to\infty} \mathcal{C}_n e_0(x) = \lim_{n\to\infty} e^{\mu x - \mu b_n} \left(e^{\frac{\mu b_n}{n}} + 1 - e^{\frac{\mu x}{n}} \right)^n = 1, \tag{2.3}$$

$$\lim_{n\to\infty} n (\mathcal{C}_n e_0(x) - 1) = \lim_{n\to\infty} n \left(e^{\mu x - \mu b_n} \left(e^{\frac{\mu b_n}{n}} + 1 - e^{\frac{\mu x}{n}} \right)^n - 1 \right) = \mu^2 x, \tag{2.4}$$

and

$$\lim_{n\to\infty} n^2 C_n(\exp^4_{\mu,x}; x) = 0.$$

3 Quantitative Results

All concepts mentioned below can be found in [7] more generally. We denote by $C_\mu[0,\infty)$ the space of continous functions $f \in C[0,\infty)$ with the property that exists $M > 0$ such that $|f(x)| \le Me^{\mu x}$, for every $x \in [0, b_n]$. This space endowed with norm

$$\|f\|_\mu = \sup_{x\in[0,b_n]} \frac{|f(x)|}{e^{\mu x}}.$$

Also,

$$C^k_\mu[0,\infty) := \left\{ f : f \in C_\mu[0,\infty) \text{ and } \lim_{x\to\infty} \frac{|f(x)|}{e^{\mu x}} = k,\ k \text{ is constant.} \right\}.$$

For $f \in C^k_\mu[0,\infty)$ we use the following modulus of continuity:

$$\Omega_\mu(f;\delta) = \sup_{\substack{x,t\in[0,b_n] \\ |e^{\mu t}-e^{\mu x}|\le\delta}} \frac{|f(x)-f(t)|}{[|e^{\mu t}-e^{\mu x}|+1]\,e^{\mu x}}.$$

In [7], the authors proved the most general form of the following lemmas.

In the following, we give the main properties of the modulus of continuity.

Lemma 3 ([7]) *If* $f \in C_\mu[0,\infty)$ *and* $\lambda > 0$, *then*

$$\Omega_\mu(f;\lambda\delta) \le (1+\lambda)(1+\delta)\,\Omega_\mu(f;\delta).$$

holds for every $\delta > 0$.

Lemma 4 ([7]) *For* $\delta > 0$, $f \in C_\mu[0,\infty)$ *and* $x, t \in [0, b_n]$, *the inequality*

$$|f(t)-f(x)| \le 2e^{\mu x}(1+\delta)^2\left(1+\frac{\left(e^{\mu x}-e^{\mu t}\right)^2}{\delta^2}\right)\Omega_\mu(f;\delta)$$

holds.

Lemma 5 ([7]) *For any* $f \in C^k_\mu[0,\infty)$, *we have*

$$\lim_{\delta\to 0}\Omega_\mu(f;\delta) = 0.$$

Quantitative approximation theorems for sequences of linear positive operators play an important role not only in approximating functions, but also in estimating the error of the approximation. One of the most important convergence results in approximation theory is the Voronovskaya theorem. Roughly speaking, it is obtained to describe the rate of pointwise convergence.

Moreover, the other results presented in this paper are a quantitative-Voronovskaya type and a Grüss–Voronovskaya type theorems for the new operators. For more details, see [1]. Recently, Gal and Gonska obtained a Voronovskaya type theorem with the aid of Grüss inequality for Bernstein operators in [8] and called it Grüss–Voronovskaya type theorem. In this paper, we extend some of these results for our operators $\mathcal{C}_n$.

First, in the following theorem, we give quantitative type theorem for our operator $\mathcal{C}_n$:

Theorem 1 *For* $f \in C_{\mu}^{k}[0,\infty)$ *and* $x \in [0, b_n]$, *we have*

$$\begin{aligned}|\mathcal{C}_n f(x) - f(x)| &\leq 8e^{\mu x}(1 + \mathcal{C}_n e_0(x))\left(1 + e^{\mu x}\right)\Omega_{\mu}\left(f; \sqrt{(\mathcal{C}_n e_0(x) - 1)}\right) \\ &\quad + f(x)\,|(\mathcal{C}_n e_0(x) - 1)|\,.\end{aligned}$$

Proof Suppose that $\delta < 1$. Using Lemma 3, 4 and (2.2) , we have

$$\begin{aligned}&|\mathcal{C}_n f(x) - f(x)| \\ &\quad\leq 2e^{\mu x}(1+\delta)^2\left(\mathcal{C}_n e_0(x) + \frac{1}{\delta^2}\mathcal{C}_n\left(\exp^2_{\mu,x}; x\right)\right)\Omega_{\mu}(f;\delta) + f(x)\,|(\mathcal{C}_n e_0(x) - 1)| \\ &\quad\leq 8e^{\mu x}(1 + \mathcal{C}_n e_0(x))\,\Omega_{\mu}\left(f; \sqrt{\mathcal{C}_n\left(\exp^2_{\mu,x}; x\right)}\right) + f(x)\,|(\mathcal{C}_n e_0(x) - 1)| \\ &\quad\leq 8e^{\mu x}(1 + \mathcal{C}_n e_0(x))\left(1 + e^{\mu x}\right)\Omega_{\mu}\left(f; \sqrt{(\mathcal{C}_n e_0(x) - 1)}\right) + f(x)\,|(\mathcal{C}_n e_0(x) - 1)|\,.\end{aligned}$$

□

We have that our operator has a different approach charecteristics

Remark 1 If in the previous theorem, we assume

$$\delta^2 = \lambda_n(x) := (\mathcal{C}_n e_0(x) - 1)\,,$$

then the estimate reads as

$$|\mathcal{C}_n f(x) - f(x)| \leq f(x)\,\lambda_n(x) + 8e^{\mu x}(1 + \mathcal{C}_n e_0(x))\left(1 + e^{\mu x}\right)\Omega_{\mu}\left(f; \sqrt{(\mathcal{C}_n e_0(x) - 1)}\right).$$

Hence, velocity of convergence of $\mathcal{C}_n f(x)$ to $f(x)$ is managed by the velocity of convergence of $\mathcal{C}_n e_0(x)$ to $e_0(x) = 1$, or equivalently, the one of $\lambda_n(x)$ to 0, and this is given by the undermentioned limit, that can be easily computed by elementary calculus.

$$\lim_{n\to\infty} n\left(\mathcal{C}_n e_0(x) - 1\right) = \lim_{n\to\infty} n\lambda_n(x)$$
$$= \lim_{n\to\infty} n\left(e^{\mu x-\mu b_n}\left(e^{\frac{\mu b_n}{n}} + 1 - e^{\frac{\mu x}{n}}\right)^n - 1\right) = \mu^2 x.$$

Now, we state quantitative-Voronovskaya type theorem for $\mathcal{C}_n$:

Theorem 2 *If* $f \in C_\mu^k[0,\infty)$ *and* $x \in (0, b_n)$, *then we get*

$$\left|\mathcal{C}_n f(x) - f(x) - \left(\mathcal{C}_n e_0(x) - 1\right)\left(f(x) - \frac{3}{2}\mu^{-1} f'(x) + \frac{1}{2}\mu^{-2} f''(x)\right)\right|$$
$$\leq 8e^{\mu x}\mathcal{C}_n\left(\exp^2_{\mu,x}; x\right)\Omega_\mu\left(\left(f\circ\log_\mu\right)''(\exp_\mu); \sqrt{\frac{\mathcal{C}_n\left(\exp^4_{\mu,x}; x\right)}{\mathcal{C}_n\left(\exp^2_{\mu,x}; x\right)}}\right).$$

Proof By Taylor's theorem, we have

$$f(t) = \left(f\circ\log_\mu\right)\left(e^{\mu t}\right)$$
$$= \left(f\circ\log_\mu\right)\left(e^{\mu x}\right) + \left(f\circ\log_\mu\right)'\left(e^{\mu x}\right)\exp_{\mu,x}(t) + \frac{1}{2}\left(f\circ\log_\mu\right)''\left(e^{\mu x}\right)\exp^2_{\mu,x}(t)$$
$$+ h(x,t)\exp^2_{\mu,x}(t),$$

where

$$h_x(t) := h(x,t) = \frac{\left(f\circ\log_\mu\right)''(\exp_\mu)(\xi) - \left(f\circ\log_\mu\right)''(\exp_\mu)(x)}{2}$$

with ξ a number between x and t. Applying the operator $\mathcal{C}_n$ to both side of above inequality, we get

$$\mathcal{C}_n f(x) = \mathcal{C}_n e_0(x) f(x) + \left(f\circ\log_\mu\right)'\left(e^{\mu x}\right)\mathcal{C}_n\left(\exp_{\mu,x}; x\right) + \frac{1}{2}\left(f\circ\log_\mu\right)''\left(e^{\mu x}\right)\mathcal{C}_n\left(\exp^2_{\mu,x}; x\right)$$
$$+ \mathcal{C}_n\left(h_x \exp^2_{\mu,x}; x\right).$$

Using Lemma 4 and the fact that $\left|e^{\mu\xi} - e^{\mu x}\right| \leq \left|e^{\mu t} - e^{\mu x}\right|$, then we can write

$$|h(x,t)| \leq e^{\mu x}(1+\delta)^2\left(1 + \frac{\left(e^{\mu\xi} - e^{\mu x}\right)^2}{\delta^2}\right)\Omega_\mu\left(\left(f\circ\log_\mu\right)''(\exp_\mu); \delta\right)$$
$$\leq e^{\mu x}(1+\delta)^2\left(1 + \frac{\left(e^{\mu t} - e^{\mu x}\right)^2}{\delta^2}\right)\Omega_\mu\left(\left(f\circ\log_\mu\right)''(\exp_\mu); \delta\right).$$

Suppose that $\delta < 1$. Thus, we can write

$$|h(x,t)| \leq 4e^{\mu x}\left(1+\frac{\left(e^{\mu t}-e^{\mu x}\right)^2}{\delta^2}\right)\Omega_\mu\left(\left(f\circ\log_\mu\right)''\left(\exp_\mu\right);\delta\right).$$

Multiplying this relation with $\exp^2_{\mu,x}$ and applying the operator $\mathcal{C}_n$, we get

$$\mathcal{C}_n\left(h_x\exp^2_{\mu,x};x\right) \leq 4e^{\mu x}\left(\mathcal{C}_n\left(\exp^2_{\mu,x};x\right)+\frac{1}{\delta^2}\mathcal{C}_n\left(\exp^4_{\mu,x};x\right)\right)\Omega_\mu\left(\left(f\circ\log_\mu\right)''\left(\exp_\mu\right);\delta\right). \tag{3.1}$$

Using (2.1) and (2.2), we get

$$\begin{aligned}\mathcal{C}_n f(x)-f(x) &= f(x)\left(\mathcal{C}_n e_0(x)-1\right)+\left(f\circ\log_\mu\right)'\left(e^{\mu x}\right)e^{\mu x}\left(1-\mathcal{C}_n e_0(x)\right)\\ &\quad+\frac{1}{2}\left(f\circ\log_\mu\right)''\left(e^{\mu x}\right)e^{2\mu x}\left(\mathcal{C}_n e_0(x)-1\right)\\ &\quad+\mathcal{C}_n\left(h_x\exp^2_{\mu,x};x\right).\end{aligned}$$

We know that, since

$$\left(f\circ\tau^{-1}\right)'=\left(f'\circ\tau^{-1}\right)\left(\tau^{-1}\right)'$$

and

$$\left(\tau^{-1}\right)'(\tau(t))=\frac{1}{\tau'(t)},$$

we have

$$\left(f\circ\tau^{-1}\right)'(\tau(t))=\frac{f'(t)}{\tau'(t)}.$$

Also since

$$\left(f\circ\tau^{-1}\right)''=\left(f''\circ\tau^{-1}\right)\left(\left(\tau^{-1}\right)'\right)^2+\left(f'\circ\tau^{-1}\right)\left(\tau^{-1}\right)''$$

and

$$\frac{d}{dt}\left(\left(\tau^{-1}\right)'(\tau(t))\right)=\left(\tau^{-1}\right)''(\tau(t))\,\tau'(t)=-\frac{\tau''(t)}{\left(\tau'(t)\right)^2},$$

we get

$$\left(f\circ\tau^{-1}\right)''(\tau(t))=\frac{f''(t)}{\left(\tau'(t)\right)^2}-f'(t)\frac{\tau''(t)}{\left(\tau'(t)\right)^3}.$$

Therefore, since

$$\left(f\circ\log_\mu\right)'\left(e^{\mu x}\right)=e^{-\mu x}\mu^{-1}f'(x)$$

and

$$\left(f\circ\log_\mu\right)''\left(e^{\mu x}\right)=e^{-2\mu x}\left(\mu^{-2}f''(x)-\mu^{-1}f'(x)\right),$$

we can write

$$\left|\mathcal{C}_n f(x) - f(x) - (\mathcal{C}_n e_0(x) - 1)\left(f(x) - \frac{3}{2}\mu^{-1} f'(x) + \frac{1}{2}\mu^{-2} f''(x)\right)\right|$$
$$\leq 4e^{\mu x}\left(\mathcal{C}_n\left(\exp^2_{\mu,x}; x\right) + \frac{1}{\delta^2}\mathcal{C}_n\left(\exp^4_{\mu,x}; x\right)\right)\Omega_\mu\left(\left(f \circ \log_\mu\right)''\left(\exp_\mu\right); \delta\right)$$
$$= 4e^{\mu x}\mathcal{C}_n\left(\exp^2_{\mu,x}; x\right)\left(1 + \frac{1}{\delta^2}\frac{\mathcal{C}_n\left(\exp^4_{\mu,x}; x\right)}{\mathcal{C}_n\left(\exp^2_{\mu,x}; x\right)}\right)\Omega_\mu\left(\left(f \circ \log_\mu\right)''\left(\exp_\mu\right); \delta\right).$$

Choosing $\delta = \sqrt{\frac{\mathcal{C}_n(\exp^4_{\mu,x};x)}{\mathcal{C}_n(\exp^2_{\mu,x};x)}}$, we have desired result. □

Later, we express quantitative-Grüss–Voronovskaya type theorem for $\mathcal{C}_n$:

Theorem 3 *If $f, g \in C^k_\mu[0,\infty)$, then for all $x \in [0, b_n]$ and $n \in \mathbb{N}$ we have*

$$n\left|\mathcal{C}_n(fg)(x) - \mathcal{C}_n f(x)\mathcal{C}_n g(x) - x f'(x) g'(x)(\mathcal{C}_n e_0(x) - 1) + \mu^2 x f(x) g(x)(\mathcal{C}_n e_0(x) - 1)\right|$$
$$\leq \mathcal{G}_n(\mathcal{C}_n, (fg); x) + \|f\|_\mu e^{\mu x}\mathcal{G}_n(\mathcal{C}_n, g; x) + \|g\|_\mu e^{\mu x}\mathcal{G}_n(\mathcal{C}_n, f; x) + n I_n(f) I_n(g),$$

where

$$\mathcal{G}_n(\mathcal{C}_n, f; x) := 8e^{\mu x}\mathcal{C}_n\left(\exp^2_{\mu,x}; x\right)\Omega_\mu\left(\left(f \circ \log_\mu\right)''\left(\exp_\mu\right); \sqrt{\frac{\mathcal{C}_n\left(\exp^4_{\mu,x}; x\right)}{\mathcal{C}_n\left(\exp^2_{\mu,x}; x\right)}}\right)$$

and

$$I_n(f) := \frac{\left\|\left(f \circ \log_\mu\right)''\right\|_\mu e^{\mu x}}{2}\left\{\mathcal{C}_n\left(\exp^2_{\mu,x}; x\right) + \sqrt{\mathcal{C}_n\left(\exp^4_{\mu,x}; x\right)}\right\} + 2\mu^{-1} f'(x)\left|1 - \mathcal{C}_n e_0(x)\right|.$$

Also, $\mathcal{G}_n(\mathcal{C}_n, g; x)$, $\mathcal{G}_n(\mathcal{C}_n, (fg); x)$, and $I_n(g)$ are the analogous one.

Proof For $x \in [0,\infty)$ and $n \in \mathbb{N}$, it is easily seen that we can write

$$\mathcal{C}_n(fg)(x) - \mathcal{C}_n f(x)\mathcal{C}_n g(x) - x f'(x) g'(x)(\mathcal{C}_n e_0(x) - 1) + \mu^2 x g(x) f(x)(\mathcal{C}_n e_0(x) - 1)$$
$$= \left[\mathcal{C}_n(fg)(x) - (fg)(x) - (\mathcal{C}_n e_0(x) - 1)\left(\mu^2 x (fg)(x) - \frac{3}{2}\mu x (fg)'(x) + \frac{1}{2}x (fg)''(x)\right)\right]$$
$$-f(x)\left[\mathcal{C}_n g(x) - g(x) - (\mathcal{C}_n e_0(x) - 1)\left(\mu^2 x g(x) - \frac{3}{2}\mu x g'(x) + \frac{1}{2}x g''(x)\right)\right]$$
$$-g(x)\left[\mathcal{C}_n f(x) - f(x) - (\mathcal{C}_n e_0(x) - 1)\left(\mu^2 x f(x) - \frac{3}{2}\mu x f'(x) + \frac{1}{2}x f''(x)\right)\right]$$
$$+[g(x) - \mathcal{C}_n g(x)][\mathcal{C}_n f(x) - f(x)]$$
$$= I_1 + I_2 + I_3 + I_4.$$

So, we get

$$\left|\mathcal{C}_n(fg)(x) - \mathcal{C}_n f(x)\,\mathcal{C}_n g(x) - xf'(x)\,g'(x)\,(\mathcal{C}_n e_0(x) - 1) + \mu^2 x g(x) f(x)(\mathcal{C}_n e_0(x) - 1)\right|$$
$$\leq |I_1| + |I_2| + |I_3| + |I_4|.$$

By Theorem 2, we have the estimates

$$|I_1| \leq 8e^{\mu x}\mathcal{C}_n\left(\exp^2_{\mu,x}; x\right)\Omega_\mu\left(\left((fg)\circ\log_\mu\right)''(\exp_\mu); \sqrt{\frac{\mathcal{C}_n\left(\exp^4_{\mu,x}; x\right)}{\mathcal{C}_n\left(\exp^2_{\mu,x}; x\right)}}\right),$$

$$|I_2| \leq \|f\|_\mu\, 8e^{2\mu x}\mathcal{C}_n\left(\exp^2_{\mu,x}; x\right)\Omega_\mu\left(\left(g\circ\log_\mu\right)''(\exp_\mu); \sqrt{\frac{\mathcal{C}_n\left(\exp^4_{\mu,x}; x\right)}{\mathcal{C}_n\left(\exp^2_{\mu,x}; x\right)}}\right)$$

and

$$|I_3| \leq \|g\|_\mu\, 8e^{2\mu x}\mathcal{C}_n\left(\exp^2_{\mu,x}; x\right)\Omega_\mu\left(\left(f\circ\log_\mu\right)''(\exp_\mu); \sqrt{\frac{\mathcal{C}_n\left(\exp^4_{\mu,x}; x\right)}{\mathcal{C}_n\left(\exp^2_{\mu,x}; x\right)}}\right).$$

On the other hand, since $f \in C^k_\mu[0,\infty)$ we write

$$\mathcal{C}_n(f; x) - f(x) = \left(f\circ\log_\mu\right)'(e^{\mu x})\,\mathcal{C}_n\left(\exp_{\mu,x}; x\right) + \frac{1}{2}\mathcal{C}_n\left(\left(f\circ\log_\mu\right)''\left(e^{\mu\xi}\right)\exp^2_{\mu,x}; x\right)$$

and so we get

$$|\mathcal{C}_n(f; x) - f(x)| \leq \mu^{-1}f'(x)\,|1 - \mathcal{C}_n e_0(x)| + \frac{1}{2}\mathcal{C}_n\left(\left(f\circ\log_\mu\right)''\left(e^{\mu\xi}\right)\exp^2_{\mu,x}; x\right)$$
$$\leq \mu^{-1}f'(x)\,|1 - \mathcal{C}_n e_0(x)| + \left\|\left(f\circ\log_\mu\right)''\right\|_\mu \frac{1}{2}\mathcal{C}_n\left(e^{\mu\xi}\exp^2_{\mu,x}; x\right)$$

where ξ is a number between t and x. If $t < \xi < x$, then $e^{\mu\xi} \leq e^{\mu x}$. In this case, we have

$$|\mathcal{C}_n(f; x) - f(x)| \leq \frac{\left\|\left(f\circ\log_\mu\right)''\right\|_\mu e^{\mu x}}{2}\mathcal{C}_n\left(\exp^2_{\mu,x}; x\right) + \mu^{-1}f'(x)\,|1 - \mathcal{C}_n e_0(x)|$$

or if $x < \xi < t$, then $e^{\mu\xi} \leq e^{\mu t}$. In this case, with the help of Hölder's inequality, we get

$$
\begin{aligned}
|\mathcal{C}_n(f;x)-f(x)| &\le \frac{\left\|\left(f\circ\log_\mu\right)''\right\|_\mu}{2}\mathcal{C}_n\left(\exp_\mu\exp^2_{\mu,x};x\right)+\mu^{-1}f'(x)\left|1-\mathcal{C}_ne_0(x)\right| \\
&\le \frac{\left\|\left(f\circ\log_\mu\right)''\right\|_\mu}{2}\mathcal{C}_n\left(\exp^2_\mu;x\right)^{\frac{1}{2}}\mathcal{C}_n\left(\exp^4_{\mu,x};x\right)^{\frac{1}{2}}+\mu^{-1}f'(x)\left|1-\mathcal{C}_ne_0(x)\right| \\
&= \frac{\left\|\left(f\circ\log_\mu\right)''\right\|_\mu e^{\mu x}}{2}\sqrt{\mathcal{C}_n\left(\exp^4_{\mu,x};x\right)}+\mu^{-1}f'(x)\left|1-\mathcal{C}_ne_0(x)\right|.
\end{aligned}
$$

Hence, we gain for two cases of ξ that

$$
\begin{aligned}
|\mathcal{C}_n(f;x)-f(x)| \le{}& \frac{\left\|\left(f\circ\log_\mu\right)''\right\|_\mu e^{\mu x}}{2}\left\{\mathcal{C}_n\left(\exp^2_{\mu,x};x\right)+\sqrt{\mathcal{C}_n\left(\exp^4_{\mu,x};x\right)}\right\} \\
&+2\mu^{-1}f'(x)\left|1-\mathcal{C}_ne_0(x)\right| := I_n(f).
\end{aligned}
$$

A similar reasoning yields $|\mathcal{C}_n(g;x)-g(x)|\le I_n(g)$. Therefore we get

$$
\begin{aligned}
& n\Big|\mathcal{C}_n(fg)(x)-\mathcal{C}_nf(x)\mathcal{C}_ng(x)-xf'(x)g'(x)\left(\mathcal{C}_ne_0(x)-1\right) \\
&\qquad +\mu^2xg(x)f(x)\left(\mathcal{C}_ne_0(x)-1\right)\Big| \\
&\le 8e^{\mu x}\mathcal{C}_n\left(\exp^2_{\mu,x};x\right)\Omega_\mu\left(\left((fg)\circ\log_\mu\right)''\left(\exp_\mu\right);\sqrt{\frac{\mathcal{C}_n\left(\exp^4_{\mu,x};x\right)}{\mathcal{C}_n\left(\exp^2_{\mu,x};x\right)}}\right) \\
&\quad +\|f\|_\mu 8e^{2\mu x}\mathcal{C}_n\left(\exp^2_{\mu,x};x\right)\Omega_\mu\left(\left(g\circ\log_\mu\right)''\left(\exp_\mu\right);\sqrt{\frac{\mathcal{C}_n\left(\exp^4_{\mu,x};x\right)}{\mathcal{C}_n\left(\exp^2_{\mu,x};x\right)}}\right) \\
&\quad +\|g\|_\mu 8e^{2\mu x}\mathcal{C}_n\left(\exp^2_{\mu,x};x\right)\Omega_\mu\left(\left(f\circ\log_\mu\right)''\left(\exp_\mu\right);\sqrt{\frac{\mathcal{C}_n\left(\exp^4_{\mu,x};x\right)}{\mathcal{C}_n\left(\exp^2_{\mu,x};x\right)}}\right) \\
&\quad +nI_n(f)I_n(g),
\end{aligned}
$$

as desired. □

Theorem 4 *For each $n\in\mathbb{N}$ and $x\in[0,\infty)$, we have*

$$
\lim_{n\to\infty}\left(\frac{\mathcal{C}_nf}{\exp_\mu}\right)'(x)=\left(\frac{f}{\exp_\mu}\right)'(x).
$$

Proof Using (1.5), we obtain

$$\left(\frac{\mathcal{C}_n f}{\exp_\mu}\right)'(a_n(x)) = \left(C_n\left(\frac{f}{\exp_\mu}\right)(x)\right)'$$
$$= \left[\sum_{k=0}^{n}\left(\frac{f}{\exp_\mu}\right)\left(\frac{kb_n}{n}\right)p_{n,k}(a_n(x))\right]'$$
$$= \frac{a_n'(x)}{a_n(x)\left(1-\frac{a_n(x)}{b_n}\right)}\sum_{k=0}^{n}\left(\frac{f}{\exp_\mu}\right)\left(\frac{kb_n}{n}\right)$$
$$\times p_{n,k}(a_n(x))\frac{n}{b_n}\left(\frac{kb_n}{n}-a_n(x)\right). \tag{3.2}$$

First, we take into account the case $x = 0$.

From (3.2), we have

$$\left(\frac{\mathcal{C}_n f}{\exp_\mu}\right)'(a_n(x)) = -\left(\frac{f}{\exp_\mu}\right)(0)\,na_n'(x)\left(1-\frac{a_n(x)}{b_n}\right)^{n-1}$$
$$+\left(\frac{f}{\exp_\mu}\right)\left(\frac{b_n}{n}\right)na_n'(x)\left(1-n\frac{a_n(x)}{b_n}\right)\left(1-\frac{a_n(x)}{b_n}\right)^{n-2}$$
$$\sum_{k=2}^{n}\left(\frac{f}{\exp_\mu}\right)\left(\frac{kb_n}{n}\right)\binom{n}{k}a_n'(x)\left(k-n\frac{a_n(x)}{b_n}\right)\left(\frac{a_n(x)}{b_n}\right)^{k-1}$$
$$\times\left(1-\frac{a_n(x)}{b_n}\right)^{n-k-1}.$$

For $x = 0$, because of $a_n(x) = 0$, we get

$$\left(\frac{\mathcal{C}_n f}{\exp_\mu}\right)'(0) = -na_n'(x)\left(\frac{f}{\exp_\mu}\right)(0) + na_n'(x)\left(\frac{f}{\exp_\mu}\right)\left(\frac{b_n}{n}\right)$$
$$= a_n'(x)\frac{\left(\frac{f}{\exp_\mu}\right)\left(\frac{b_n}{n}\right)-\left(\frac{f}{\exp_\mu}\right)(0)}{\frac{1}{n}-0}.$$

If the limit of both sides is taken above equality, then we obtain

$$\lim_{n\to\infty}\left(\frac{\mathcal{C}_n f}{\exp_\mu}\right)'(0) = \left(\frac{f}{\exp_\mu}\right)'(0).$$

Now, let's $x > 0$.

We consider the following function:

$$\lambda_x(t)=\frac{\left(\frac{f}{\exp_\mu}\circ\log_\mu\right)\left(e^{\mu t}\right)-\left(\frac{f}{\exp_\mu}\circ\log_\mu\right)\left(e^{\mu x}\right)}{e^{\mu t}-e^{\mu x}}-\left(\frac{f}{\exp_\mu}\circ\log_\mu\right)'\left(e^{\mu x}\right).$$

In that case, $\lim\limits_{t\to x}\lambda_x(t)=0$. We get

$$\left(\frac{f}{\exp_\mu}\circ\log_\mu\right)\left(e^{\mu t}\right)=\left(\frac{f}{\exp_\mu}\circ\log_\mu\right)\left(e^{\mu x}\right)+\left(\frac{f}{\exp_\mu}\circ\log_\mu\right)'\left(e^{\mu x}\right)\left(e^{\mu t}-e^{\mu x}\right)$$
$$+\lambda_x(t)\left(e^{\mu t}-e^{\mu x}\right).$$

If $\frac{kb_n}{n}$ is changed instead of t, then we have

$$\left(\frac{f}{\exp_\mu}\circ\log_\mu\right)\left(e^{\mu\frac{kb_n}{n}}\right)=\left(\frac{f}{\exp_\mu}\circ\log_\mu\right)\left(e^{\mu x}\right)+\left(\frac{f}{\exp_\mu}\circ\log_\mu\right)'\left(e^{\mu x}\right)\left(e^{\mu\frac{kb_n}{n}}-e^{\mu x}\right)$$
$$+\lambda_x(t)\left(e^{\mu\frac{kb_n}{n}}-e^{\mu x}\right).$$

If this equality is written in (3.2), then we attain

$$\begin{aligned}\left(\frac{C_nf}{\exp_\mu}\right)'(x)&=\frac{a_n'(x)}{a_n(x)\left(1-\frac{a_n(x)}{b_n}\right)}\left[\left(\frac{f}{\exp_\mu}\right)(x)\sum_{k=0}^{n}p_{n,k}(a_n(x))\frac{n}{b_n}\left(\frac{kb_n}{n}-a_n(x)\right)\right.\\&\quad+\frac{n}{b_n}\frac{\left(\frac{f}{\exp_\mu}\right)'(x)}{\mu e^{\mu x}}\sum_{k=0}^{n}\left(e^{\mu\frac{kb_n}{n}}-e^{\mu x}\right)p_{n,k}(a_n(x))\left(\frac{kb_n}{n}-a_n(x)\right)\\&\quad\left.+\frac{n}{b_n}\sum_{k=0}^{n}\lambda_x(t)\left(e^{\mu\frac{kb_n}{n}}-e^{\mu x}\right)p_{n,k}(a_n(x))\left(\frac{kb_n}{n}-a_n(x)\right)\right]\\&=\frac{a_n'(x)}{a_n(x)\left(1-\frac{a_n(x)}{b_n}\right)}\frac{n}{b_n}\left[\left(\frac{f}{\exp_\mu}\right)(x)C_n(t-a_n(x);a_n(x))\right.\\&\quad+\frac{\left(\frac{f}{\exp_\mu}\right)'(x)}{\mu e^{\mu x}}C_n\left(\left(e^{\mu t}-e^{\mu x}\right)(t-a_n(x));a_n(x)\right)\\&\quad\left.+C_n\left(\lambda_x(t)\left(e^{\mu t}-e^{\mu x}\right)(t-a_n(x));a_n(x)\right)\right].\end{aligned}$$

We know

$$C_n(t-a_n(x);a_n(x))=0.$$

We can write

$$\begin{aligned}C_n\left(\left(e^{\mu t}-e^{\mu x}\right)(t-a_n(x));a_n(x)\right)&=C_n\left(te^{\mu t}-te^{\mu x}-a_n(x)e^{\mu t}+a_n(x)e^{\mu x};a_n(x)\right)\\&=C_n\left(te^{\mu t};a_n(x)\right)-e^{\mu x}C_n(t;a_n(x))\\&\quad-a_n(x)C_n\left(e^{\mu t};a_n(x)\right)+a_n(x)e^{\mu x}C_n(1;a_n(x)).\end{aligned}$$

Because

$$C_n\left(te^{\mu t};a_n(x)\right)=a_n(x)\,e^{\frac{\mu b_n+\mu x(n-1)}{n}},$$

$$C_n\left(e^{\mu t};a_n(x)\right)=e^{\mu x},$$

$$\lim_{n\to\infty}a_n(x)=x$$

and

$$\lim_{n\to\infty}a_n'(x)=\lim_{n\to\infty}b_n\frac{\frac{\mu}{n}e^{\frac{\mu x}{n}}}{e^{\frac{\mu b_n}{n}}-1}=1,$$

we have

$$\lim_{n\to\infty}\frac{a_n'(x)}{a_n(x)\left(1-\frac{a_n(x)}{b_n}\right)}\frac{n}{b_n}\frac{1}{\mu e^{\mu x}}C_n\left(\left(e^{\mu t}-e^{\mu x}\right)(t-a_n(x));a_n(x)\right)=1.$$

Now, we use Hölder inequality:

$$0\le C_n\left(\lambda_x(t)\left(e^{\mu t}-e^{\mu x}\right)(t-a_n(x));a_n(x)\right)\le$$
$$\left(C_n\left(\lambda_x^2(t);a_n(x)\right)\right)^{\frac{1}{2}}\left(C_n\left(\left(e^{\mu t}-e^{\mu x}\right)^2;a_n(x)\right)\right)^{\frac{1}{2}}\left(C_n\left((t-a_n(x))^2;a_n(x)\right)\right)^{\frac{1}{2}}.$$

From Korovkin theorem, we know

$$\lim_{n\to\infty}C_n\left(\lambda_x^2(t);a_n(x)\right)=\lambda_x^2(x)=0.$$

As

$$\lim_{n\to\infty}C_n\left(\left(e^{\mu t}-e^{\mu x}\right)^2;a_n(x)\right)=0$$

and

$$\lim_{n\to\infty}C_n\left((t-a_n(x))^2;a_n(x)\right)=0,$$

we obtain desired result. □

4 Variation Detracting Property of Bernstein–Chlodowsky Operators

The first study about the variation detracting property and the convergence in variation of a sequence of linear positive operators was come out by Lorentz (1953). He proved that B_n have

$$V_{[0,1]}[B_n f] \leq V_{[0,1]}[f]$$

and it is called the variation detracting property.

The main purpose of this section is to confirm the variation detracting property and convergence in the variation seminorm for the Bernstein–Chlodowsky operators. We firstly give the definitions related to variation detracting property.

Definition 1 ([11]) The least upper bound of the set of all possible sums V is called the total variation of the function $f(x)$ on $[a, b]$ and is designated by $V_{[a,b]}[f]$.

Definition 2 ([2]) The class of all functions of bounded variation on I is called BV space and denoted by $BV(I)$. This space can be endowed both with seminorm $|.|_{BV(I)}$ and with a norm, $\|.\|_{BV(I)}$, where

$$|f|_{BV(I)} := V_I[f] \quad , \quad \|f\|_{BV(I)} := V_I[f] + |f(a)| \, ,$$

$f \in BV(I)$, a being any fixed point of I.

Definition 3 ([3]) Let $I \subseteq \mathbb{R}$ be a fixed integral, and $V_I[f]$ the total variation of the function $f : I \to \mathbb{R}$. The class of all bounded functions of bounded variation on I endowed with the seminorm

$$\|f\|_{TV(I)} := V_I[f]$$

is called TV space and is denoted by $TV(I)$.

Definition 4 ([11]) Let $f(x)$ be a finite function defined on the closed interval $[a, b]$. Suppose that for every $\epsilon > 0$, there exists a $\delta > 0$ such that

$$\left| \sum_{k=1}^{n} \{f(b_k) - f(a_k)\} \right| < \epsilon$$

for all numbers $a_1, b_1, \ldots, a_n, b_n$ such that $a_1 < b_1 \leq a_2 < b_2 \leq \cdots \leq a_n < b_n$ and

$$\sum_{k=1}^{n} (b_k - a_k) < \delta.$$

Then the function $f(x)$ is said to be absolutely continuous. The class of all absolutely continuous function on $[a, b]$ is denoted by $AC[a, b]$.

Now, we give the variation detracting property of the Bernstein–Chlodowsky operators:

Theorem 5 *If* $f \in TV[0, b_n]$, *then* $V_{[0,b_n]}\left[\frac{C_n f}{\exp_\mu}\right] \leq V_{[0,b_n]}\left[\frac{f}{\exp_\mu}\right]$.

Proof As $\frac{\mathcal{C}_n f}{\exp_\mu}$ polynomials are differentiable and their derivatives are integrable, by [9, 10], the equality

$$\left\| \frac{\mathcal{C}_n f}{\exp_\mu} \right\|_{TV[0,b_n]} = V_{[0,b_n]} \left[\frac{\mathcal{C}_n f}{\exp_\mu} \right] = \int_0^{b_n} \left| \frac{d}{dx} \frac{\mathcal{C}_n}{\exp_\mu} (f; x) \right| dx$$

is implemented. From (1.5), we can write

$$\begin{aligned} V_{[0,b_n]} \left[\frac{\mathcal{C}_n f}{\exp_\mu} \right] &= \int_0^{b_n} \left| \frac{d}{dx} \frac{\mathcal{C}_n}{\exp_\mu} (f; x) \right| dx \\ &= \int_0^{b_n} \left| \frac{d}{dx} \left[\frac{\mathcal{C}_n}{\exp_\mu} (f; a_n (x)) \right] \right| dx. \end{aligned}$$

By Theorem 3.13 in [6], we get

$$\begin{aligned} V_{[0,b_n]} \left[\frac{\mathcal{C}_n f}{\exp_\mu} \right] &= \int_0^{b_n} \left| \frac{n}{b_n} \sum_{k=0}^{n-1} p_{n-1,k} (a_n (x)) \Delta_{\frac{b_n}{n}} \frac{f}{\exp_\mu} \left(\frac{k}{n} b_n \right) \right| a_n' (x) \, dx \\ &\leq \frac{n}{b_n} \sum_{k=0}^{n-1} \int_0^{b_n} \left| p_{n-1,k} (a_n (x)) \Delta_{\frac{b_n}{n}} \frac{f}{\exp_\mu} \left(\frac{k}{n} b_n \right) \right| a_n' (x) \, dx \\ &= \frac{n}{b_n} \sum_{k=0}^{n-1} \left| \Delta_{\frac{b_n}{n}} \frac{f}{\exp_\mu} \left(\frac{k}{n} b_n \right) \right| \int_0^{b_n} p_{n-1,k} (a_n (x)) \, a_n' (x) \, dx. \end{aligned}$$

If $\frac{a_n(x)}{b_n} = y$ is changed, then we have

$$V_{[0,b_n]} \left[\frac{\mathcal{C}_n f}{\exp_\mu} \right] \leq n \sum_{k=0}^{n-1} \binom{n-1}{k} \left| \Delta_{\frac{b_n}{n}} \frac{f}{\exp_\mu} \left(\frac{k}{n} b_n \right) \right| \int_0^1 y^k (1-y)^{n-k-1} \, dy.$$

Now, let's consider the integral on the left side of the inequality. From definition of Beta function, we obtain

$$\begin{aligned} V_{[0,b_n]} \left[\frac{\mathcal{C}_n f}{\exp_\mu} \right] &\leq n \sum_{k=0}^{n-1} \binom{n-1}{k} \left| \Delta_{\frac{b_n}{n}} \frac{f}{\exp_\mu} \left(\frac{k}{n} b_n \right) \right| \frac{1}{n \binom{n-1}{k}} \\ &= \sum_{k=0}^{n-1} \left| \Delta_{\frac{b_n}{n}} \frac{f}{\exp_\mu} \left(\frac{k}{n} b_n \right) \right| \end{aligned}$$

$$\leq \sup \sum_{k=0}^{n-1} \left| \Delta_{\frac{b_n}{n}} \frac{f}{\exp_\mu} \left(\frac{k}{n} b_n \right) \right|$$
$$= \sup \sum_{k=0}^{n-1} \left| \frac{f}{\exp_\mu} \left(\frac{k+1}{n} b_n \right) - \frac{f}{\exp_\mu} \left(\frac{k}{n} b_n \right) \right|$$
$$= V_{[0,b_n]} \left[\frac{f}{\exp_\mu} \right] = \left\| \frac{f}{\exp_\mu} \right\|_{TV[0,b_n]}.$$

□

Theorem 6 *Let $f \in TV\,[0, b_n]$. There holds*

$$\lim_{n\to\infty} \left\| \frac{\mathcal{C}_n f}{\exp_\mu} - \frac{f}{\exp_\mu} \right\|_{TV[0,\infty)} = 0 \iff \frac{f}{\exp_\mu} \in AC\,[0, b_n]\,.$$

Proof Since $\frac{f}{\exp_\mu}$ and $\frac{\mathcal{C}_n f}{\exp_\mu} \in AC\,[0, b_n]$, then $\frac{\mathcal{C}_n f}{\exp_\mu} - \frac{f}{\exp_\mu} \in AC\,[0, b_n]$. By Theorem 3.13 and Remark 3.20 in [6], it is written

$$\lim_{n\to\infty} \left\| \frac{\mathcal{C}_n f}{\exp_\mu} - \frac{f}{\exp_\mu} \right\|_{TV[0,\infty)} = \lim_{n\to\infty} \int_0^\infty \left| \left(\frac{\mathcal{C}_n f}{\exp_\mu} \right)' (x) - \left(\frac{f}{\exp_\mu} \right)' (x) \right| dx.$$

From Theorem 4, it can be seen easily that $\left(\frac{\mathcal{C}_n f}{\exp_\mu} \right)' (x) \longrightarrow \left(\frac{f}{\exp_\mu} \right)' (x)$ as $n \to \infty$. Therefore,

$$\lim_{n\to\infty} \left\| \frac{\mathcal{C}_n f}{\exp_\mu} - \frac{f}{\exp_\mu} \right\|_{TV[0,\infty)} = 0.$$

Conversely, let $\lim_{n\to\infty} \left\| \frac{\mathcal{C}_n f}{\exp_\mu} - \frac{f}{\exp_\mu} \right\|_{TV[0,\infty)} = 0$. This means that $\frac{\mathcal{C}_n f}{\exp_\mu} \longrightarrow \frac{f}{\exp_\mu}$ in TV space. Therefore $\frac{f}{\exp_\mu}$ is in AC because of AC is closed. □

Acknowledgements The first author of this paper would like to thank TUBITAK (The Scientific and Technological Research Council of Turkey) for their financial supports during his Ph.D. studies.

References

1. T. Acar, A. Aral, I. Raşa, The new forms of Voronovskaya's theorem in weighted spaces. Positivity **20**, 25–40 (2016)
2. O. Agratini, On the variation detracting property of a class of operators. Appl. Math. Lett. **19**, 1261–1264 (2006)

3. C. Bardaro, P.L. Butzer, R.L. Stens, G. Vinti, Convergence in variation and rates of approximation for Bernstein-type polynomials and singular convolution integrals. Munich. Anal. **23**(4), 299–346 (2003)
4. P.L. Butzer, On the extension of Bernstein polynomials to the in nite interval. Proc. Am. Math. Soc. **5**, 547–553 (1954)
5. P.L. Butzer, H. Karslı, Voronovskaya-type theorems for derivatives of the Bernstein–Chlodovsky polynomials and the Szász-Mirakyan operator. Comment. Math. **49**(1), 3358 (2009)
6. R. Demiray, Convergence in the variation seminorm of Bernstein and Bernstein Chlodovsky polynomials (Supervisor: Harun Karslı), Master thesis, The Near East University, 2013
7. A.D. Gadjiev, A. Aral, The estimates of approximation by using a new type of weighted modulus of continuity. Comput. Math. Appl. **54**(1), 127–135 (2007)
8. S.G. Gal, H. Gonska, Grüss and Grü ss-Voronovskaya-type estimates for some Bernstein-type polynomials of real and complex variables (2014), arXiv:1401.6824v1
9. R. Kannan, K.C. King, *Advanced Analysis on the Real Line* (Springer, New York, 1996), p. 259
10. H. Karsli, Ö.Ö. Yılık, F.T. Yeş ildal, Convergence of the Bernstein–Durrmeyer operators in variation seminorm. Results Math. **72**(3), 1257–1270 (2017)
11. I.P. Natanson, *Theory of Functions of a Real Variable*, vol. 1–2 (Ungar, New York, 1955–1964)

Iterative Approximation of Common Fixed Points in Kasahara Spaces

Alexandru-Darius Filip and Voichiţa Adriana Radu

Abstract Let $(X, \to, d)$ be a Kasahara space and $f, g : X \to X$ be two operators. Let $x_0 \in X$ and $x_1 := f(x_0), x_2 := g(x_1), \ldots, x_{2n} := g(x_{2n-1}), x_{2n+1} := f(x_{2n}), \ldots$ The aim of this paper is to give conditions on the pair (f, g) such that the sequence $(x_n)_{n\in\mathbb{N}}$ converges with respect to $\to$ to a common fixed point of f and g.

Keywords Fixed point · Common fixed point · Kasahara space · Generalized Kasahara space · Matrix convergent to zero

2010 MSC 47H10 · 54H25

1 Introduction

In the mathematical literature, there are various common fixed point theorems, most of them are given for self-mappings defined on a metric space (X, d) satisfying some contractive conditions. Some of these theorems can be found in the work of Rus [14], Rus, Petruşel and Petruşel [17] and the references therein, Aliouche and Popa [1], Chandok [3], Cho and Bae [4], Pant and Bisht [11], Wang and Guo [20].

There are also some common fixed point results given in a more general setting, more precisely, in these results the metric space (X, d) is replaced by the L-space $(X, \to)$ endowed with a functional $d : X \times X \to \mathbb{R}_+$ which does not necessarily satisfy all of the metric axioms. In this sense, see Kasahara [7, 9, 10].

In 2010, Rus introduced in [16] the notion of Kasahara space. Let $(X, \to, d)$ be a Kasahara space and $f, g : X \to X$ be two operators. Let $x_0 \in X$ and $x_1 := f(x_0)$,

A.-D. Filip · V. A. Radu (✉)
Department of Statistics-Forecasts-Mathematics, Babes-Bolyai University, FSEGA, Cluj-Napoca, Romania
e-mail: voichita.radu@econ.ubbcluj.ro; voichita.radu@gmail.com

A.-D. Filip
e-mail: darius.filip@econ.ubbcluj.ro; darius.fsegamath@gmail.com

N. Deo et al. (eds.), *Mathematical Analysis I: Approximation Theory*, Springer Proceedings in Mathematics & Statistics 306,
https://doi.org/10.1007/978-981-15-1153-0_12

$x_2 := g(x_1), \ldots, x_{2n} := g(x_{2n-1}), x_{2n+1} := f(x_{2n}), \ldots$ The aim of this paper is to give conditions on the pair (f, g) such that the sequence $(x_n)_{n\in\mathbb{N}}$ converges with respect to $\to$ to a common fixed point of f and g.

2 Basic Notions and Notations

Let X be a nonempty set and $f : X \to X$ be an operator. The set of all fixed points of f is denoted by

$$F_f := \{x \in X | x = f(x)\}.$$

If $f, g : X \to X$ are two operators on X, then an element $x^* \in X$ is a common fixed point for f and g if and only if

$$x^* \in F_f \cap F_g.$$

We recall next the notions of L-space, Kasahara space and generalized Kasahara space.

Definition 2.1 (*Fréchet* [6], *Rus* [15]) Let X be a nonempty set. Let

$$s(X) := \left\{(x_n)_{n\in\mathbb{N}} \mid x_n \in X,\ n \in \mathbb{N}\right\}.$$

Let $c(X) \subset s(X)$ be a subset of $s(X)$ and $Lim : c(X) \to X$ be an operator. By definition, the triple $(X, c(X), Lim)$ is called an L-space if the following conditions are satisfied:

(i) If $x_n = x$, for all $n \in \mathbb{N}$, then $(x_n)_{n\in\mathbb{N}} \in c(X)$ and $Lim(x_n)_{n\in\mathbb{N}} = x$.
(ii) If $(x_n)_{n\in\mathbb{N}} \in c(X)$ and $Lim(x_n)_{n\in\mathbb{N}} = x$, then for all subsequences $(x_{n_i})_{i\in\mathbb{N}}$ of $(x_n)_{n\in\mathbb{N}}$ we have that $(x_{n_i})_{i\in\mathbb{N}} \in c(X)$ and $Lim(x_{n_i})_{i\in\mathbb{N}} = x$.

By definition, an element $(x_n)_{n\in\mathbb{N}}$ of $c(X)$ is a convergent sequence and $x = Lim(x_n)_{n\in\mathbb{N}}$ is the limit of this sequence and we shall write

$$x_n \to x \text{ as } n \to \infty.$$

We denote an L-space by $(X, \to)$.

Example 2.1 Let (X, d) be a metric space. Let $\overset{d}{\to}$ be the convergence structure induced by d on X. Then $(X, \overset{d}{\to})$ is an L-space.

In general, an L-space is any set endowed with a structure implying a notion of convergence for sequences. Other examples of L-spaces are: Hausdorff topological spaces, generalized metric spaces in Perov' sense (i.e. $d(x, y) \in \mathbb{R}^m_+$), generalized metric spaces in Luxemburg' sense (i.e. $d(x, y) \in \mathbb{R}_+ \cup \{+\infty\}$), K-metric spaces (i.e. $d(x, y) \in K$, where K is a cone in an ordered Banach space), gauge spaces, 2-metric spaces, D-R-spaces, probabilistic metric spaces, syntopogenous spaces.

Remark 2.1 Let $(X, \to)$ be an L-space and $f : X \to X$ be a self-operator on X. Let us consider the following set (also called the *graph* of f):

$$Graph(f) := \{(x, y) \in X \times X \mid f(x) = y\}.$$

Then f has closed graph with respect to $\to$ if and only if $Graph(f)$ is closed with respect to $\to$, i.e. for any sequences $(x_n)_{n\in\mathbb{N}}$ and $(y_n)_{n\in\mathbb{N}}$ of X satisfying

(i) $x_n \to x \in X$ as $n \to \infty$;
(ii) $y_n \to y \in X$ as $n \to \infty$;
(iii) $f(x_n) = y_n$, for all $n \in \mathbb{N}$

we get that $f(x) = y$.

Definition 2.2 (*Rus* [16]) Let $(X, \to)$ be an L-space and $d : X \times X \to \mathbb{R}_+$ be a functional. The triple $(X, \to, d)$ is a Kasahara space if and only if we have the following compatibility condition between $\to$ and d:

$$x_n \in X, \sum_{n\in\mathbb{N}} d(x_n, x_{n+1}) < +\infty \Rightarrow (x_n)_{n\in\mathbb{N}} \text{ converges in } (X, \to). \quad (1)$$

Example 2.2 (Kasahara [8]) Let X denote the closed interval $[0, 1]$ and $\to$ be the usual convergence structure on $\mathbb{R}$. Let $d : X \times X \to \mathbb{R}_+$ be defined by

$$d(x, y) = \begin{cases} |x - y|, & \text{if } x \neq 0 \text{ and } y \neq 0 \\ 1, & \text{otherwise .} \end{cases}$$

Then $(X, \to, d)$ is a Kasahara space.

Definition 2.3 (*Rus* [16]) Let $(X, \to)$ be an L-space, $(G, +, \leq, \overset{G}{\to})$ be an L-space ordered semigroup with unity, 0 be the least element in $(G, \leq)$ and $d_G : X \times X \to G$ be an operator. The triple $(X, \to, d_G)$ is a generalized Kasahara space if and only if we have the following compatibility condition between $\to$ and d_G:

$$x_n \in X, \sum_{n\in\mathbb{N}} d_G(x_n, x_{n+1}) < +\infty \Rightarrow (x_n)_{n\in\mathbb{N}} \text{ converges in } (X, \to). \quad (2)$$

Notice that by the inequality with the symbol $+\infty$ in the compatibility condition (2), we mean that the series $\sum_{n\in\mathbb{N}} d_G(x_n, x_{n+1})$ is convergent in $(G, +, \overset{G}{\to})$.

For the case of generalized Kasahara space, we will consider $G := \mathbb{R}^m_+$.

Example 2.3 (*Rus* [16]) Let $\rho : X \times X \to \mathbb{R}^m_+$ be a generalized complete metric on a set X. Let $x_0 \in X$ and $\lambda \in \mathbb{R}^m_+$ with $\lambda \neq 0$. Let $d_\lambda : X \times X \to \mathbb{R}^m_+$ be defined by

$$d_\lambda(x, y) := \begin{cases} \rho(x, y), & \text{if } x \neq x_0 \text{ and } y \neq x_0 \\ \lambda, & \text{if } x = x_0 \text{ or } y = x_0. \end{cases}$$

Then $(X, \overset{\rho}{\rightarrow}, d_\lambda)$ is a generalized Kasahara space.

We mention that if $\alpha, \beta \in \mathbb{R}^m_+$, $\alpha = (\alpha_1, \alpha_2, \dots, \alpha_m)$, $\beta = (\beta_1, \beta_2, \dots, \beta_m)$ and $c \in \mathbb{R}_+$, then by $\alpha \leq \beta$ (respectively $\alpha < \beta$), we mean that $\alpha_i \leq \beta_i$ (respectively $\alpha_i < \beta_i$), for all $i = \overline{1, m}$ and by $\alpha \leq c$ we mean that $\alpha_i \leq c$, for all $i = \overline{1, m}$.

Throughout this paper we denote by $\mathcal{M}_{m,m}(\mathbb{R}_+)$ the set of all $m \times m$ matrices with positive elements, by Θ the zero $m \times m$ matrix and by I_m the identity $m \times m$ matrix. If $A \in \mathcal{M}_{m,m}(\mathbb{R}_+)$, then the symbol A^τ stands for the transpose matrix of A. Notice also that, for the sake of simplicity, we will make an identification between row and column vectors in $\mathbb{R}^m$.

A matrix $A \in \mathcal{M}_{m,m}(\mathbb{R}_+)$ is said to be convergent to zero if and only if $A^n \to \Theta$ as $n \to \infty$ (see [19]). Regarding this class of matrices we have the following classical result in matrix analysis (see [2] (Lemma 3.3.1, p. 55), [12], [13] (p. 37), [19] (p. 12)). More considerations can be found in [18].

Theorem 2.1 *Let $A \in \mathcal{M}_{m,m}(\mathbb{R}_+)$. The following statements are equivalent:*

(i) $A^n \to \Theta$, *as* $n \to \infty$;
(ii) the eigenvalues of A lies in the open unit disc, i.e. $|\lambda| < 1$, *for all* $\lambda \in \mathbb{C}$ *with* $det(A - \lambda I_m) = 0$;
(iii) the matrix $I_m - A$ *is non-singular and*

$$(I_m - A)^{-1} = I_m + A + A^2 + \cdots + A^n + \cdots ;$$

(iv) the matrix $(I_m - A)$ *is non-singular and* $(I_m - A)^{-1}$ *has nonnegative elements;*
(v) the matrices Aq *and* $q^\tau A$ *converges to zero for each* $q \in \mathbb{R}^m$.

We recall also a useful tool for proving the uniqueness of a fixed point in a Kasahara space (see Kasahara [8], Rus [16]).

Lemma 2.1 (Kasahara's lemma) *Let $(X, \rightarrow, d_G)$ be a generalized Kasahara space. Then:*

$$x, y \in X,\ d_G(x, y) = d_G(y, x) = 0 \implies x = y.$$

3 Main Results

In this section, we present our main common fixed point results obtained in Kasahara spaces and then in generalized Kasahara spaces. The uniqueness of the common fixed point is also discussed.

Theorem 3.1 *Let $(X, \to, d)$ be a Kasahara space, where $d : X \times X \to \mathbb{R}_+$ is a functional and let $f, g : X \to X$ be two operators with closed graph with respect to $\to$. Suppose that there exists $\alpha, \beta \in \mathbb{R}_+$ with $\alpha\beta < 1$ such that:*

(i) $d(f(x), g(f(x))) \le \alpha d(x, f(x))$, *for all* $x \in X$;
(ii) $d(g(x), f(g(x))) \le \beta d(x, g(x))$, *for all* $x \in X$.

Then

(1) $F_f \cap F_g \neq \emptyset$;
(2) the sequence $(x_n)_{n\in\mathbb{N}}$ *defined by* $x_0 \in X$, $x_{2n+1} = f(x_{2n})$, $x_{2n+2} = g(x_{2n+1})$, *for all* $n \in \mathbb{N}$, *converges to a common fixed point of* f *and* g.

Proof We consider the sequence $(x_n)_{n\in\mathbb{N}}$ defined by $x_0 \in X$, $x_{2n+1} = f(x_{2n})$, $x_{2n+2} = g(x_{2n+1})$, for all $n \in \mathbb{N}$ and we prove by induction that

$$d(x_{2n}, x_{2n+1}) \le (\alpha\beta)^n d(x_0, x_1), \text{ for all } n \in \mathbb{N} \tag{3}$$

$$d(x_{2n+1}, x_{2n+2}) \le \alpha(\alpha\beta)^n d(x_0, x_1), \text{ for all } n \in \mathbb{N}. \tag{4}$$

For $n = 0$ the relations (3) and (4) hold.

We assume that (3) and (4) hold for a fixed $n \in \mathbb{N}$ and we prove that

$$d(x_{2n+2}, x_{2n+3}) \le (\alpha\beta)^{n+1} d(x_0, x_1) \tag{5}$$

$$d(x_{2n+3}, x_{2n+4}) \le \alpha(\alpha\beta)^{n+1} d(x_0, x_1). \tag{6}$$

We have

$$\begin{aligned} d(x_{2n+2}, x_{2n+3}) &= d(g(x_{2n+1}), f(g(x_{2n+1}))) \\ &\le \beta d(x_{2n+1}, g(x_{2n+1})) = \beta d(f(x_{2n}), g(f(x_{2n}))) \\ &\le \alpha\beta d(x_{2n}, f(x_{2n})) \le \alpha\beta(\alpha\beta)^n d(x_0, x_1) \end{aligned}$$

and hence (5) is proved.

On the other hand

$$\begin{aligned} d(x_{2n+3}, x_{2n+4}) &= d(x_{2(n+1)+1}, g(x_{2(n+1)+1})) = d(f(x_{2(n+1)}), g(f(x_{2(n+1)}))) \\ &\le \alpha d(x_{2(n+1)}, f(x_{2(n+1)})) = \alpha d(x_{2n+2}, x_{2n+3}) \\ &\le \alpha(\alpha\beta)^{n+1} d(x_0, x_1), \end{aligned}$$

so (6) is proved.

Now we have the following estimations:

$$\begin{aligned}\sum_{n\in\mathbb{N}} d(x_n, x_{n+1}) &= \sum_{n\in\mathbb{N}} d(x_{2n}, x_{2n+1}) + \sum_{n\in\mathbb{N}} d(x_{2n+1}, x_{2n+2})\\ &\leq \sum_{n\in\mathbb{N}} (\alpha\beta)^n d(x_0, x_1) + \sum_{n\in\mathbb{N}} \alpha(\alpha\beta)^n d(x_0, x_1)\\ &= \frac{1}{1-\alpha\beta} d(x_0, x_1) + \frac{\alpha}{1-\alpha\beta} d(x_0, x_1) = \frac{1+\alpha}{1-\alpha\beta} d(x_0, x_1).\end{aligned}$$

Since $(X, \to, d)$ is a Kasahara space and $\sum_{n\in\mathbb{N}} d(x_n, x_{n+1}) < +\infty$, it follows that the sequence $(x_n)_{n\in\mathbb{N}}$ converges in $(X, \to)$. So, there exists $x^* \in X$ such that $x_n \to x^*$ as $n \to \infty$.

On the other hand, $x_{2n} \to x^*$ and $x_{2n+1} = f(x_{2n}) \to x^*$ as $n \to \infty$. Since f has closed graph in $(X, \to)$, we get that $x^* \in F_f$. By a similar way of proof, we have that $x^* \in F_g$. So $F_f \cap F_g \neq \emptyset$. □

Remark 3.1 In Theorem 3.1, item (2) also holds if only f has closed graph with respect to $\to$ and $d(x, y) = 0 \Leftrightarrow x = y$, for all $x, y \in X$.

Indeed, if f has closed graph then we get the existence of $x^* \in F_f$ as shown in the proof of Theorem 3.1. So $x^* = f(x^*)$ and we get further that

$$d(x^*, g(x^*)) = d(f(x^*), g(f(x^*))) \leq \alpha d(x^*, f(x^*)) = 0$$

and hence $x^* \in F_g$.

Remark 3.2 In Theorem 3.1, if the functional d satisfies $d(x, y) = d(y, x)$, for all $x, y \in X$ and $\beta < 1$, then we have the uniqueness of common fixed point.

Indeed, let $x^* \in F_f \cap F_g$ and $y^* \in F_g$ such that $x^* \neq y^*$. Then

$$d(y^*, x^*) = d(g(y^*), f(x^*)) = d(g(y^*), f(g(x^*))) \leq \beta d(y^*, x^*).$$

Since $\beta < 1$ it follows that $d(y^*, x^*) = 0$. By Kasahara's lemma we get $x^* = y^*$.

Hence $F_f \cap F_g = \{x^*\}$.

Corollary 3.1 *Let $(X, \to, d)$ be a Kasahara space, where $d : X \times X \to \mathbb{R}_+$ is a functional satisfying:*

(i) $d(x, x) = 0$, for all $x \in X$;
(ii) $d(x, y) = d(y, x)$, for all $x, y \in X$.

Let $f, g : X \to X$ be two operators having closed graph with respect to $\to$. If there exists $\alpha \in [0, \frac{1}{2}[$ such that

$$d(f(x), g(y)) \leq \alpha[d(x, f(x)) + d(y, g(y))], \text{ for all } x, y \in X$$

then

(1) $F_f \cap F_g = \{x^*\}$;
(2) *the sequence* $(x_n)_{n\in\mathbb{N}}$ *defined by* $x_0 \in X$, $x_{2n+1} = f(x_{2n})$, $x_{2n+2} = g(x_{2n+1})$, *for all* $n \in \mathbb{N}$, *converges to* x^* *with respect to* $\to$.

Proof Let $x \in X$ and $y = f(x) \in X$. Then we have

$$d(f(x), g(f(x))) \leq \alpha[d(x, f(x)) + d(f(x), g(f(x)))], \text{ for all } x \in X$$

such as

$$d(f(x), g(f(x))) \leq \frac{\alpha}{1-\alpha} d(x, f(x)),$$

for all $x \in X$.

Let $\delta = \dfrac{\alpha}{1-\alpha}$. Thus, there exists $\delta \in [0, 1[$ such that $d(f(x), g(f(x))) \leq \delta d(x, f(x))$, for all $x \in X$. Hence, the assumption (i) of Theorem 3.1 is satisfied.

On the other hand, we have

$$d(g(x), f(g(x))) \leq \alpha[d(x, g(x)) + d(g(x), f(g(x)))], \text{ for all } x \in X$$

such that

$$d(g(x), f(g(x))) \leq \frac{\alpha}{1-\alpha} d(x, g(x)),$$

for all $x \in X$. The assumption (ii) of Theorem 3.1 is satisfied for $\beta := \dfrac{\alpha}{1-\alpha} \in [0, 1[$.

Applying Theorem 3.1 and taking into account the Remark 3.2 the conclusions follow. □

We give next our common fixed point results in generalized Kasahara spaces.

Theorem 3.2 *Let* $(X, \to, d)$ *be a generalized Kasahara space, where* $d : X \times X \to \mathbb{R}^m_+$ *is a functional and let* $f, g : X \to X$ *be two operators with closed graph with respect to* $\to$. *Suppose that there exist* $A, B \in \mathcal{M}_{m,m}(\mathbb{R}_+)$ *such that:*

(i) $d(f(x), g(f(x))) \leq Ad(x, f(x))$, *for all* $x \in X$;
(ii) $d(g(x), f(g(x))) \leq Bd(x, g(x))$, *for all* $x \in X$.

If the matrix BA *is convergent to zero, then*

(1) $F_f \cap F_g \neq \emptyset$;
(2) *the sequence* $(x_n)_{n\in\mathbb{N}}$ *defined by* $x_0 \in X$, $x_{2n+1} = f(x_{2n})$, $x_{2n+2} = g(x_{2n+1})$, *for all* $n \in \mathbb{N}$, *converges to a common fixed point of* f *and* g.

Proof Let $(x_n)_{n\in\mathbb{N}}$ be the sequence defined by $x_0 \in X$, $x_{2n+1} = f(x_{2n})$, $x_{2n+2} = g(x_{2n+1})$, for all $n \in \mathbb{N}$ and we prove by induction that

$$d(x_{2n}, x_{2n+1}) \leq (BA)^n d(x_0, x_1), \text{ for all } n \in \mathbb{N} \tag{7}$$

$$d(x_{2n+1}, x_{2n+2}) \leq A(BA)^n d(x_0, x_1), \text{ for all } n \in \mathbb{N}. \tag{8}$$

For $n = 0$ the relations (7) and (8) hold. We assume that (7) and (8) hold for a fixed $n \in \mathbb{N}$ and we prove that

$$d(x_{2n+2}, x_{2n+3}) \leq (BA)^{n+1} d(x_0, x_1) \tag{9}$$

$$d(x_{2n+3}, x_{2n+4}) \leq A(BA)^{n+1} d(x_0, x_1). \tag{10}$$

We have

$$\begin{aligned} d(x_{2n+2}, x_{2n+3}) &= d(g(x_{2n+1}), f(g(x_{2n+1}))) \\ &\leq Bd(x_{2n+1}, g(x_{2n+1})) = Bd(f(x_{2n}), g(f(x_{2n}))) \\ &\leq BAd(x_{2n}, f(x_{2n})) \leq BA(BA)^n d(x_0, x_1) \end{aligned}$$

and hence (9) is proved.
On the other hand

$$\begin{aligned} d(x_{2n+3}, x_{2n+4}) &= d(x_{2(n+1)+1}, g(x_{2(n+1)+1})) = d(f(x_{2(n+1)}), g(f(x_{2(n+1)}))) \\ &\leq Ad(x_{2(n+1)}, f(x_{2(n+1)})) = Ad(x_{2n+2}, x_{2n+3}) \\ &\leq A(BA)^{n+1} d(x_0, x_1), \end{aligned}$$

so (10) is proved.

Now we have the following estimations:

$$\begin{aligned} \sum_{n\in\mathbb{N}} d(x_n, x_{n+1}) &= \sum_{n\in\mathbb{N}} d(x_{2n}, x_{2n+1}) + \sum_{n\in\mathbb{N}} d(x_{2n+1}, x_{2n+2}) \\ &\leq \sum_{n\in\mathbb{N}} (BA)^n d(x_0, x_1) + \sum_{n\in\mathbb{N}} A(BA)^n d(x_0, x_1) \\ &= (I_m - BA)^{-1} d(x_0, x_1) + A(I_m - BA)^{-1} d(x_0, x_1) \\ &= (I_m + A)(I_m - BA)^{-1} d(x_0, x_1). \end{aligned}$$

Since $(X, \to, d)$ is a generalized Kasahara space and $\sum_{n\in\mathbb{N}} d(x_n, x_{n+1}) < +\infty$, it follows that the sequence $(x_n)_{n\in\mathbb{N}}$ converges in $(X, \to)$. So, there exists $x^* \in X$ such that $x_n \to x^*$ as $n \to \infty$.

On the other hand, $x_{2n} \to x^*$ and $x_{2n+1} = f(x_{2n}) \to x^*$ as $n \to \infty$. Since f has closed graph in $(X, \to)$, we get that $x^* \in F_f$. By a similar way of proof, we have that $x^* \in F_g$. So $F_f \cap F_g \neq \emptyset$. □

Remark 3.3 In Theorem 3.2, item (2) also holds if only f has closed graph with respect to $\to$ and $d(x, y) = 0 \in \mathbb{R}^m_+ \Leftrightarrow x = y$, for all $x, y \in X$.

Remark 3.4 In Theorem 3.2, if the functional d satisfies $d(x, y) = d(y, x)$, for all $x, y \in X$ and B is a matrix convergent to zero, then we have the uniqueness of the common fixed point.

Indeed, let $x^* \in F_f \cap F_g$ and $y^* \in F_g$ such that $x^* \neq y^*$. Then

$$d(y^*, x^*) = d(g(y^*), f(x^*)) = d(g(y^*), f(g(x^*))) \leq Bd(y^*, x^*)$$

i.e. $(I_m - B)d(y^*, x^*) \leq 0 \in \mathbb{R}^m_+$. Since B converges towards zero, the matrix $(I_m - B)$ is non-singular and $(I_m - B)^{-1}$ has positive elements (see Theorem 2.1, item (iv)). It follows that $d(y^*, x^*) = 0$. By Kasahara's lemma we get $x^* = y^*$. Hence $F_f \cap F_g = \{x^*\}$.

We give next a similar result to Corollary 3.1 in generalized Kasahara spaces. To achieve our goal, let us consider the following set:

$$\mathcal{M}^{\Delta}_{m,m}(\mathbb{R}_+) := \left\{ Q = \begin{pmatrix} q_{11} & q_{12} & \dots & q_{1m} \\ 0 & q_{22} & \dots & q_{2m} \\ \vdots & \vdots & & \vdots \\ 0 & 0 & \dots & q_{mm} \end{pmatrix} \in \mathcal{M}_{m,m}(\mathbb{R}_+) \;\middle|\; \max_{i=\overline{1,m}} q_{ii} < \frac{1}{2} \right\}.$$

Then we have:

Lemma 3.1 (Filip [5]) *If $Q \in \mathcal{M}^{\Delta}_{m,m}(\mathbb{R}_+)$ then*

(1) the matrix Q is convergent to zero;
(2) the matrix $(I_m - Q)^{-1}Q$ is convergent to zero.

Corollary 3.2 *Let $(X, \to, d)$ be a Kasahara space, where $d : X \times X \to \mathbb{R}^m_+$ is a functional satisfying:*

(i) $d(x, x) = 0 \in \mathbb{R}^m_+$, for all $x \in X$;
(ii) $d(x, y) = d(y, x)$, for all $x, y \in X$.

Let $f, g : X \to X$ be two operators having closed graph with respect to $\to$.
If there exists $A \in \mathcal{M}^{\Delta}_{m,m}(\mathbb{R}_+)$ such that

$$d(f(x), g(y)) \leq A[d(x, f(x)) + d(y, g(y))], \text{ for all } x, y \in X$$

then

(1) $F_f \cap F_g = \{x^*\}$;
(2) the sequence $(x_n)_{n\in\mathbb{N}}$ defined by $x_0 \in X$, $x_{2n+1} = f(x_{2n})$, $x_{2n+2} = g(x_{2n+1})$, for all $n \in \mathbb{N}$, converges to x^* with respect to $\to$.

Proof Let $x \in X$ and $y = f(x) \in X$. Then we have

$$d(f(x), g(f(x))) \leq A[d(x, f(x)) + d(f(x), g(f(x)))], \text{ for all } x \in X$$

then

$$d(f(x), g(f(x))) \leq (I_m - A)^{-1} A d(x, f(x)),$$

for all $x \in X$.

Let $\Lambda = (I_m - A)^{-1} A$. Thus, there exists $\Lambda \in \mathcal{M}^{\Delta}_{m,m}(\mathbb{R}_+)$ such that

$$d(f(x), g(f(x))) \leq \Lambda d(x, f(x)),$$

for all $x \in X$. Hence, the assumption (i) of Theorem 3.2 is satisfied.

On the other hand, we have

$$d(g(x), f(g(x))) \leq A[d(x, g(x)) + d(g(x), f(g(x)))], \text{ for all } x \in X$$

so

$$d(g(x), f(g(x))) \leq (I_m - A)^{-1} A d(x, g(x)),$$

for all $x \in X$. The assumption (ii) of Theorem 3.2 is satisfied for the matrix $B := (I_m - A)^{-1} A \in \mathcal{M}^{\Delta}_{m,m}(\mathbb{R}_+)$.

Since the matrix $B\Lambda \in \mathcal{M}^{\Delta}_{m,m}(\mathbb{R}_+)$ we can apply Theorem 3.2 and taking into account the Remark 3.4 the conclusions follow. □

References

1. A. Aliouche, V. Popa, Common fixed point theorems for occasionally weakly compatible mappings in compact metric spaces. Acta Univ. Apulensis **28**, 203–214 (2011)
2. G. Allaire, *Numerical Linear Algebra* (Springer, New York, 2008)
3. S. Chandok, Some common fixed point theorems for generalized nonlinear contractive mappings. Comput. Math. Appl. **62**(10), 3692–3699 (2011)
4. S.-H. Cho, J.-S. Bae, Common fixed point theorems for mappings satisfying property (E.A) on cone metric spaces. Math. Comput. Model. **53**(5–6), 945–951 (2011)
5. A.-D. Filip, Perov's fixed point theorem for multivalued mappings in generalized Kasahara spaces. Stud. Univ. Babeş-Bolyai Math. **56**(3), 19–28 (2011)
6. M. Fréchet, *Les espaces abstraits* (Gauthier-Villars, Paris, 1928)
7. S. Kasahara, Common fixed point theorems for mappings in L-spaces. Math. Semin. Notes **3**, 203–212 (1975)
8. S. Kasahara, On some generalizations of the Banach contraction theorem. Publ. RIMS Kyoto Univ. **12**, 427–437 (1976)
9. S. Kasahara, Common fixed point theorems in certain L-spaces. Math. Semin. Notes **5**, 173–178 (1977)
10. S. Kasahara, On some common fixed point theorems for contractive mappings. Math. Semin. Notes **6**, 57–75 (1978)

11. R.P. Pant, R.K. Bisht, Common fixed point theorems under a new continuity condition. Ann. Univ. Ferrara **58**(1), 127–141 (2012)
12. R. Precup, The role of the matrices that are convergent to zero in the study of semilinear operator systems. Math. Comput. Model. **49**, 703–708 (2009)
13. I.A. Rus, *Principii şi Aplicaţii ale Teoriei Punctului Fix* (Dacia, Cluj-Napoca, 1979)
14. I.A. Rus, *Generalized Contractions and Applications* (Cluj University Press, Cluj-Napoca, 2001)
15. I.A. Rus, Picard operators and applications. Sci. Math. Jpn. **58**, 191–219 (2003)
16. I.A. Rus, Kasahara spaces. Sci. Math. Jpn. **72**, 101–110 (2010)
17. I.A. Rus, A. Petruşel, G. Petruşel, *Fixed Point Theory* (Cluj University Press, Cluj-Napoca, 2008)
18. M. Turinici, Finite dimensional vector contractions and their fixed points. Stud. Univ. Babeş-Bolyai Math. **35**(1), 30–42 (1990)
19. R.S. Varga, *Matrix Iterative Analysis* (Springer, Berlin, 2000)
20. S. Wang, B. Guo, Distance in cone metric spaces and common fixed point theorems. Appl. Math. Lett. **24**(10), 1735–1739 (2011)

Fixed Point Theorem in Fuzzy Metric Space Via α-Series Contraction

Vizender Sihag, Dinesh and Vinod

Abstract Starting from the setting of fuzzy metric spaces, in the present article, the authors utilize new concept of α-series contraction to establish fixed point theorems for a sequence of mappings. These results unify, extend, and complement some theorems on fuzzy metric spaces existing in the literature. The established results are supported by an illustrative example and finally by furnishing an application in product space.

Keywords Fuzzy metric space · α-series and product spaces

MSC 47H10 · 54H25

1 Introduction

In the study of real analysis, the concept of metric space is important and in fixed point theory, Banach Contraction Principle is one of pivotal results. Until now many results in fixed point theory are extended by improving the contractive conditions involved. In the same direction, many efforts have been done by various authors, one of them is Rhoadas [9], who made a comparison of more than hundred types of contractive conditions. Following the same tradition Sihag et al. [11] gave new notion of $alpha$-series and established fixed point theorem in G-metric space. Utilising this new concept, the present paper proves common fixed point theorem for a sequence

---

V. Sihag (✉)
Faculty of Mathematics, Directorate of Distance Education, GJUS&T, Hisar, India
e-mail: vsihag3@gmail.com

Dinesh
Department of Mathematics, GJUS&T, Hisar, India
e-mail: dineshreddu3@gmail.com

Vinod
Faculty of Computer Science, GJUS&T, Hisar, India
e-mail: vinodgoyal.gjust@gmail.com

N. Deo et al. (eds.), *Mathematical Analysis I: Approximation Theory*,
Springer Proceedings in Mathematics & Statistics 306,
https://doi.org/10.1007/978-981-15-1153-0_13

of mappings in setting of fuzzy metric space, which generalizes Banach Contraction Principle in fuzzy metric space by Grabie et al. [2] and further by Vasuki et al. [14].

The notion of fuzzy metric space was introduced by different authors (see [2, 7]) in different ways. Further using these different concepts various authors [1, 3, 7] proved theorem which assures the existence of fixed point.

2 Preliminaries

In what follows, we collect some relevant definitions, results, examples for our further use.

Definition 2.1 ([15]) A fuzzy set A in X is a function with domain X and values in $[0, 1]$.

Definition 2.2 A continuous t-norm (in sense of Schweizer and Sklar [10]) is a binary operation T on $[0, 1]$ satisfying the following conditions:

(i) T is a commutative and associative;
(ii) $T(a, 1) = a$ for all $a \in [0, 1]$;
(iii) $T(a, b) \leq T(c, d)$ whenever $a \leq c$ and $b \leq d$, for all $a, b, c, d \in [0, 1]$;
(iv) The mapping $T : [0, 1] \times [0, 1] \to [0, 1]$ is continuous.

Remark 2.3 The following are classical example of continuous t-norm

(i) $T_M(a, b) = \min\{a, b\}$, minimum t-norm.
(ii) $T_H(a, b) = \begin{cases} 0 & \text{if } a = b = 0, \\ \dfrac{ab}{a + b + ab} & \text{otherwise,} \end{cases}$ Hamacher product.
(iii) $T_P(a, b) = ab$, product t-norm.
(iv) $T_N(a, b) = \begin{cases} \min\{a, b\} & \text{if } a + b > 1, \\ 0 & \text{otherwise,} \end{cases}$ Nilpotent minimum.
(v) $T_L(a, b) = \max\{a + b - 1, 0\}$, Lukasiewict t-norm.
(vi) $T_D(a, b) = \begin{cases} b & \text{if } a = 1, \\ a & \text{if } b = 1, \\ 0 & \text{otherwise,} \end{cases}$ Drastic t-norm.

The minimum t-norm is point wise largest t-norm and the drastic t-norm is point wise smallest t-norm; that is, $T_M(a, b) = T(a, b) = T_D(a, b)$ for any t-norm t with $a, b \in [0, 1]$.

Kramosil and Michalek in [13] generalized the concept of probabilistic metric space given by Menger to the fuzzy framework as follows.

Definition 2.4 A fuzzy metric space (in sense of Kramosil and Michalek [13]) is a triple $(X, M, *)$, where X is a nonempty set, $*$ is a continuous t-norm, and M is a fuzzy set on $X^2 \times [0, \infty)$ such that the following axioms holds:

(FM-1) $M(x, y, 0) = 0$ $(x, y \in X)$;
(FM-2) $M(x, y, t) = 1$ for all $t > 0$ iff $x = y$;
(FM-3) $M(x, y, t) = M(y, x, t)$ $(x, y \in X, t > 0)$;
(FM-4) $M(x, y, \cdot) : [0, \infty) \to [0, 1]$ is left continuous for all $x, y \in X$;
(FM-5) $M(x, z, t + s) \geq M(x, y, t) * M(y, z, s)$ for all $x, y, z \in X$ and $s, t > 0$.

We will refer to these spaces as KM-fuzzy metric spaces.

Lemma 2.5 ([5]) *For every $x, y \in X$, the mapping $M(x, y, \cdot)$ is nondecreasing on $(0, \infty)$.*

In order to introduce a Hausdorff topology on the fuzzy metric spaces, George and Veeramani in [6] modified in a slight but appealing way the notion of fuzzy metric spaces of Kramosil and Michalek.

Definition 2.6 A fuzzy metric space (in sense of George and Veeramani [7]) is a triple $(X, M, *)$, where X is a nonempty set, $*$ is a continuous t-norm, and M is a fuzzy set on $X^2 \times (0, \infty)$ such that the following axioms holds:

(GV-1) $M(x, y, t) > 0(x, y \in X)$;
(GV-2) $M(x, y, t) = 1$ for all $t > 0$ iff $x = y$;
(GV-3) $M(x, y, t) = M(y, x, t)$ $(x, y \in X, t > 0)$;
(GV-4) $M(x, y, \cdot) : (0, \infty) \to (0, 1]$ is continuous for all $x, y \in X$;
(GV-5) $M(x, z, t + s) \geq M(x, y, t) * M(y, z, s)$ for all $x, y, z \in X$ and $s, t > 0$.

Notice that condition (GV-5) is a fuzzy version of triangular inequality. The value $M(x, y, t)$ can be thought of as degree of nearness between x and y with respect to t and from axiom (GV-2) we can relate the value 0 and 1 of a fuzzy metric to the notions of ∞ and 0 of classical metric, respectively.

We will refer to these spaces as GV-fuzzy metric spaces.

Definition 2.7 ([8]) A fuzzy metric M on X is said to be stationary if M does not depend on t, i.e., the function $M_{x,y}(t) = M(x, y, t)$ is constant.

Definition 2.8 ([10]) If $(X, M, *)$ is a KM-fuzzy metric space and $\{x_n\}$, $\{y_n\}$ are sequences in X such that $x_n \to x$, $y_n \to y$, then $M(x_n, y_n, t) \to M(x, y, t)$ for every continuity point t of $M(x, y, \cdot)$.

We can fuzzify example of metric space into fuzzy metric spaces in a normal way.

Example 2.9 ([8]) Let (X, d) be metric space and $g : R^+ \to R^+$ is an increasing continuous function. For $m > 0$, we define the function M by

$$M(x, y, t) = \frac{g(t)}{g(t) + m \cdot d(x, y)} \tag{2.1}$$

Then for $a * b = a \cdot b$, $(X, M, *)$ is a GV-fuzzy metric space on X.

As a particular case if we take $g(t) = t^n$ where $n \in N$ and $m = 1$. Then (2.1) becomes

$$M(x, y, t) = \frac{t^n}{t^n + d(x, y)} \tag{2.2}$$

Then for $a * b = T_M(a, b)$, $(X, M, *)$ is a GV-fuzzy metric space on X.

If we take $n = 1$, in (2.2), the well-known fuzzy metric space is obtained.

On the other hand, if we take g as a constant function in (2.1), i.e., $g(t) = k > 0$ and $m = 1$, we obtain

$$M(x, y, t) = \frac{k}{k + d(x, y)}$$

And so $(X, M, *)$ is a stationary GV-fuzzy metric space for $a * b = a \cdot b$ but, in general, (X, M, T_M) is not.

In what follows, Sihag [11], introduced the following definition of α-series, which will be utilized in proving main result.

Definition 2.10 Let $\{a_n\}$ be a sequence of nonnegative real numbers. We say that a series $\sum_{n=1}^{+\infty} a_n$ is an α-series, if there exist $0 < \alpha < 1$ and $n_\alpha \in N$ such that $\sum_{i=1}^{k} a_i \leq \alpha k$ for each $k \geq n_\alpha$.

Remark 2.11 Each convergent series of nonnegative real terms is an α-series. However, there are also divergent series that are α-series. For example, $\sum_{n=1}^{+\infty} \frac{1}{n}$ is an α-series.

3 Main Result

Let us start this section with the following theorem:

Theorem 3.1 *Let $\{T_n\}$ be a sequence of self mappings of a complete fuzzy metric space $(X, M, \star)$ such that*

$$\begin{aligned} M(T_i(x), T_j(y), t) \geq{}& \beta_{i,j}[M(x, T_i(x), t) + M(y, T_j(y), t)] \\ &+ \gamma_{i,j} M(x, y, t) \end{aligned} \tag{3.1}$$

for $x, y, z \in X$ with $x \neq y$, $0 \leq \beta_{i,j}, \gamma_{i,j} < 1$, $i, j = 1, 2, \ldots$.

If $\sum_{i=1}^{+\infty} \left(\frac{\beta_{i,i+1} + \gamma_{i,i+1}}{1 - \beta_{i,i+1}} \right)$ is an α-series, then $\{T_n\}$ has a unique common fixed point in X.

Proof For any $x_0 \in X$, we can consider the sequence $x_n = T_n(x_{n-1})$, $n = 1, 2, \ldots$. By (3.1), we have

$$\begin{aligned}
M(x_1, x_2, t) &= M(T_1(x_0), T_2(x_1), t) \\
&\geq \beta_{1,2}[M(x_0, T_1(x_0), t) + M(x_1, T_2(x_1), t)] \\
&\quad + \gamma_{1,2} M(x_0, x_1, t) \\
&= \beta_{1,2}[M(x_0, x_1, t) + M(x_1, x_2, t)] + \gamma_{1,2} M(x_0, x_1, t).
\end{aligned}$$

It follows that

$$(1 - \beta_{1,2}) M(x_1, x_2, t) \geq (\beta_{1,2} + \gamma_{1,2}) M(x_0, x_1, t),$$

or equivalently,

$$M(x_1, x_2, t) \geq \left(\frac{\beta_{1,2} + \gamma_{1,2}}{1 - \beta_{1,2}} \right) M(x_0, x_1, t).$$

Also, we get

$$\begin{aligned}
M(x_2, x_3, t) &= M(T_2(x_1), T_3(x_2), t) \\
&\geq \left(\frac{\beta_{2,3} + \gamma_{2,3}}{1 - \beta_{2,3}} \right) M(x_1, x_2, t) \\
&\geq \left(\frac{\beta_{2,3} + \gamma_{2,3}}{1 - \beta_{2,3}} \right) \left(\frac{\beta_{1,2} + \gamma_{1,2}}{1 - \beta_{1,2}} \right) M(x_0, x_1, t).
\end{aligned}$$

Repeating the above reasoning, we obtain

$$M(x_n, x_{n+1}, t) \geq \prod_{i=1}^{n} \left(\frac{\beta_{i,i+1} + \gamma_{i,i+1}}{1 - \beta_{i,i+1}} \right) M(x_0, x_1, t). \tag{3.2}$$

Moreover, for $p > 0$ and by repeated use of $(G5)$, we have

$$\begin{aligned}
M(x_n, x_{n+p}, t) &\geq M(x_n, x_{n+1}, t) + M(x_{n+1}, x_{n+2}, t) \\
&\quad + \cdots + M(x_{n+p-1}, x_{n+p}, t) \\
&\geq \prod_{i=1}^{n} \left(\frac{\beta_{i,i+1} + \gamma_{i,i+1}}{1 - \beta_{i,i+1}} \right) M(x_0, x_1, t) \\
&\quad + \prod_{i=1}^{n+1} \left(\frac{\beta_{i,i+1} + \gamma_{i,i+1}}{1 - \beta_{i,i+1}} \right) M(x_0, x_1, t) \\
&\quad + \cdots + \\
&\quad + \prod_{i=1}^{n+p-1} \left(\frac{\beta_{i,i+1} + \gamma_{i,i+1}}{1 - \beta_{i,i+1}} \right) M(x_0, x_1, t) \\
&= \sum_{k=0}^{p-1} \prod_{i=1}^{n+k} \left(\frac{\beta_{i,i+1} + \gamma_{i,i+1}}{1 - \beta_{i,i+1}} \right) M(x_0, x_1, t)
\end{aligned}$$

$$= \sum_{k=n}^{n+p-1} \prod_{i=1}^{k} \left(\frac{\beta_{i,i+1} + \gamma_{i,i+1}}{1 - \beta_{i,i+1}} \right) M(x_0, x_1, t).$$

Let α and n_α as in Definition 2.10, then, for $n \geq n_\alpha$, it follows that

$$\begin{aligned} M(x_n, x_{n+p}, xt) &\geq \sum_{k=n}^{n+p-1} \left[\frac{1}{k} \sum_{i=1}^{k} \left(\frac{\beta_{i,i+1} + \gamma_{i,i+1}}{1 - \beta_{i,i+1}} \right) \right]^k M(x_0, x_1, t) \\ &\geq \left(\sum_{k=n}^{n+p-1} \alpha^k \right) M(x_0, x_1, t) \\ &\leq \frac{\alpha^n}{1 - \alpha} M(x_0, x_1, t). \end{aligned} \tag{3.3}$$

Now, letting the limit as $n \to +\infty$, we deduce that $M(x_n, x_{n+p}, t) \to 0$. Thus $\{x_n\}$ is a Cauchy sequence and, by completeness of X, converges to u (say) in X.

For any positive integer m, we have

$$\begin{aligned} M(x_n, T_m(u), t) &= M(T_n(x_{n-1}), T_m(u), t) \\ &\geq \beta_{n,m}[M(x_{n-1}, x_n, t) + M(u, T_m(u), t)] \\ &+ \gamma_{n,m} M(x_{n-1}, u, t). \end{aligned}$$

Letting $n \to +\infty$, we obtain

$$\begin{aligned} M(u, T_m(u), t) &\geq \beta_{n,m}[M(u, u, t) + M(u, T_m(u), t)] + \gamma_{n,m} M(u, u, t) \\ &= \beta_{n,m} M(u, T_m(u), t), \end{aligned}$$

and so as $\beta_{n,m} < 1$, it follows that $M(u, T_m(u), t) = 0$, that is $T_m(u) = u$. Then, u is a common fixed point of $\{T_m\}$. Finally, we prove uniqueness of the common fixed point u. To this aim, let us suppose that v is another common fixed point of $\{T_m\}$, i.e., $T_m(v) = v$. Then, using (3.1), we have

$$\begin{aligned} M(u, v, t) &= M(T_m(u), T_m(v), t) \\ &\geq \beta_{n,m}[M(u, T_m(u), t) + M(v, T_m(v), t)] + \gamma_{n,m} M(u, v, T) \\ &= \beta_{n,m}[G(u, u, t) + M(v, v, t)] + \gamma_{n,m} M(u, v, t). \end{aligned}$$

This implies that $M(u, v, t) \geq \gamma_{n,m} M(u, v, t)$, which yields that $u = v$ as $\gamma_{n,m} < 1$. So, u is the unique common fixed point of $\{T_m\}$.

As particular case of Theorem 3.1, we state the following corollary.

Corollary 3.2 *Let $\{T_n\}$ be a sequence of self mappings of a complete fuzzy metric space $(X, M, \star)$ such that*

$$M(T_i(x), T_j(y), t) \geq \beta_{i,j}[M(x, T_i(x), t) + M(y, T_j(y), t)] \tag{3.4}$$

for $x, y, z \in X$ *with* $x \neq y$, $0 \leq \beta_{i,j} < 1$, $i, j = 1, 2, \ldots$.

If $\sum_{i=1}^{+\infty}\left(\frac{\beta_{i,i+1}}{1-\beta_{i,i+1}}\right)$ *is an* α*-series, then* $\{T_n\}$ *has a unique common fixed point in* X.

Example 3.3 Let $X = [0, 1]$ and $M(x, y, t) = \min\left\{\frac{t}{t+|x-y|}, \frac{t}{t+|y-z|}, \frac{t}{t+|z-x|}\right\}$. Clearly, $(X, M, \star)$ is a complete fuzzy metric space. Define also $\beta_{i,j} = \frac{1}{1+2^i}$ for all $i, j = 1, 2, \ldots$ and $T_i(x) = \frac{x}{4^i}$ for all $x \in X$ and $i = 1, 2, \ldots$.

Assume $i < j$ and $x > y \geq z$, so that we have

$$M(T_i(x), T_j(y), t) = \frac{t}{t + \left|\frac{x}{4^i} - \frac{y}{4^j}\right|}$$

and

$$M(x, T_i(x), t) + M(y, T_j(y), t) = \frac{t}{t + \left|x - \frac{x}{4^i}\right|} + \frac{t}{\left|y - \frac{z}{4^j}\right|}.$$

Therefore condition (3.4) is satisfied for all $x, y, z \in X$ with $x \neq y$. Moreover, the series

$$\sum_{i=1}^{+\infty}\left(\frac{\beta_{i,i+1}}{1-\beta_{i,i+1}}\right) = \sum_{i=1}^{+\infty}\frac{1}{2^i}$$

is an α-series with $\alpha = 1/2$. Then, by Corollary 3.2, $\{T_n\}$ has a unique common fixed point $0 \in X$.

Following the same lines of the proof of Theorem 3.1, one can prove the next theorem.

Theorem 3.4 *Let* $\{T_n\}$ *be a sequence of self mappings of a complete fuzzy metric space* $(X, M, \star)$ *such that*

$$M(T_i^p(x), T_j^p(y), t) \geq \beta_{i,j}[M(x, T_i^p(x), t) + M(y, T_j^p(x), t) + \gamma_{i,j}G(x, y, z)]$$

for $x, y, z \in X$ *with* $x \neq y$, $0 \leq \beta_{i,j}, \gamma_{i,j} < 1$, $i, j = 1, 2, \ldots$.

If $\sum_{i=1}^{+\infty}\left(\frac{\beta_{i,i+1}+\gamma_{i,i+1}}{1-\beta_{i,i+1}}\right)$ *is an* α*-series, then* $\{T_n\}$ *has a unique common fixed point in* X.

Further in the next section, finally we establish an application of our result in product space.

4 Application

Let $\{T_n^{'}\}$ be a sequence of self mappings of a complete fuzzy metric space $X \times X = X^2$, i.e.,$(X^2, M, *)$ such that

$$M\{T_i^{'}(x,y), T_j^{'}(x^{'},y^{'}), t\} \geq \beta_{ij}\{M((x,y), T_i^{'}(x,y), t) + M((x^{'},y^{'}), T_i^{'}(x^{'},y^{'}), t)\} + \gamma_{ij} M((x,y),(x^{'},y^{'}), t) \tag{4.1}$$

for all $(x,y),(x^{'},y^{'}) \in X^2$ with $(x,y) \neq (x^{'},y^{'})$, $0 \leq \beta_{ij}, \gamma_{ij} < 1$, $i, j = 1, 2, \ldots$ if $\sum_{i=1}^{\infty}(\frac{\beta_{i,i+1},\gamma_{i,i+1}}{1-\beta_{i,i+1}})$ is an α-series then sequence $\{T_n^{'}\}$ has a unique fixed point in X.

Proof Fix $y = y^{'} \in X$. Let $T_i, T_j : X \to X$ be such that

$$T_i(x) = T_i^{'}(x,y)$$

$$T_j(x^{'}) = T_j^{'}(x^{'},y^{'}), \forall x, x^{'} \in X$$

then condition (4.1) reduces to (3.1).

Then by our main result $\{T_n^{'}\}$ has a unique fixed point $(z(y), y)$, that is

$$T_n(z(y)) = T_n^{'}(z(y), y) = (z(y), y) = z(y).$$

5 Future Scope

Some further generalizations are possible, the self-mapping can be replaced by multivalued, bivariate, trivariate, or n-variate mapping and coupled, tripled or n-tuple common fixed point theorem can be proved.

References

1. M. Abbas, B. Damjanovic, R. Lazovic, Fuzzy common fixed point theorems for generalized contractive mapping. Appl. Math. Lett. **23**, 1325–1330 (2010)
2. A. George, P. Veeramani, On some results on fuzzy metric spaces. Fuzzy Sets Syst. **64**, 395–399 (1994)
3. M. Grabiec, Fixed points in fuzzy metric spaces. Fuzzy Sets Syst. **27**, 385–389 (1988)
4. V. Gregori, S. Morillas, A. Sapena, Examples of fuzzy metrics and application. Fuzzy Sets Syst. **170**, 95–111 (2011)
5. V. Gregory, S. Romaguera, Characterizing compliable fuzzy metric spaces. Fuzzy Sets Syst. **144**, 411–420 (2004)
6. O. Kaleva, S. Seikkala, On fuzzy metric spaces. Fuzzy Sets Syst. **12**, 215–229 (1984)

7. O. Kramosil, J. Michalek, Fuzzy metrics and statistical metric spaces. Kybernetika **15**, 326–334 (1975)
8. S.N. Mishra, N. Sharma, S.L. Singh, Common fixed points of maps on fuzzy metric space. Int. J. Math. Sci. **17**, 253–258 (1994)
9. B.E. Rhoades, A comparison of various definition of contractive mappings. Trans. Am. Math. Soc. **226**(2), 257–290 (1977)
10. B. Schweizer, A. Sklar, *Probabilistic Metric Spaces* (Elsevier, Amesterdam, North Holland, 1983)
11. V. Sihag, R.K. Vats, C. Vetro, A fixed point theorem in G-metric spaces via α-series. Quest. Math. **37**, 1–6 (2014)
12. B. Singh, M.S. Chauhan, Common fixed point of compatible maps in fuzzy metric spaces. Fuzzy Sets Syst. **115**, 470–475 (2000)
13. P.V. Subrahmanyam, Common fixed point theorem in fuzzy metric spaces. Inform. Sci. **83**, 109–112 (1995)
14. R. Vasuki, A common fixed point theorem in fuzzy metric spaces spaces. Fuzzy Sets Syst. **97**, 375–397 (1998)
15. L.A. Zadeh, Fuzzy sets. Inform. Control **8**, 103–112 (1965)

The Unique Common Fixed Point Theorem for Four Mappings Satisfying Common Limit in the Range Property in b-Metric Space

Anushri A. Aserkar and Manjusha P. Gandhi

Abstract In the present paper a unique common fixed point theorem has been established in b-metric space for four weakly compatible mappings in pairs, satisfying common limit range property. We have proved this theorem without using the condition of completeness of the b-metric space. The result is an extension and generalization of many results available in metric space. A suitable example is also discussed to validate the result.

Keywords b-metric space · CLR property · Weakly compatible

1 Introduction

The most remarkable result in fixed point theory is the Banach contraction principle. Over the century or so researchers have extended and generalized it in different directions and spaces [1–11]. Bhaktin [12] in 1989 and Czerwik [13, 14] in 1993 came up with the concept of b-metric space which is one of the important generalizations of metric spaces.

Numerous researchers have proved fixed point theorems for multiple mappings. In some cases commutative property between the maps are required to obtain a common fixed point. Sessa [15] initiated the term weak commutative mappings. Jungck [16] generalized the notion of weak commutative condition by introducing the concept of compatible maps and then in [17] he introduced weakly compatible maps.

In 2002, Aamri and Moutawakil [1] put forward the idea of E.A., which is a true generalization of non-compatible maps in metric spaces. Many researchers [18–20] used this theory to establish unique common fixed point theorems. In 2011, the concept of common limit in the range (CLR) property for a pair of self-mappings in

---

A. A. Aserkar (✉)
Rajiv Gandhi College of Engineering and Research, Nagpur, India
e-mail: aserkar_aaa@rediffmail.com

M. P. Gandhi
Yeshwantrao Chavan College of Engineering, Nagpur, India
e-mail: manjusha_g2@rediffmail.com

N. Deo et al. (eds.), *Mathematical Analysis I: Approximation Theory*,
Springer Proceedings in Mathematics & Statistics 306,
https://doi.org/10.1007/978-981-15-1153-0_14

Fuzzy metric space was initiated by Sintunavarat and Kumam [21]. This property is outstanding because it does not require the closed subspaces.

In the present paper a unique common fixed point theorem has been established in b-metric for four weakly compatible mappings in pairs satisfying common limit range property. This theorem has been proved without using the condition of completeness of the b-metric space. The result is an extension and generalization of [22–25] available in metric space and b-metric spaces. A suitable example is also discussed to support this result.

2 Preliminary

Some basic definitions are necessary to discuss before we start the main theorems.

2.1 b-Metric Space [12, 13]

Let X be a (nonempty) set and $s \geq 1$ be a given real number. A function $d : X \times X \to [0, \infty)$ is a b-metric on X if the following conditions hold:

(i) $d(x, y) = 0$ if and only if $x = y$
(ii) $d(x, y) = d(y, x)$ for any $x, y \in X$
(iii) $d(x, y) \leq s\{d(x, z) + d(z, y)\}$ for any $x, y, z \in X$

Then the pair (X, d) is called a b-metric space (or metric type space).

2.2 .

Let (X, d) be a b-metric space.

(i) The sequence $\{x_n\}$ converges to $x \in X$ if and only if $\lim_{n\to\infty} d(x_n, x) = 0$;
(ii) The sequence $\{x_n\}$ is Cauchy if and only if $\lim_{m,n\to\infty} d(x_n, x_m) = 0$;
(iii) The space is complete if and only if every Cauchy sequence in X is convergent.

2.3 Common Limit in the Range Property [21]

Suppose that (X, d) is a metric space. Two mappings $F, Q : X \to X$ satisfies the common limit in the range of Q property (CLR_Q) if $\lim_{n\to\infty} Fx_n = \lim_{n\to\infty} Qx_n = Qx \in QX$ for some $x \in X$.

Wu et al. [26] extended the property to b-metric space for three mappings.

Let $F, P, Q : X \to X$ be three self-mappings of a b-metric space (X, d). The pair F, Q satisfies the common limit in the range of P property (CLR_P) if there exists a sequence $\{x_n\} \subseteq X$ and a point $x \in X$ such that $\lim_{n\to\infty} Fx_n = \lim_{n\to\infty} Qx_n = Px \in PX$.

Particularly, if $Q = P$ then the pair F, Q satisfies the (CLR_Q)-property.

2.4 *Altering Distance Function [27]*

A function $\xi : [0, +\infty) \to [0, +\infty)$ is called an altering distance function if it satisfies:

(i) ξ is continuous and non-decreasing.
(ii) $\xi(t) = 0$ if and only if $t = 0$.
Let Ψ denote the set of all continuous functions $\xi, \psi : [0, +\infty) \to [0, +\infty)$ and $s \geq 1$ be a given real number such that
(iii) $s\xi(t) \leq \xi(t) - \eta(t)$ if and only if $t = 0$.

2.5 *Weakly Compatible Map [17]*

Let P and Q be two self-maps defined on a set X, then P and Q are said to be weakly compatible if they commute at coincidence points. That is if $Px = Qx$ for some $x \in X$, then $PQx = QPx$.

3 Main Theorem

We have established the following theorem in b-metric space for four weakly compatible mappings in pair, which satisfy common limit range property.

Theorem 3.1 *Let (X, d) be a b-metric space with $s \geq 1$ and $F, G, P, Q : X \to X$. Suppose that $\xi, \eta \in \Psi$ and $L \geq 0$ such that*

(i) *F, Q satisfies CLR_p and G, P satisfies CLR_Q property.*

(ii) $s\xi(d(Fx, Gy)) \leq \xi(M(x, y)) - \eta(M(x, y)) + \text{LN}(x, y)$ (1)

$$\text{where } M(x, y) = \max\left\{\begin{array}{l} d(Py, Qx), \frac{d(Qx, Fx)*d(Py, Gy)}{1+d(Fx,Gy)}, \frac{(d(Py, Fx))^2+(d(Qx, Gy))^2}{d(Py, Fx)+d(Qx, Gy)}, \\ \frac{d(Qx, Fx)*d(Qx, Gy)+d(Py, Gy)*d(Py, Fx)}{d(Qx, Gy)+d(Py, Fx)} \end{array}\right\}$$

and $N(x, y) = \min\{d(Qx, Fx), d(Qx, Gy), d(Py, Fx), d(Py, Gy)\}$ *for all* $x, y \in X$.

(iii) *The pairs* (F, Q) *and* (G, P) *are weakly compatible.*

Then F, G, P, Q *have a unique common fixed point.*

Proof As F, Q satisfies CLR_p and G, P satisfies CLR_Q property, there exists sequences $\{x_n\}$ and $\{y_n\}$ in X and $r, t \in X$ such that

$$\lim_{n\to\infty} Fx_n = \lim_{n\to\infty} Qx_n = Pr \in PX \text{ and } \lim_{n\to\infty} Gy_n = \lim_{n\to\infty} Py_n = Qt \in QX$$

Putting $x = x_n$ and $y = y_n$ in (1),

$$s\xi(d(Fx_n, Gy_n)) \le \xi(M(x_n, y_n)) - \eta(M(x_n, y_n)) + \mathrm{LN}(x_n, y_n)$$

where

$$M(x_n, y_n) = \max\left\{ \begin{array}{l} d(Py_n, Qx_n), \frac{d(Qx_n, Fx_n)*d(Py_n, Gy_n)}{1+d(Fx_n, Gy_n)}, \frac{(d(Py_n, Fx_n))^2+(d(Qx_n, Gy_n))^2}{d(Py_n, Fx_n)+d(Qx_n, Gy_n)}, \\ \frac{d(Qx_n, Fx_n)*d(Qx_n, Gy_n)+d(Py_n, Gy_n)*d(Py_n, Fx_n)}{d(Qx_n, Gy_n)+d(Py_n, Fx_n)} \end{array} \right\}$$

and $N(x_n, y_n) = \min\{d(Qx_n, Fx_n), d(Qx_n, Gy_n), d(Py_n, Fx_n), d(Py_n, Gy_n)\}$
Taking $\lim_{n\to\infty}$ to both sides, we get

$$\lim_{n\to\infty} s\xi(d(Fx_n, Gy_n)) \le \lim_{n\to\infty} \xi(M(x_n, y_n)) - \lim_{n\to\infty} \eta(M(x_n, y_n)) + \lim_{n\to\infty} \mathrm{LN}(x_n, y_n)$$

where

$$\lim_{n\to\infty} M(x_n, y_n) = \lim_{n\to\infty} \max\left\{ \begin{array}{l} d(Py_n, Qx_n), \frac{d(Qx_n, Fx_n)*d(Py_n, Gy_n)}{1+d(Fx_n, Gy_n)}, \frac{(d(Py_n, Fx_n))^2+(d(Qx_n, Gy_n))^2}{d(Py_n, Fx_n)+d(Qx_n, Gy_n)}, \\ \frac{d(Qx_n, Fx_n)*d(Qx_n, Gy_n)+d(Py_n, Gy_n)*d(Py_n, Fx_n)}{d(Qx_n, Gy_n)+d(Py_n, Fx_n)} \end{array} \right\}$$

$$= \max\left\{ \begin{array}{l} d(Qt, Pr), \frac{d(Pr, Pr)*d(Qt, Qt)}{1+d(Pr, Qt)}, \frac{(d(Qt, Pr))^2+(d(Pr, Qt))^2}{d(Qt, Pr)+d(Pr, Qt)}, \\ \frac{d(Pr, Pr)*d(Pr, Qt)+d(Qt, Qt)*d(Qt, Pr)}{d(Pr, Qt)+d(Qt, Pr)} \end{array} \right\} = d(Qt, Pr)$$

and

$$\lim_{n\to\infty} N(x_n, y_n) = \lim_{n\to\infty} \min\{d(Qx_n, Fx_n), d(Qx_n, Gy_n), d(Py_n, Fx_n), d(Py_n, Gy_n)\}$$
$$= \min\{d(Pr, Pr), d(Pr, Qy), d(Qy, Pr), d(Qy, Qy)\} = 0.$$

$$\therefore s\xi(d(Pr, Qt)) \le \xi(d(Pr, Qt)) - \eta(d(Pr, Qt))$$
$$\therefore d(Pr, Qt) = 0 \Rightarrow Pr = Qt \tag{2}$$

Putting $x = x_n$ and $y = r$ in (1),

$$s\xi(d(Fx_n, Gr)) \leq \xi(M(x_n, r)) - \eta(M(x_n, r)) + \mathrm{LN}(x_n, r)$$

where

$$M(x_n, r) = \max\left\{ \begin{array}{l} d(Pr,\ Qx_n),\ \frac{d(Qx_n, Fx_n) * d(Pr, Gr)}{1 + d(Fx_n, Gr)},\ \frac{(d(Pr, Fx_n))^2 + (d(Qx_n, Gr))^2}{d(Pr, Fx_n) + d(Qx_n, Gr)}, \\ \frac{d(Qx_n, Fx_n) * d(Qx_n, Gr) + d(Pr, Gr) * d(Pr, Fx_n)}{d(Qx_n, Gr) + d(Pr, Fx_n)} \end{array} \right\}$$

and $N(x_n, r) = \min\{d(Qx_n, Fx_n), d(Qx_n, Gr), d(Pr, Fx_n), d(Pr, Gr)\}$
Taking $\lim_{n\to\infty}$ to both sides, we get

$$\lim_{n\to\infty} s\xi(d(Fx_n, Gr)) \leq \lim_{n\to\infty} \xi(M(x_n, r)) - \lim_{n\to\infty} \eta(M(x_n, r)) + \lim_{n\to\infty} \mathrm{LN}(x_n, r)$$

where

$$\lim_{n\to\infty} M(x_n, r) = \lim_{n\to\infty} \max\left\{ \begin{array}{l} d(Pr,\ Qx_n),\ \frac{d(Qx_n, Fx_n) * d(Pr, Gr)}{1 + d(Fx_n, Gr)},\ \frac{(d(Pr, Fx_n))^2 + (d(Qx_n, Gr))^2}{d(Pr, Fx_n) + d(Qx_n, Gr)}, \\ \frac{d(Qx_n, Fx_n) * d(Qx_n, Gr) + d(Pr, Gr) * d(Pr, Fx_n)}{d(Qx_n, Gr) + d(Pr, Fx_n)} \end{array} \right\}$$

$$= \max\left\{ \begin{array}{l} d(Pr,\ Pr),\ \frac{d(Pr, Pr) * d(Pr, Gr)}{1 + d(Pr, Gr)}, \\ \frac{(d(Pr, Pr))^2 + (d(Pr, Gr))^2}{d(Pr, Pr) + d(Pr, Gr)}, \\ \frac{d(Pr, Pr) * d(Pr, Gr) + d(Pr, Gr) * d(Pr, Pr)}{d(Pr, Gr) + d(Pr, Pr)} \end{array} \right\} = d(Pr,\ Gr)$$

and

$$\lim_{n\to\infty} N(x_n, r) = \lim_{n\to\infty} \min\{d(Qx_n, Fx_n), d(Qx_n, Gr), d(Pr, Fx_n), d(Pr, Gr)\}$$
$$= \min\{d(Pr, Pr), d(Pr, Gr), d(Pr, Pr), d(Pr, Gr)\} = 0$$

$$\therefore s\xi(d(Pr, Gr)) \leq \xi(d(Pr, Gr)) - \eta(d(Pr, Gr))$$
$$\therefore d(Pr, Gr) = 0 \Rightarrow Pr = Gr \tag{3}$$

Similarly, it may be proved that

$$Ft = Qt \tag{4}$$

∴ from (2), (3) and (4) we get

$$\therefore Ft = Qt = Pr = Gr \tag{5}$$

Let $Ft = Qt = Pr = Gr = u$
$\because F, Q$ are weakly compatible,

$$Fu = FQt = QFt = Qu$$

Putting $x = u$, $y = r$ in (1), we get

$$s\xi(d(Fu, Gr)) \le \xi(M(u,r)) - \eta(M(u,r)) + LN(u,r)$$

where

$$M(u,r) = \max\left\{\begin{array}{l} d(Pr,\ Qu), \frac{d(Qu,Fu)*d(Pr,Gr)}{1+d(Fu,Gr)}, \frac{(d(Pr,Fu))^2+(d(Qu,Gr))^2}{d(Pr,Fu)+d(Qu,Gr)}, \\ \frac{d(Qu,Fu)*d(Qu,Gr)+d(Pr,Gr)*d(Pr,Fu)}{d(Qu,Gr)+d(Pr,Fu)} \end{array}\right\}$$

$$= \max\left\{\begin{array}{l} d(u,\ Fu), \frac{d(Fu,Fu)*d(u,u)}{1+d(Fu,u)}, \frac{(d(u,Fu))^2+(d(Fu,u))^2}{d(u,Fu)+d(Fu,u)}, \\ \frac{d(Fu,Fu)*d(Fu,u)+d(u,u)*d(u,Fu)}{d(Fu,u)+d(u,Fu)} \end{array}\right\} = d(Fu,\ u)$$

and

$$\begin{aligned} N(u,r) &= \min\{d(Qu,Fu), d(Qu,Gr), d(Ps,Fu), d(Pr,Gr)\} \\ &= \min\{d(Fu,Fu), d(Fu,u), d(u,Fu), d(u,u)\} = 0 \end{aligned}$$

$$\begin{aligned} & s\xi(d(Fu,u)) \le \xi(d(Fu,u)) - \eta(d(Fu,u)) \\ & \therefore d(Fu,u) = 0 \Rightarrow Fu = u \\ & \therefore Fu = Qu = u \end{aligned}$$

Similarly we may prove that $Gu = Pu = u$

$$\therefore Fu = Qu = Pu = Gu = u.$$

Uniqueness: Let if possible there are two fixed points u, u^*, i.e., $Fu = Gu = Pu = Qu = u$ and $Fu^* = Gu^* = Pu^* = Qu^* = u^*$.

Putting $x = u$, $y = u^*$ in (1), we get

$$s\xi\big(d(Fu, Gu^*)\big) \le \xi\big(M(u,u^*)\big) - \eta\big(M(u,u^*)\big) + \mathrm{LN}(u,u^*)$$

where

$$M(u,u^*) = \max\left\{\begin{array}{l} d(Pu^*,\ Qu), \frac{d(Qu,Fu)\times d(Pu^*,Gu^*)}{1+d(Fu,Gu^*)}, \frac{(d(Pu^*,Fu))^2+(d(Qu,Gu^*))^2}{d(Pu^*,Fu)+d(Qu,Gu^*)}, \\ \frac{d(Qu,Fu)\times d(Qu,Gu^*)+d(Pu^*,Gu^*)\times d(Pu^*,Fu)}{d(Qu,Gu^*)+d(Pu^*Fu)} \end{array}\right\}$$

$$= \max\left\{\begin{array}{l} d(u^*,\ u), \frac{d(u,u)\times d(u^*,u^*)}{1+d(u,u^*)}, \frac{(d(u^*,u))^2+(d(u,u^*))^2}{d(u^*,u)+d(u,u^*)}, \\ \frac{d(u,u)\times d(u,u^*)+d(u^*,u^*)\times d(r^*,r)}{d(u,u^*)+d(u^*,u)} \end{array}\right\} = d(u, u*)$$

and

$$N(u,u^*) = \min\{d(Qu,Fu), d(Qu,Gu^*), d(Pu^*,Fu), d(Pu^*,Gu^*)\}$$
$$= \min\{d(u,u),\ d(u,u^*),\ d(u^*,u),\ d(u^*,u^*)\} = 0$$

$$s\xi\big(d(Fu,Gu^*)\big) \le \xi\big(M(u,u^*)\big) - \eta\big(M(u,u^*)\big) + \mathrm{LN}(u,u^*)$$
$$\therefore s\xi\big(d(u,u^*)\big) \le \xi\big(d(u,u^*)\big) - \eta\big(d(u,u^*)\big)$$
$$\therefore d(u,u^*) = 0 \Rightarrow u = u^*$$

Thus the fixed point is unique.

Corollary 3.1 *Let (X,d) be a b-metric space with $s \ge 1$ and $F, G, P, Q : X \to X$. Suppose that $\xi, \eta \in \Psi$ such that*

(i) *F, Q satisfies CLR_p and G, P satisfies CLR_Q property.*
(ii) $s\xi(d(Fx,Gy)) \le \xi(M(x,y)) - \eta(M(x,y))$

where $M(x,y) = \max\left\{\begin{array}{l} d(Py,Qx), \frac{d(Qx,Fx)*d(Py,Gy)}{1+d(Fx,Gy)}, \frac{(d(Py,Fx))^2+(d(Qx,Gy))^2}{d(Py,Fx)+d(Qx,Gy)}, \\ \frac{d(Qx,Fx)*d(Qx,Gy)+d(Py,Gy)*d(Py,Fx)}{d(Qx,Gy)+d(Py,Fx)} \end{array}\right\}$

for all $x, y \in X$.

(iii) *The pairs (F,Q) and (G,P) are weakly compatible.*

Then F, G, P, Q have a unique common fixed point.

Proof By substituting $L = 0$ in Theorem 3.1, F, G, P, Q have a unique common fixed point.

Corollary 3.2 *Let (X,d) be a b-metric space with $s \ge 1$ and $F, G, P, Q : X \to X$. Suppose that $0 \le k < 1$ such that*

(i) *F, Q satisfies CLR_p and G, P satisfies CLR_Q property.*
(ii) $sd(Fx,Gy) \le k\ M(x,y)$

where $M(x,y) = \max\left\{\begin{array}{l} d(Py,\ Qx), \frac{d(Qx,\ Fx)*d(Py,\ Gy)}{1+d(Fx,\ Gy)}, \frac{(d(Py,\ Fx))^2+(d(Qx,\ Gy))^2}{d(Py,\ Fx)+d(Qx,\ Gy)}, \\ \frac{d(Qx,\ Fx)*d(Qx,\ Gy)+d(Py,\ Gy)*d(Py,\ Fx)}{d(Qx,\ Gy)+d(Py,\ Fx)} \end{array}\right\}$

for all $x, y \in X$.

(iii) *The pairs (F,Q) and (G,P) are weakly compatible.*

Then F, G, P, Q have a unique common fixed point.

Proof Let $\xi(t) = t$,$\eta(t) = (1-k)t$ and $L = 0$. Then by Corollary 3.1, F, G, P, Q have a unique common fixed point.

Corollary 3.3 *Let (X,d) be a b-metric space with $s \ge 1$ and $F, G, P, Q : X \to X$. Suppose that $0 \le k < \frac{1}{4}$ and $L \ge 0$ such that*

(i) *F, Q satisfies CLR_p and G, P satisfies CLR_Q property.*

$sd(Fx, Gy) \leq$

(ii) $$k\left\{\begin{array}{l} d(Py, Qx) + \frac{d(Qx,Fx)*d(Py,Gy)}{1+d(Fx,Gy)} + \frac{(d(Py, Fx))^2+(d(Qx, Gy))^2}{d(Py, Fx)+d(Qx, Gy)} \\ + \frac{d(Qx, Fx)*d(Qx, Gy)+d(Py, Gy)*d(Py, Fx)}{d(Qx, Gy)+d(Py, Fx)} \end{array}\right\} + \mathrm{LN}(x, y)$$

where $N(x, y) = \min\{d(Qx, Fx), d(Qx, Gy), d(Py, Fx), d(Py, Gy)\}$ *for all* $x, y \in X$.

(iii) *The pairs* (F, Q) *and* (G, P) *are weakly compatible.*

Then F, G, P, Q *have a unique common fixed point.*

Proof Obviously

$$k\left\{\begin{array}{l} d(Py, Qx) + \frac{d(Qx, Fx)*d(Py, Gy)}{1+d(Fx, Gy)} + \frac{(d(Py, Fx))^2+(d(Qx, Gy))^2}{d(Py, Fx)+d(Qx, Gy)} \\ + \frac{d(Qx, Fx)*d(Qx, Gy)+d(Py, Gy)*d(Py, Fx)}{d(Qx, Gy)+d(Py, Fx)} \end{array}\right\} + \mathrm{LN}(x, y)$$

$$\leq 4k\max\left\{\begin{array}{l} d(Py, Qx), \frac{d(Qx, Fx)*d(Py, Gy)}{1+d(Fx, Gy)}, \frac{(d(Py, Fx))^2+(d(Qx, Gy))^2}{d(Py, Fx)+d(Qx, Gy)}, \\ \frac{d(Qx, Fx)*d(Qx, Gy)+d(Py, Gy)*d(Py, Fx)}{d(Qx, Gy)+d(Py, Fx)} \end{array}\right\} + \mathrm{LN}(x, y)$$

Let $\xi(t) = t$ and $\eta(t) = (1 - 4k)t$. Then by Theorem 3.1, F, G, P, Q have a unique common fixed point.

Corollary 3.4 *Let* (X, d) *be a b-metric space with* $s \geq 1$ *and* $F, G, P, Q : X \to X$. *Suppose that* $\eta \in \Psi$ *and* $L \geq 0$ *such that*

(i) F, Q *satisfies* CLR_p *and* G, P *satisfies* CLR_Q *property.*

(ii) $sd(Fx, Gy) \leq M(x, y) - \eta(M(x, y)) + \mathrm{LN}(x, y)$

where $$M(x, y) = \max\left\{\begin{array}{l} d(Py, Qx), \frac{d(Qx, Fx)*d(Py, Gy)}{1+d(Fx, Gy)}, \frac{(d(Py, Fx))^2+(d(Qx, Gy))^2}{d(Py, Fx)+d(Qx, Gy)}, \\ \frac{d(Qx, Fx)*d(Qx, Gy)+d(Py, Gy)*d(Py, Fx)}{d(Qx, Gy)+d(Py, Fx)} \end{array}\right\}$$

and $N(x, y) = \min\{d(Qx, Fx), d(Qx, Gy), d(Py, Fx), d(Py, Gy)\}$ *for all* $x, y \in X$.

(iii) *The pairs* (F, Q) *and* (G, P) *are weakly compatible.*

Then F, G, P, Q *have a unique common fixed point.*

Proof Let $\xi(t) = t$. Then by Theorem 3.1, F, G, P, Q have a unique common fixed point.

Example Let $X = [1, 6]$ and $d : X \times X \to [0, \infty)$ be defined by $d(x, y) = |x - y|$. We define mappings

$$Fx = \begin{cases} 3, & x \leq 3, \\ 4, & x > 3. \end{cases} \quad Gx = \begin{cases} 4, & x < 3, \\ \frac{x+3}{2}, & x \geq 3. \end{cases} \quad Qx = \begin{cases} 6 - x, & x \leq 3, \\ 5, & x > 3. \end{cases} \quad \text{and}$$

$$Px = \begin{cases} 6, & x < 3, \\ \frac{2x+3}{3}, & x \geq 3. \end{cases}$$

Obviously, d is a b-metric with $s = 2$.

Let $\xi(t) = \frac{t}{2}, \eta(t) = \frac{t}{10}, L = 10$.

To prove that F, Q satisfies CLR$_p$ and G, P satisfies CLR$_Q$ property, consider sequences $\{x_n\}$ and $\{y_n\}$ defined by $x_n = 3 - \frac{1}{n}$ and $y_n=3 + \frac{1}{n}$.

We have $\lim_{n\to\infty} Fx_n = \lim_{n\to\infty} Qx_n = 3 = P3 \in PX$ and $\lim_{n\to\infty} Gy_n = \lim_{n\to\infty} Py_n = 3 = Q3 \in QX$

It is easily proved that F, Q and G, P are weakly compatible.

Case-I If $x < 3$, $y < 3$ then we have

$Fx = 3,\ Gx = 4,\ Px = 6$ and $Qx = 6 - x$

$$\text{L.H.S.} = s\xi(d(Fx, Gy)) = 2 \times \frac{1}{2}|3-4| = 1$$
$$\text{R.H.S.} = \xi(M(x, y)) - \eta(M(x, y)) + \text{LN}(x, y)$$

where

$$M(x, y) = \max\left\{\begin{array}{l} d(Py, Qx), \frac{d(Qx, Fx)*d(Py, Gy)}{1+d(Fx, Gy)}, \frac{(d(Py, Fx))^2+(d(Qx, Gy))^2}{d(Py, Fx)+d(Qx, Gy)}, \\ \frac{d(Qx, Fx)*d(Qx, Gy)+d(Py, Gy)*d(Py, Fx)}{d(Qx, Gy)+d(Py, Fx)} \end{array}\right\}$$

$$= \max\left\{\begin{array}{l} |6-(6-x)|, \frac{|6-x-3| \times |6-4|}{1+|3-4|}, \frac{|6-3|^2+|(6-x)-4|^2}{|6-3|+|(6-x)-4|}, \\ \frac{|6-x-3| \times |(6-x)-4| + |6-4| \times |6-3|}{|(6-x)-4|+|6-3|} \end{array}\right\} \text{Let } y = 2.$$

$$= \max\{2, 1, 3, 3\} = 3$$

$$N(x, y) = \min\{d(Qx, Fx), d(Qx, Gy), d(Py, Fx), d(Py, Gy)\}$$
$$= \min\{|6-x-3|, |6-x-4|, |6-3|, |6-4|\} = \min\{1, 0, 3, 2\} = 0$$

Therefore we obtain

$$\text{R.H.S.} = \xi(M(x, y)) - \eta(M(x, y)) + \text{LN}(x, y)$$
$$= \frac{3}{2} - \frac{3}{10} + 10 \times 0 = 1.2$$
$$\therefore \text{L.H.S.} \le \text{R.H.S.}$$

Case-II Let $x = 3$, $y = 3$

$$Fx = 3,\ Gx = 3,\ Px = 3 \text{ and } Qx = 3$$

L.H.S. = R.H.S. = 0.

Case-III Let $x > 3$, $y > 3$,

$$Fx = 4 \quad Gx = \frac{x+3}{2} \quad Qx = 5 \quad \text{and } Px = \frac{2x+3}{3}$$

$$\text{L.H.S.} = s\xi(d(Fx, Gy)) = 2 \times \frac{1}{2}\left|4 - \left(\frac{x+3}{2}\right)\right| = 0.5 \quad \text{(For calculation } y = 4)$$

$$\text{R.H.S.} = \xi(M(x, y)) - \eta(M(x, y)) + \text{LN}(x, y)$$

where

$$M(x, y) = \max\left\{\begin{array}{l} d(Py,\ Qx),\ \frac{d(Qx, Fx) \times d(Py, Gy)}{1 + d(Fx, Gy)},\ \frac{(d(Py, Fx))^2 + (d(Qx, Gy))^2}{d(Py, Fx) + d(Qx, Gy)}, \\ \frac{d(Qx, Fx) \times d(Qx, Gy) + d(Py, Gy) \times d(Py, Fx)}{d(Qx, Gy) + d(Py, Fx)} \end{array}\right\}$$

$$= \max\left\{\left|\frac{11}{3} - 5\right|, \frac{1 \times \frac{1}{6}}{1 + \frac{1}{2}}, \frac{\left|\frac{11}{3} - 4\right|^2 + \left|5 - \frac{7}{2}\right|^2}{\left|\frac{11}{3} - 4\right| + \left|5 - \frac{7}{2}\right|}, \frac{1 \times \left|5 - \frac{7}{2}\right| + \frac{1}{6} \times \left|\frac{11}{3} - 4\right|}{\left|5 - \frac{7}{2}\right| + \left|\frac{11}{3} - 4\right|}\right\}$$

$$= \max\{1.33, 0.111, 1.29, 0.85\} = 1.33$$

and

$$N(x, y) = \min\{d(Qx, Fx), d(Qx, Gy), d(Py, Fx), d(Py, Gy)\}$$
$$= \min\{1, 1.5, 0.33, 0.167\} = 0.167$$

$$\text{R.H.S.} = \xi(M(x, y)) - \eta(M(x, y)) + \text{LN}(x, y) = \frac{1.33}{2} - \frac{1.33}{10} + 10 \times 0.167 = 2.202$$

$\therefore$ L.H.S.$\leq$ R.H.S.

Thus all conditions of Theorem 3.1 are satisfied. In fact, 3 is a unique point in X such that $F(3) = G(3) = P(3) = Q(3) = 3$.

Acknowledgements The authors are thankful to the affiliated college authorities for financial support given by them.

References

1. A.F. Roldan-Lopez-de-Hierro, E. Karapınar, H.H. Alsulami, A short-note on 'Common fixed point theorems for noncompatible self-maps in generalized metric spaces'. J. Inequal. Appl. 14 (2015)
2. C. Di Bari, P. Vetro, Nonlinear quasi-contractions of Ciric type. Fixed Point Theory **13**, 453–459 (2012)
3. F. He, X.Y. Nan, A unified view on common fixed point theorems for Ciric quasi-contraction maps. Fixed Point TheoryAppl. 17 (2015)
4. H. Aydi, C. Vetro, W. Sintunavarat, P. Kumam, Coincidence and fixed points for contractions and cyclical contractions in partial metric spaces. Fixed Point Theory Appl. 18 (2012)

5. H. Aydi, E. Karapınar, B. Samet, Fixed points for generalized (α,ψ)-contractions on generalized metric spaces. J. Inequal. Appl. 16 (2014)
6. H.K. Nashine, B. Samet, C. Vetro, Fixed point theorems in partially ordered metric spaces and existence results for integral equations. Numer. Func. Anal. Optim. **33**, 1304–1320 (2012)
7. L.B. Ciric, A generalization of Banach's contraction principle. Proc. Amer. Math. Soc. **45**, 267–273 (1974)
8. M. Jovanovic, Z. Kadelburg, S. Radenovic, Common fixed point results in metric-type spaces. Fixed Point Theory Appl. 15 (2010)
9. W.S. Du, A note on cone metric fixed point theory and its equivalence. Nonlinear Anal. **72**, 2259–2261 (2010)
10. Z.Z. Yang, Common fixed point theorems for non-compatible self-maps in generalized metric spaces. J. Inequal. Appl. 12 (2014)
11. Z.Z. Yang, H. Sadati, S. Sedghi, N. Shobe, Common fixed point theorems for non-compatible self-maps in b-metric spaces. J. Nonlinear Sci. Appl. **8**, 1022–1031 (2015)
12. I.A. Bakhtin, The contraction principle in quasimetric spaces. Funct. Anal. **30**, 26–37 (1989)
13. S. Czerwik, Contraction mappings in-metric spaces. Acta Math. Inform. Univ. Ostrav. **1**, 5–11 (1993)
14. S. Czerwik, Nonlinear set-valued contraction mappings in b-metric spaces. AttiSemin. Mat. Fis. Univ. Modena **46**(2), 263–276 (1998)
15. S. Sessa, On a weak commutativity condition of mappings in fixed point consideration. Publ. Inst. Math. **32**, 149–153 (1982)
16. G. Jungck, Compatible mappings and common fixed points. Int. J. Math. Math. Sci. **9**, 771–779 (1986)
17. G. Jungck, Common fixed points for non-continuous non-self maps on non-metric spaces. Far East J. Math. Sci. **4**(2), 199–215 (1996)
18. A.H. Soliman, M. Imdad, M. Hasan, Proving unified common fixed point theorems via common property (E.A) in symmetric spaces. Commun. Korean Math. Sci. **25**(4), 629–645 (2010)
19. A. Kaewcharoen, Common fixed point theorems for contractive mappings satisfying Φ-maps in G-metric spaces. Banach J. Math. Anal. **6**(1), 101–111 (2012)
20. D. Gopal, M. Imdad, C. Vetro, Impact of common property (E.A) on fixed point theorems in fuzzy metric spaces. Fixed Point Theory Appl. Article ID 297360, 14 (2011)
21. W. Sintunavarat, P. Kumam, Common fixed point theorems for a pair of weakly compatible mappings in fuzzy metric spaces. J. Appl. Math. Article ID 637958, 14 (2011)
22. J.R Roshan, V. Parvaneh, S. Sedghi, N. Shobkolaei, W. Shatanawi, Common fixed points of almost generalized (ψ,ϕ)-contractive mappings in ordered b-metric spaces. Fixed Point Theory Appl. Article ID 159 (2013). https://doi.org/10.1186/1687-1812-2013-159
23. M. Abbas, Common fixed point for Lipschitzian mapping satisfying rational contractive conditions. Ital. J. Pure Appl. Math. **27**, 141–146 (2010)
24. M. Imdad, A. Ahmad, Four mappings with a common fixed point. Univ.uNovomSadu Zb.Rad.Priod.mat.Fak.Ser.Mat. **24**, 23–30 (1994)
25. N. Hussain, V. Parvaneh, J.R. Roshan, Z. Kadelburg, Fixed points of cyclic weakly (ψ, ϕ, L, A, B)-contractive mappings in ordered b-metric spaces with applications. Fixed Point Theory Appl. (2013). http://www.fixedpointtheoryandapplications.com/content/2013/1/256
26. B. Wu, F. He, T. Xu, Common fixed point theorems for Ciric type mappings in b-metric spaces without any completeness assumption. J. Nonlinear Sci. Appl. **10**, 3180–3190, (2017). ISSN 2008-1898
27. M.S. Khan, M. Swaleh, S. Sessa, Fixed point theorems by altering distances between the points. Bull. Aust. Math. Soc. **30**, 1–9 (1984)
28. M. Aamri, DEl Moutawakil, Some new common fixed point theorems under strict contractive conditions. J. Math. Anal. Appl. **270**, 181–188 (2002)

Radius Estimates for Three Leaf Function and Convex Combination of Starlike Functions

Shweta Gandhi

Abstract We study radii problems for the class $\mathcal{S}^*_{3\mathcal{L}}$ consisting of normalized analytic functions f in the unit disk with $zf'(z)/f(z)$ subordinate to $1 + 4z/5 + z^4/5$ and the class associated with convex combination of linear and exponential functions.

Keywords Starlike functions · Coefficient bounds · Growth · Radius problems

2010 Mathematics Subject Classification 30C45 · 30C50 · 30C80

1 Introduction

Let $\mathcal{A}$ denote the class of analytic functions in the unit disk $\mathbb{D} := \{z : |z| < 1\}$ whose Taylor's series expansion is given by $f(z) = z + \sum_{k=2} a_k z^k$. In particular for $n = 1$ let $\mathcal{A} := \mathcal{A}_1$. Let $\mathcal{S}$ be the subclass of $\mathcal{A}$ consisting of univalent functions. An analytic function f is said to be subordinate to an analytic function g, denoted by $f \prec g$, if there exists a Schwarz function w defined on $\mathbb{D}$ such that $f(z) = g(w(z))$ for all $z \in \mathbb{D}$. When g is a univalent function, this definition is equivalent to $f(0) = g(0)$ and $f(\mathbb{D}) \subset g(\mathbb{D})$. Among the various subclasses of $\mathcal{S}$, class of starlike functions and convex functions are prominently studied. In 1992 Ma and Minda integrated various subclasses of starlike and convex function using the subordination theory.

For a univalent functions φ normalized by $\varphi(0) = 1$ and $\varphi'(0) > 0$ with $\mathrm{Re}(\varphi(z)) > 0$ and $\varphi(\mathbb{D})$ is symmetric with respect to real axis and starlike with respect to 1, Ma and Minda [15] studied distortion and growth properties of the classes

$$\mathcal{S}^*(\varphi) = \left\{ f \in \mathcal{A} : \frac{zf'(z)}{f(z)} \prec \varphi(z) \right\} \quad \text{and} \quad \mathcal{K}(\varphi) = \left\{ f \in \mathcal{A} : 1 + \frac{zf''(z)}{f'(z)} \prec \varphi(z) \right\}.$$

S. Gandhi (✉)
Department of Mathematics, Miranda House, University of Delhi, Delhi 110 007, India
e-mail: gandhishwetagandhi@gmail.com

N. Deo et al. (eds.), *Mathematical Analysis I: Approximation Theory*,
Springer Proceedings in Mathematics & Statistics 306,
https://doi.org/10.1007/978-981-15-1153-0_15

By restricting values of φ to lie in some specific regions like half plane, disks, sectors, parabolas, cardioid, lemniscate of Bernoulli, booth lemniscate in the right-half plane of $\mathbb{C}$, various interesting subclasses of starlike and convex functions can be obtained. For example, $\mathcal{S}^*((1+Az)/(1+Bz)) =: \mathcal{S}^*[A,B]$ [13] and $\mathcal{K}((1+Az)/(1+Bz)) =: \mathcal{K}[A,B]$ $(-1 \leq B < A \leq 1)$ are the familiar classes of Janowski starlike and Janowski convex functions, respectively. For $0 \leq \alpha < 1$, $\mathcal{S}^*(\alpha) := \mathcal{S}^*[1-2\alpha,-1]$ and $\mathcal{K}(\alpha) := \mathcal{K}[1-2\alpha,-1]$ denote the classes of starlike and convex functions of order α, respectively which were introduced in [27]. The classes $\mathcal{S}^* := \mathcal{S}^*(0)$ and $\mathcal{K} := \mathcal{K}(0)$ are the well-known classes of starlike and convex functions, respectively. Similarly, the classes $\mathcal{S}^*_e := \mathcal{S}^*(e^z)$, $\mathcal{S}^*_{RL} := \mathcal{S}^*\left(\sqrt{2}-(\sqrt{2}-1)\sqrt{(1-z)/(1+2(\sqrt{2}-1)z)}\right)$ and $\mathcal{S}^*_L := \mathcal{S}^*(\sqrt{1+z})$ were introduced by Mendiratta et al. [19, 20], Sokół and Stankiewicz [35], respectively. Various other results related to above classes can be found in [3–5, 22, 30–34]. Raina and Sokol [24] introduced the class $\mathcal{S}^*_q = \mathcal{S}^*(z+\sqrt{1+z^2})$, which was further studied by Gandhi and Ravichandran [10]. We call the function $\phi_{3\mathcal{L}}(z) = 1+4z/5+z^4/5$ as three leaf function. This function is univalent, starlike with respect to $\phi(0)=1$ and its image of the unit disk is symmetric with respect to real axis. We investigate various geometric and analytical properties of the class $\mathcal{S}^*(1+4z/5+z^4/5) = \mathcal{S}^*_{3\mathcal{L}}$. In 1952 and 1953, Rahmanov [23] studied various properties of convex combination of functions belonging to several well-known classes of functions. Campbell [6] in his survey article provides various results related to combination of univalent functions as well as of locally univalent functions.

Cho et al. [7] determined radii of convexity, starlikeness, lemniscate starlikeness, and close to convexity for the convex combination of identity map and a normalized convex function. Recently, Khatter et al. [14] investigated various geometrical properties of the convex combination of constant function $f(z)=1$ with e^z and $\sqrt{1+z}$. Motivated by these results, we have considered the convex combination of two starlike functions, namely, e^z and $1+z$. For $0 \leq k \leq 1$, define $\phi_{EL}(z) = ke^z + (1-k)(1+z)$. In this paper, we investigate various geometric properties of the class $\mathcal{S}^*_{EL} := \mathcal{S}^*(\phi_{EL})$ consisting of analytical functions satisfying $zf'(z)/f(z) \prec \phi_{EL}(z)$, $(z \in \mathbb{D})$.

2 $\mathcal{S}^*_{3\mathcal{L}}$-Radii for Several Classes

Let $\mathcal{M}$ and $\mathcal{M}'$ be subsets of $\mathcal{A}$ then the $\mathcal{M}$-radius in $\mathcal{M}'$ denoted by $\mathcal{R}_{\mathcal{M}}(\mathcal{M}')$ is the largest $\rho_0 \in (0,1)$ such that for every $f \in \mathcal{M}'$, the function $F(z) = f(\rho z)/\rho \in \mathcal{M}$, whenever $0 < \rho < \rho_0$. In this section, sharp $\mathcal{S}^*_{3\mathcal{L}}$-radius for certain well-known classes of functions are obtained. The first two results of this section determines $\mathcal{S}^*_{3\mathcal{L}}$ radius for the class $\mathcal{S}^*[A,B]$ for the cases $B \geq 0$ and $B < 0$. For $-1 \leq B < A \leq 1$, let $\mathcal{P}[A,B]$ be the class of analytic functions p of the form $p(z) = 1 + a_1 z + a_2 z^2 + \cdots$ satisfying $p(z) \prec (1+Az)/(1+Bz)$ for all $z \in \mathbb{D}$. Let $\mathcal{P}[1-2\alpha,-1] = \mathcal{P}(\alpha)$

$(0 \leq \alpha < 1)$ and $\mathcal{P}(0) = \mathcal{P}$. We require the following results and definitions to prove our result.

Lemma 2.1 ([26, Lemma 2.1, p. 267], [28]) *For $p \in \mathcal{P}(\alpha)$ and $|z| = r$, we have*

$$\left| p(z) - \frac{1+(1-2\alpha)r^2}{1-r^2} \right| \leq \frac{2(1-\alpha)r}{1-r^2}, \quad \textit{and} \quad \left| \frac{zp'(z)}{p(z)} \right| \leq \frac{2r(1-\alpha)}{(1-r)(1+(1-2\alpha)r)}.$$

More generally, for $p \in \mathcal{P}[A, B]$, we have

$$\left| p(z) - \frac{1-ABr^2}{1-B^2r^2} \right| \leq \frac{(A-B)r}{1-B^2r^2}.$$

We find the largest disk with given center on real axis which is contained inside $\phi_{3\mathcal{L}}(\mathbb{D})$. This helps us to find condition for which several well-known classes can be associated with the class $\mathcal{S}^*_{3\mathcal{L}}$.

Lemma 2.2 *For $2/5 < a \leq 2$, let r_a be given by*

$$r_a = \begin{cases} a - 2/5 & \textit{if } \ 2/5 < a \leq 1 \\ \sqrt{(a-7/5)^2 + a/5} & \textit{if } \ 1 \leq a < 51/35 \\ 2 - a & \textit{if } \ 51/35 \leq a < 2. \end{cases}$$

If $\varphi(z) = 1 + \frac{4}{5}z + \frac{1}{5}z^4$, then $\{w : |w - a| < r_a\} \subset \varphi(\mathbb{D})$.

Proof Let $\phi_{3\mathcal{L}}(z) = 1 + 4z/5 + z^4/5$. Then parametric form of any point on the boundary of $\phi_{3\mathcal{L}}(\mathbb{D})$ is given by $w = \phi_{3\mathcal{L}}(e^{it})$. The symmetry of the curve $w = \phi_{3\mathcal{L}}(e^{it})$ with respect to real axis permits us to consider interval $0 \leq t \leq \pi$. The parametric equation of $w = \phi_{3\mathcal{L}}(e^{it})$ becomes

$$w = \phi_{3\mathcal{L}}(e^{it}) = 1 + \frac{4}{5}\cos t + \frac{1}{5}\cos 4t + i\left(\frac{4}{5}\sin t + \frac{1}{5}\sin 4t\right).$$

Let $z(t)$ denote the square of the distance from any point $(a, 0)$ on real axis to the points on the curve $w = \phi_{3\mathcal{L}}(e^{it})$. Then we have

$$z(t) = \frac{1}{25}\left(42 + 25a^2 + (-50 - 40\cos t - 10\cos 4t)a + 40\cos t + 8\cos 3t + 10\cos 4t\right).$$

It can be easily seen that

$$z'(t) = \frac{1}{25}\left((a-1)40\sin t + (a-1)40\sin 4t - 24\sin 3t\right).$$

When $a = 1$, we have $z'(t) = \frac{-24}{25}\sin 3t$ and $z'(t) = 0$ at $0, \pi/3$ and $2\pi/3$. The minimum value turns out to be $z(\pi)$. When $a \neq 1$, we have $z'(t) = \frac{8}{25}\sin t\left(\cos t - \frac{1}{2}\right)$ $(\cos t - r_{a_1})(\cos t - r_{a_2})$ where r_{a_1}, r_{a_2} are roots of the equation

$$40(a-1)\cos^2 t + (20a-32)\cos t + (-10a+4) = 0.$$

For $-1 < r_{a_1}, r_{a_2} < 1$, we have the conditions and range for a as required. By comparing the values of $z(t)$ at critical points of these range we have the desired result. □

Theorem 2.3 *The $\mathcal{S}^*_{3\mathcal{L}}$-radius for the class $\mathcal{S}^*[A, B]$ is given by*

$$\mathcal{R}_{\mathcal{S}^*_{3\mathcal{L}}}(\mathcal{S}^*[A, B]) = \frac{3}{5A-2B} \quad 0 \le B < A \le 1.$$

Proof Let $f \in S^*[A, B]$, then we have $zf'(z)/f(z) \prec (1+Az)/(1+Bz)$ which implies $zf'/f \in P[A, B]$. By Lemma 2.1 we have

$$\left|\frac{zf'(z)}{f(z)} - \frac{(1-ABr^2)}{(1-B^2r^2)}\right| \le \frac{(A-B)r}{1-B^2r^2} \quad (|z| = r < 1).$$

Since $B \ge 0$, we have $(1-ABr^2)/(1-B^2r^2) \le 1$. In view of Lemma 2.2 $f \in \mathcal{S}^*_{3\mathcal{L}}$ if $((A-B)r)/(1-B^2r^2) \le (1-ABr^2)/(1-B^2r^2) - 2/5$.

On solving this inequality for r, we get $r \le 3/(5A-2B)$. Also the function

$$f(z) = \begin{cases} z(1+Bz)^{\frac{A-B}{B}} & \text{if } B \neq 0 \\ ze^{Az} & \text{if } B = 0. \end{cases}$$

belongs to $\mathcal{S}^*[A, B]$ and $zf'(z)/f(z) = (1+Az)/(1+Bz)$ and at the point $z = -3/(5A-2B)$, the function $zf'(z)/f(z)$ assumes the value $2/5$. This shows that radius obtained is sharp. □

Theorem 2.4 *Let $-1 \le B < A \le 1$ with $B < 0$. Let*

$$R_1 = \min\left(1, \frac{4}{\sqrt{51B^2 - 35AB}}\right), \quad R_2 = \min\left(1, \frac{1}{A-2B}\right)$$

and

$$R_3 = \min\left(1, \frac{3}{\sqrt{25A^2 - 65AB + 49B^2}}\right).$$

Then $\mathcal{S}^*_{3\mathcal{L}}$ radius of $\mathcal{S}^*[A, B]$ is given by

$$\mathcal{R}_{\mathcal{S}^*_{3\mathcal{L}}}(\mathcal{S}^*[A, B]) = \begin{cases} R_2 & \text{if } R_2 \le R_1 \\ R_3 & \text{if } R_2 > R_1. \end{cases}$$

Proof Let $f \in \mathcal{S}^*[A, B]$, then by Lemma 2.1, we have

$$\left|\frac{zf'(z)}{f(z)} - \frac{1-ABr^2}{1-B^2r^2}\right| \le \frac{(A-B)r}{1-B^2r^2}.$$

We determine numbers R_1, R_2, and R_3. We have $r \leq R_1$ if and only if $(1 - ABr^2)/(1 - B^2r^2) \leq 51/35$. This yields $r \leq 4/\sqrt{51B^2 - 35AB}$. We determine R_2 such that $r \leq R_2$ if and only if

$$\frac{(A-B)r}{1-B^2r^2} \leq \sqrt{\left(\frac{1-ABr^2}{1-B^2r^2} - \frac{7}{5}\right)^2 + \frac{1}{5}\left(\frac{1-ABr^2}{1-B^2r^2}\right)}.$$

On computing this inequality we get $3/\sqrt{25A^2 - 65AB + 49B^2}$. We determine R_3 such that $r \leq R_3$ if and only if $(A-B)r/(1-B^2r^2) \leq 2 - (1-ABr^2)/(1-B^2r^2)$. A simple calculation yields $r \leq 1/(A-2B)$. A similar argument as in previous section gives the desired result. □

Theorem 2.5 *The $S^*_{3\mathcal{L}}$ radius of the class $\mathcal{S}^*_L$ is given by $\mathcal{R}_{S^*_{3\mathcal{L}}}(\mathcal{S}^*_L) = \frac{21}{25}$.*

Proof If $f \in \mathcal{S}^*_L$. Then $zf'(z)/f(z) \prec \sqrt{1+z}$. For $|z|=r$, we get $\left|zf'(z)/f(z) - 1\right| \leq |\sqrt{1+z} - 1| \leq 1 - \sqrt{1-r}$. From Lemma 2.2, it follows that if $1 - \sqrt{1-r} \leq 3/5$ then $f \in \mathcal{S}^*_{\phi_{3\mathcal{L}}}$. This gives $r \leq 21/25$ and this bound is best possible as seen by the function in this class given by

$$f(z) = \frac{4z\exp(2\sqrt{1+z} - 2)}{(1+\sqrt{1+z})^2}.$$

The above radius is sharp because at the point $z = -21/25$, we have $zf'(z)/f(z) = \sqrt{1+z} = 2/5$. □

Theorem 2.6 *The $S^*_{3\mathcal{L}}$ radius of the class $\mathcal{S}^*_{RL}$ is given by $\mathcal{R}_{S^*_{3\mathcal{L}}}(\mathcal{S}^*_{RL}) = 3\left(\frac{1169+444\sqrt{2}}{6439}\right)$.*

Proof Suppose that $f \in \mathcal{S}^*_{RL}$. Then

$$\frac{zf'(z)}{f(z)} \prec \sqrt{2} - \left(\sqrt{2}-1\right)\sqrt{\frac{1-z}{1+2\left(\sqrt{2}-1\right)z}}$$

and it is easy to deduce that

$$\left|\frac{zf'(z)}{f(z)} - 1\right| \leq 1 - \left(\sqrt{2} - \left(\sqrt{2}-1\right)\sqrt{\frac{1+r}{1-2\left(\sqrt{2}-1\right)r}}\right)$$

for $|z| = r$. By applying Lemma 2.2, we obtain $f \in \mathcal{S}^*_{\phi_{3\mathcal{L}}}$ provided

$$1 - \left(\sqrt{2} - \left(\sqrt{2}-1\right)\sqrt{\frac{1+r}{1-2\left(\sqrt{2}-1\right)r}}\right) \leq \frac{3}{5}$$

which by simple calculation shows that $r \leq 3(1169 + 444\sqrt{2})/6439$. The result is sharp for the function f given by

$$\frac{zf'(z)}{f(z)} = \sqrt{2} - (\sqrt{2} - 1)\sqrt{\frac{1 - z}{1 + 2(\sqrt{2} - 1)z}}.$$

□

There are various studies [5, 17, 18, 20] on classes of functions in the class $\mathcal{A}$, which are characterized by the ratio of functions f and g belonging to particular subclasses of $\mathcal{A}$. Recall that $\mathcal{W}$ denote the class of functions $f \in \mathcal{A}$ satisfying $\operatorname{Re}(f(z)/z) > 0$ for all $z \in \mathbb{D}$. Set of functions $f \in \mathcal{A}$ such that $f(z)/g(z) \in \mathcal{P}$ for some $g \in \mathcal{W}$ is denoted by $\mathcal{F}_1$ and the set of functions $f \in \mathcal{A}$ satisfying the inequality $|f(z)/g(z) - 1| < 1$ $(z \in \mathbb{D})$ for some $g \in \mathcal{W}$ is denoted by $\mathcal{F}_2$.

Theorem 2.7 *The $S^*_{3\mathcal{L}}$ radius of the class $\mathcal{W}$ is given by $\mathcal{R}_{\mathcal{S}^*_{3\mathcal{L}}}(\mathcal{W}) = \frac{\sqrt{34}-5}{3}$.*

Proof From Lemma 2.1, we get $|zf'(z)/f(z) - 1| \leq 2r/(1 - r^2)$. Lemma 2.2 shows that this disk lies inside $\phi_{3\mathcal{L}}(\mathbb{D})$ provided $2r/(1 - r^2) \leq 3/5$. This gives $r \leq (\sqrt{34} - 5)/3$. The function $f(z) = z(1 + z)/(1 - z) \in \mathcal{W}$ and at the point $z = -(\sqrt{34} - 5)/(3)$, we get $zf'(z)/f(z) = 2/5$ which shows that this bound is the best possible. □

Theorem 2.8 *The $S^*_{3\mathcal{L}}$ radius of the class $\mathcal{F}_1$ is given by $\mathcal{R}_{\mathcal{S}^*_{3\mathcal{L}}}(\mathcal{F}_1) = \frac{\sqrt{109}-10}{3}$.*

Proof Let $f \in \mathcal{F}_1$ and the functions $p, q : \mathbb{D} \to \mathbb{C}$ defined by

$$p(z) = \frac{g(z)}{z} \quad \text{and} \quad q(z) = \frac{f(z)}{g(z)}.$$

Then p, q belong to $\mathcal{P}$ and in view of Lemma 2.1, we get

$$\left|\frac{zf'(z)}{f(z)} - 1\right| \leq \left|\frac{zp'(z)}{p(z)}\right| + \left|\frac{zq'(z)}{q(z)}\right| \leqslant \frac{4r}{1 - r^2} \quad (|z| = r < 1).$$

By using Lemma 2.2, $f \in \mathcal{S}^*_{\phi_{3\mathcal{L}}}$ provided $4r/(1 - r^2) \leq 3/5$ which gives $r \leq (\sqrt{109} - 10)/3$. For sharpness, consider the function

$$f_0(z) = \frac{z(1 + z)^2}{(1 - z)^2} \quad \text{with} \quad g_0(z) = \frac{z(1 + z)}{(1 - z)}.$$

It is easy to see that $f_0 \in \mathcal{F}_1$ and at the point $z = (\sqrt{109} - 10)/3$, a simple computation shows that

$$\frac{zf_0'(z)}{f_0(z)} = \frac{1 + 4z - z^2}{1 - z^2} = \frac{2}{5}.$$

□

Recall that the class $\mathcal{M}(\beta)$, $\beta > 1$ is given by $\mathcal{M}(\beta) = \Big\{f \in A : \operatorname{Re} \frac{zf'(z)}{f(z)} < \beta, z \in \mathbb{D}\}$.

Theorem 2.9 *The $S^*_{3\mathcal{L}}$ radius of the class $\mathcal{M}(\beta)$ is given by $\mathcal{R}_{S^*_{3\mathcal{L}}}(\mathcal{M}(\beta)) = \frac{3}{10\beta-7}$.*

Proof For $f \in \mathcal{M}(\beta)$, then from Lemma 2.1 shows that

$$\left|\frac{zf'(z)}{f(z)} - \frac{1+(1-2\beta)r^2}{1-r^2}\right| \le \frac{2(\beta-1)r}{1-r^2} \qquad (|z|=r). \tag{2.1}$$

Clearly, the center of disk (2.1) is less than 1. Also, in view of Lemma 2.2, $f \in \mathcal{S}^*_{\phi_\mathcal{L}}$ if

$$\frac{2(\beta-1)r}{1-r^2} \le \frac{1+(1-2\beta)r^2}{1-r^2} - \frac{2}{5}.$$

This gives $r \le 3/(10\beta-7)$. Sharpness can be seen by considering the function

$$f(z) = \frac{z}{(1-z)^{2(1-\beta)}} \in \mathcal{M}(\beta).$$

One can compute that at $z = 3/(10\beta-7)$, we get $zf'(z)/f(z) = 2/5$. □

For $-\infty < \mu < \infty$, Ma and Minda [15] proved the growth, covering, rotation, and distortion theorem for the function $f \in \mathcal{S}^*(\varphi)$. Authors also obtained the sharp bounds on the functional $|a_3 - \mu a_2^2|$ which yields the sharp bounds on second and third coefficients of the function $f \in \mathcal{S}^*(\varphi)$. In 2007, Ali et al. [2] determined the sharp estimate for the Fekete–Szegö functional and sharp bound for the fourth coefficient of the function $f \in \mathcal{S}^*(\varphi)$. Determination of bounds on the coefficient a_n for $n \ge 5$ of the function $f \in \mathcal{S}^*(\varphi)$ is still an open problem. Recently, Ravichandran and Verma [25] found the sharp bound for the fifth coefficient for the functions in the class $\mathcal{S}^*_L$ and $\mathcal{S}^*_{RL}$. They proved the following lemma:

Lemma 2.10 *Let the real numbers α, β, γ and, a satisfy the inequalities $0 < \alpha < 1$, $0 < a < 1$ and*

$$\begin{aligned} 8a(1-a)\big((\alpha\beta - 2\gamma)^2 + (\alpha(a+\alpha)-\beta)^2\big) + \alpha(1-\alpha)(\beta - 2a\alpha)^2 \\ \le 4\alpha^2(1-\alpha)^2 a(1-a). \end{aligned} \tag{2.2}$$

If $p(z) = 1 + \sum_{k=1}^{\infty} c_k z^k \in \mathcal{P}$, then

$$|\gamma c_1^4 + a c_2^2 + 2\alpha c_1 c_3 - (3/2)\beta c_1^2 c_2 - c_4| \le 2.$$

As a simple application of the above lemma we have the following result.

Theorem 2.11 *If the function $f(z) = z + a_2 z^2 + a_3 z^3 + \cdots \in S^*_{\varphi_{3\mathcal{L}}}$, then $|a_5| \le 1/5$.*

3 $\mathcal{S}^*_{EL}$ Radius Estimates

In this section, we investigate the radius problems associated with the class $\mathcal{S}^*_{EL} := \mathcal{S}^*(\phi_{EL})$ consisting of functions $f \in \mathcal{A}$ satisfying $zf'(z)/f(z) \prec \phi_{EL}(z)$. These sharp radii constants are determined by first finding the largest disk with a given fixed center that contains the values of $zf'(z)/f(z)$ instead of directly estimating the real part of the expression $zf'(z)/f(z)$. This result is contained in the following lemma.

Lemma 3.1 *For* $0 \le k \le 1$ *and* $k/e \le a \le ke + 2(1-k)$, *let*

$$r_a = \begin{cases} k\left(a - \frac{1}{e}\right) + (1-k)(1-|1-a|) & \dfrac{k}{e} \le a \le \dfrac{k}{2}\left(e + \dfrac{1}{e}\right) + (1-k) \\ k(e-a) + (1-k)(1-|1-a|) & \dfrac{k}{2}\left(e + \dfrac{1}{e}\right) + (1-k) \le a \le ke + 2(1-k) \end{cases}$$

Then $\{w \in \mathbb{C} : |w - a| < r_a\} \subseteq \phi_{EL}(\mathbb{D})$.

In this section, we obtain $\mathcal{S}^*_{EL}$-radius for the class $\mathcal{S}^*[A, B]$ by distinguishing the cases $B \ge 0$ and $B \le 0$.

Theorem 3.2 *For the class* $\mathcal{S}^*[A, B]$, $\mathcal{S}^*_{EL}$*-radius is given by*

$$\mathcal{R}_{\mathcal{S}^*_{EL}}(\mathcal{S}^*[A, B]) = \frac{e-k}{Ae - Bk} \quad 0 \le B < A \le 1.$$

Proof Let $f \in \mathcal{S}^*[A, B]$, then we have

$$\frac{zf'(z)}{f(z)} \prec \frac{1+Az}{1+Bz} \quad \Rightarrow \quad \frac{zf'(z)}{f(z)} \in P[A, B]$$

and Lemma 2.1 gives

$$\left|\frac{zf'(z)}{f(z)} - \frac{1 - ABr^2}{1 - B^2r^2}\right| \le \frac{(A-B)r}{1 - B^2r^2} \quad (|z| = r < 1).$$

Since $B \ge 0$, we have$(1 - ABr^2)/(1 - B^2r^2) \le 1$ so that in view of Lemma 3.1, $f \in \mathcal{S}^*_{EL}$ if

$$\frac{(A-B)r}{1-B^2r^2} \le k\left(\frac{1-ABr^2}{1-B^2r^2} - \frac{1}{e}\right) + (1-k)\left(\frac{1-ABr^2}{1-B^2r^2}\right)$$

On solving this inequality for r, we get $r \le e - k/(Ae - Bk)$. For sharpness, we consider the function

$$f(z) = \begin{cases} z(1+Bz)^{\frac{A-B}{B}} & \text{if } B \ne 0 \\ ze^{Az} & \text{if } B = 0. \end{cases}$$

Then

$$\frac{zf'(z)}{f(z)} = \frac{1 + Az}{1 + Bz}$$

and hence $f \in \mathcal{S}^*[A, B]$. By simple calculations at $z = -(e - k)/(Ae - Bk)$, the function $zf'(z)/f(z)$ assumes the value k/e. This shows that the obtained radius is sharp. □

Now, we determine $\mathcal{S}^*_{EL}$-radius for the classes $\mathcal{S}^*_L$, $\mathcal{F}_1$, and $\mathcal{F}_2$.

Theorem 3.3 *For the class $\mathcal{S}^*_L$, the $\mathcal{S}^*_{EL}$ radius is given by*

$$\mathcal{R}_{\mathcal{S}^*_{EL}}(\mathcal{S}^*_L) = 1 - \frac{k^2}{e^2}.$$

Proof If $f \in \mathcal{S}^*_L$. Then $zf'(z)/f(z) \prec \sqrt{1+z}$. For $|z| = r$, we get

$$\left|\frac{zf'(z)}{f(z)} - 1\right| \leq |\sqrt{1+z} - 1| \leq 1 - \sqrt{1-r}.$$

From Lemma 3.1, it follows that if

$$1 - \sqrt{1-r} \leq k\left(1 - \frac{1}{e}\right) + (1-k)$$

then $f \in \mathcal{S}^*_{EL}$. This gives $r \leq 1 - k^2/e^2$ and this bound is the best possible as seen by the function in this class given by

$$f(z) = \frac{4z \exp(2\sqrt{1+z} - 2)}{(1 + \sqrt{1+z})^2} \in \mathcal{S}^*_L.$$

At the point $z = -(1 - k^2/e^2)$, we have $zf'(z)/f(z) = \sqrt{1+z} = k/e$. □

Theorem 3.4 *For the class $\mathcal{F}_1$, the $\mathcal{S}^*_{EL}$ radius is given by*

$$\mathcal{R}_{\mathcal{S}^*_{EL}}(\mathcal{F}_1) = \frac{\sqrt{4e^2 + (k-e)^2} - 2e}{e - k}.$$

Proof Let $f \in \mathcal{F}_1$ and the functions $p, q : \mathbb{D} \to \mathbb{C}$ defined by

$$p(z) = \frac{g(z)}{z} \quad \text{and} \quad q(z) = \frac{f(z)}{g(z)}.$$

Then p, q belong to $\mathcal{P}$ and in view of Lemma 2.1, we get

$$\left|\frac{zf'(z)}{f(z)} - 1\right| \leq \left|\frac{zp'(z)}{p(z)}\right| + \left|\frac{zq'(z)}{q(z)}\right| \leqslant \frac{4r}{1-r^2} \quad (|z| = r < 1).$$

By using Lemma 3.1, $f \in \mathcal{S}^*_{EL}$ provided $4r/(1-r^2) \leq k\,(1-1/e)+(1-k)$ which gives $r \leq (\sqrt{4e^2+(e-k)^2}-2e)/(e-k)$. For sharpness, consider the functions

$$f_0(z) = \frac{z(1+z)^2}{(1-z)^2} \quad \text{with} \quad g_0(z) = \frac{z(1+z)}{(1-z)}.$$

It is easy to see that $f_0 \in \mathcal{F}_1$ and at the point $z = (\sqrt{4e^2+(e-k)^2}-2e)/(e-k)$, a simple computation shows that

$$\frac{zf_0'(z)}{f_0(z)} = \frac{1-4z-z^2}{1-z^2} = \frac{k}{e}.$$

□

Theorem 3.5 *For the class $\mathcal{F}_2$, the $\mathcal{S}^*_{EL}$ radius is given by*

$$R_{\mathcal{S}^*_{EL}}(\mathcal{F}_2) = \frac{\sqrt{8e^2+(3e-2k)^2}-3e}{2(2e-k)}.$$

Proof Let $f \in \mathcal{F}_2$. Define functions $u, v : \mathbb{D} \to \mathbb{C}$ by

$$u(z) = \frac{g(z)}{z} \quad \text{and} \quad v(z) = \frac{g(z)}{f(z)}.$$

Then $u \in \mathcal{P}$ and since $|f(z)/g(z) - 1| < 1$ if and only if $\operatorname{Re}(g(z)/f(z)) > 1/2$, $v \in \mathcal{P}(1/2)$. By using Lemma 2.1 to the identity

$$\frac{zf'(z)}{f(z)} = 1 + \frac{zu'(z)}{u(z)} - \frac{zv'(z)}{v(z)}$$

we obtain

$$\left|\frac{zf'(z)}{f(z)} - 1\right| \leq \frac{2r}{1-r^2} + \frac{r}{1-r} = \frac{3r+r^2}{1-r^2}.$$

By Lemma 3.1, $f \in \mathcal{S}^*_{EL}$ provided

$$\frac{3r+r^2}{1-r^2} \leq k\left(1-\frac{1}{e}\right) + (1-k)$$

This gives $r \leq (\sqrt{8e^2+(3e-2k)^2}-3e)/2(2e-k)$. To show that this bound is the best possible, consider the function

$$f_0(z) = \frac{z(1+z)^2}{(1-z)} \quad \text{with} \quad g_0(z) = \frac{z(1+z)}{(1-z)}.$$

At the point $z = (\sqrt{8e^2+(3e-2k)^2}-3e)/(2(2e-k))$, we obtain

$$\frac{zf_0'(z)}{f_0(z)} = \frac{1-3z-2z^2}{1-z^2} = \frac{k}{e}. \qquad \square$$

References

1. R.M. Ali, V. Ravichandran, Uniformly convex and uniformly starlike functions. Math. Newslett. **21**(1), 16–30 (2011)
2. R.M. Ali, V. Ravichandran, N. Seenivasagan, Coefficient bounds for p-valent functions. Appl. Math. Comput. **187**(1), 35–46 (2007)
3. R.M. Ali, N.E. Cho, N.K. Jain, V. Ravichandran, Radii of starlikeness and convexity for functions with fixed second coefficient defined by subordination. Filomat **26**(3), 553–561 (2012)
4. R.M. Ali, N.K. Jain, V. Ravichandran, Radii of starlikeness associated with the lemniscate of Bernoulli and the left-half plane. Appl. Math. Comput. **218**(11), 6557–6565 (2012)
5. R.M. Ali, N.K. Jain, V. Ravichandran, On the radius constants for classes of analytic functions. Bull. Malays. Math. Sci. Soc. (2) **36**(1), 23–38 (2013)
6. D.M. Campbell, A survey of properties of the convex combination of univalent functions. Rocky Mt. J. Math. **5**(4), 475–492 (1975)
7. N.E. Cho, Some new criteria for p-valent meromorphic starlike functions. Nihonkai Math. J. **4**(2), 125–132 (1993)
8. N.E. Cho et al., Radius problems for starlike functions associated with the sine function. Bull. Iran. Math. Soc. **45**(1), 213–232 (2019)
9. P.L. Duren, *Univalent Functions*. Grundlehren der Mathematischen Wissenschaften, vol. 259 (Springer, New York, 1983)
10. S. Gandhi, V. Ravichandran, Starlike functions associated with a lune. Asian-Eur. J. Math. **10**(4), 1750064 (2017), 12 pp
11. A.W. Goodman, *Univalent Functions*, vol. 1–2 (Mariner, Tampa, 1983)
12. W. Janowski, Extremal problems for a family of functions with positive real part and for some related families. Ann. Pol. Math. **23**, 159–177 (1970/1971)
13. W. Janowski, Some extremal problems for certain families of analytic functions. I. Ann. Pol. Math. **28**, 297–326 (1973)
14. K. Khatter, V. Ravichandran, S. Sivaprasad Kumar, Starlike functions associated with exponential function and the lemniscate of Bernoulli. Rev. R. Acad. Cienc. Exactas Fís. Nat. Ser. A Mat. RACSAM **113**(1), 233–253 (2019)
15. W.C. Ma, D. Minda, A unified treatment of some special classes of univalent functions, in *Proceedings of the Conference on Complex Analysis (Tianjin, 1992), Conference Proceedings and Lecture Notes in Analysis. I* (International Press, Cambridge, 1992), pp. 157–169
16. T.H. MacGregor, Functions whose derivative has a positive real part. Trans. Am. Math. Soc. **104**, 532–537 (1962)
17. T.H. MacGregor, The radius of univalence of certain analytic functions. Proc. Am. Math. Soc. **14**, 514–520 (1963)
18. T.H. MacGregor, The radius of univalence of certain analytic functions. II. Proc. Am. Math. Soc. **14**, 521–524 (1963)
19. R. Mendiratta, S. Nagpal, V. Ravichandran, A subclass of starlike functions associated with left-half of the lemniscate of Bernoulli. Int. J. Math. **25**(9), 1450090 (2014), 17 pp
20. R. Mendiratta, S. Nagpal, V. Ravichandran, On a subclass of strongly starlike functions associated with exponential function. Bull. Malays. Math. Sci. Soc. **38**(1), 365–386 (2015)
21. S. Owa, H.M. Srivastava, Some generalized convolution properties associated with certain subclasses of analytic functions. JIPAM J. Inequal. Pure Appl. Math. **3**(3), Article 42 (2002), 13 pp. (electronic)

22. E. Paprocki, J. Sokół, The extremal problems in some subclass of strongly starlike functions. Zeszyty Nauk. Politech. Rzeszowskiej Mat. **20**, 89–94 (1996)
23. B.N. Rahmanov, On the theory of univalent functions. Dokl. Akad. Nauk SSSR (N.S.) **82**, 341–344 (1952)
24. R.K. Raina, J. Sokół, Some properties related to a certain class of starlike functions. C. R. Math. Acad. Sci. Paris **353**(11), 973–978 (2015). MR3419845
25. V. Ravichandran, S. Verma, Bound for the fifth coefficient of certain starlike functions. C. R. Math. Acad. Sci. Paris **353**(6), 505–510 (2015)
26. V. Ravichandran, F. Rønning, T.N. Shanmugam, Radius of convexity and radius of starlikeness for some classes of analytic functions. Complex Var. Theory Appl. **33**(1–4), 265–280 (1997)
27. M.S. Robertson, Certain classes of starlike functions. Mich. Math. J. **32**(2), 135–140 (1985)
28. G.M. Shah, On the univalence of some analytic functions. Pac. J. Math. **43**, 239–250 (1972)
29. K. Sharma, N.K. Jain, V. Ravichandran, Starlike functions associated with a cardioid. Afr. Mat. **27**(5–6), 923–939 (2016)
30. J. Sokół, On some subclass of strongly starlike functions. Demonstr. Math. **31**(1), 81–86 (1998)
31. J. Sokół, On sufficient condition to be in a certain subclass of starlike functions defined by subordination. Appl. Math. Comput. **190**(1), 237–241 (2007)
32. J. Sokół, On an application of certain sufficient condition for starlikeness. J. Math. Appl. **30**, 131–135 (2008)
33. J. Sokół, Coefficient estimates in a class of strongly starlike functions. Kyungpook Math. J. **49**(2), 349–353 (2009)
34. J. Sokół, Radius problems in the class $\mathcal{SL}^*$. Appl. Math. Comput. **214**(2), 569–573 (2009)
35. J. Sokół, J. Stankiewicz, Radius of convexity of some subclasses of strongly starlike functions. Zesz. Nauk. Politech. Rzeszowskiej Mat. **19**, 101–105 (1996)
36. B.A. Uralegaddi, M.D. Ganigi, S.M. Sarangi, Univalent functions with positive coefficients. Tamkang J. Math. **25**(3), 225–230 (1994)

First-Order Differential Subordinations for Janowski Starlikeness

Swati Anand, Sushil Kumar and V. Ravichandran

Abstract By using admissibility condition technique, certain sufficient conditions are determined so that an analytic function p defined on the open unit disk and normalized by $p(0) = 1$ satisfy the subordination $p(z) \prec (1 + Az)/(1 + Bz)$ whenever, for certain choice of ψ, the function $\psi(p(z), zp'(z))$ is subordinate to a starlike function associated with lune. Further, we obtain certain sufficient conditions for a normalized analytic function f to be in the class of Janowski starlike functions.

Keywords Differential subordination · Admissibility condition · Univalent functions · Starlike functions · Janowski starlike function · Lune

1 Introduction

Let $\mathcal{A}$ denote the class of analytic functions normalized by the condition $f(0) = 0 = f'(0) - 1$ in the unit disk $\mathbb{D} := \{z \in \mathbb{C} : |z| < 1\}$. Let $\mathcal{S} \subset \mathcal{A}$ consists of univalent functions. For a fixed nonnegative integer n, denoted by $\mathcal{H}[a, n]$, the class of functions of the form $f(z) = a + a_n z^n + a_{n+1} z^{n+1} + \cdots$. In particular, $\mathcal{H}[1, 1] = \mathcal{H}_1$. For two analytic functions f and g, f is *subordinate* to g, if there exists an analytic function ω with $\omega(0) = 0$ and $|\omega(z)| < 1$ for $z \in \mathbb{D}$ such that $f(z) = g(\omega(z))$, written as $f \prec g$. In particular, if $g \in \mathcal{S}$, then $f \prec g$ if and only if $f(0) = g(0)$ and $f(\mathbb{D}) \subseteq g(\mathbb{D})$. Let $A, B \in [-1, 1]$ be two arbitrary real numbers. The class $\mathcal{P}[A, B]$ consists of normalized analytic functions of the form $p(z) = 1 + c_1 z + c_2 z^2 + \cdots$ satisfying

---

S. Anand
Rajdhani College, University of Delhi, Delhi, India
e-mail: swati_anand01@yahoo.com

S. Kumar (✉)
Bharati Vidyapeeth's College of Engineering, Delhi 110063, India
e-mail: sushilkumar16n@gmail.com

V. Ravichandran
Department of Mathematics, National Institute of Technology, Tiruchirappalli 620 015, India
e-mail: vravi68@gmail.com

N. Deo et al. (eds.), *Mathematical Analysis I: Approximation Theory*,
Springer Proceedings in Mathematics & Statistics 306,
https://doi.org/10.1007/978-981-15-1153-0_16

$p(z) \prec (1 + Az)/(1 + Bz)$, for $z \in \mathbb{D}$. Suppose $f \in \mathcal{A}$, then f is a Janowski function if $zf'(z)/f(z) \in \mathcal{P}[A, B]$ for $z \in \mathbb{D}$. The class of Janowski functions is denoted by $S^*[A, B]$ [9]. For particular values of A and B, the class $S^*[1 - 2\alpha, -1] = \mathcal{S}^*(\alpha)$ is the class of *starlike functions of order* α. For details, see [8, 14]. In 2015, Raina and Sokół [17] studied a class $\mathcal{S}_q^* := \mathcal{S}^*(\varphi_q)$ where $\varphi_q(z) = z + \sqrt{1 + z^2}$ and discussed several other properties of the class $\mathcal{S}_q^*$. It is observed that $\mathcal{S}_q^* \subset \mathcal{A}$ such that

$$\left|\frac{zf'(z)}{f(z)} - 1\right| < 2\left|\frac{zf'(z)}{f(z)}\right|$$

which is the interior of the lune. In geometry, the lune is a concave–convex plane area which is bounded by two circular arcs of unequal radii.

In 1989, Nunokawa et al. [16] proved that if a function p satisfies the subordination $1 + zp'(z) \prec 1 + z$, then $p(z) \prec 1 + z$. Ali et al. [3] computed a condition for lower bound on β for $p(z) \prec \sqrt{1 + z}$ when $1 + \beta zp'(z)/p^j(z) \prec \sqrt{1 + z}$ $(j = 0, 1, 2)$. In [13], the bound on β is obtained such that $1 + \beta zp'(z)/p^j(z) \prec (1 + Dz)/(1 + Ez)$; $-1 < E < 1$ and $|D| \leq 1$ $(j = 0, 1, 2)$ implies $p(z)$ is subordinate to functions associated with lemniscate of Bernoulli. Recently, authors [12] obtained sharp lower bound on β so that the function p is subordinate to the functions e^z and $(1 + Az)/(1 + Bz)$ whenever $1 + \beta zp'(z)/p^j(z)$, $(j = 0, 1, 2)$ is subordinate to functions with positive real part. Ahuja et al. [1] obtained certain inclusions between the class of Carathéodory functions and the class of starlike univalent functions associated with lemniscate of Bernoulli. For related results, see [2, 4–6, 13, 18].

Motivated by the earlier discussed works, using the concept of admissibility condition, we obtain some conditions on β so that $p \in \mathcal{P}[A, B]$ whenever $p(z) + \beta zp'(z)/p^k(z)$ with $k = 0, 1$ and $1/p(z) + \beta zp'(z)/p^k(z)$ $(k = 1, 2)$ is subordinate to $\varphi_q(z)$. Further, alternate proofs are provided in which the conditions on β is determined, so that $1 + \beta zp'(z)/p^k(z) \prec \varphi_q(z)$ with $k = 0, 1, 2$ whenever the function p is subordinate to $(1 + Az)/(1 + Bz)$. As a consequence of these results, certain sufficient conditions for a function $f \in \mathcal{A}$ to be in $S^*[A, B]$ are also given.

2 Subordination Results Associated with Lune

Let p be an analytic function defined on $\mathbb{D}$ with $p(0) = 1$. Let $h \in \mathcal{S}$ and the function $\psi(r, s, t; z)$ be defined in some domain $D \subset \mathbb{C}^3 \times \mathbb{D}$. If the function p satisfies

$$\psi(p(z), zp'(z), z^2p''(z); z) \prec h(z) \tag{2.1}$$

then p is known as *solution* of the subordination relation (2.1). The univalent function q is a dominant of the solutions of the differential subordination if $p \prec q$ for all p satisfying (2.1). A dominant $\tilde{q}$ which satisfies $\tilde{q} \prec q$ for all dominant q of (2.1) is known as the *best dominant* of relation (2.1). It is unique up to rotation.

Denote by $\mathcal{Q}$ the class of all analytic and injective functions q on $\overline{\mathbb{D}} \setminus \mathbf{E}(q)$, where

$$\mathbf{E}(q) = \{\xi \in \partial\mathbb{D} : \lim_{z \to \xi} q(z) = \infty\}$$

such that $q'(\xi) \neq 0$ for $\xi \in \overline{\mathbb{D}} \setminus \mathbf{E}(q)$. Let Ω be a subset of $\mathbb{C}$, $q \in \mathcal{Q}$ and n be a positive integer. The admissibility class $\Psi_n[\Omega, q]$ consists of admissible functions $\psi : \mathbb{C}^3 \times \mathbb{D} \to \mathbb{C}$ satisfying the following condition:

$$\psi(r, s, t; z) \notin \Omega \tag{2.2}$$

whenever

$$r = q(\xi),\ s = m\,\xi\, q'(\xi) \text{ and } \operatorname{Re}\left(\frac{t}{s} + 1\right) \geq m \operatorname{Re}\left(\frac{\xi q''(\xi)}{q'(\xi)} + 1\right)$$

for $z \in \mathbb{D}$, $\xi \in \overline{\mathbb{D}} \setminus \mathbf{E}(q)$ and $m \geq n \geq 1$. When $n = 1$, let $\Psi_1[\Omega, q] = \Psi[\Omega, q]$.

Theorem 2.1 ([15, Theorem 2.3b, p. 28]) *Let the function $\psi \in \Psi_n[\Omega, q]$ with $q(0) = a$. If the function $p \in \mathcal{H}[a, n]$ satisfies*

$$\psi(p(z), zp'(z), z^2 p''(z); z) \in \Omega, \tag{2.3}$$

then $p \prec q$.

For a simply connected domain $\Omega \neq \mathbb{C}$, there is a conformal mapping of $h : \mathbb{D} \to \Omega$ satisfying $h(0) = \psi(a, 0, 0; 0)$ and $\Psi_n[\Omega, q]$ is written as $\Psi_n[h, q]$. Thus relation (2.2) can be written as

$$\psi(p(z), zp'(z), z^2 p''(z); z) \prec h(z). \tag{2.4}$$

For more details, see [10, 11].

This section deals with certain condition under which $p(z)$ is subordinate to $(1 + Az)/(1 + Bz)$ for $-1 \leq B < A \leq 1$ whenever $\psi(p(z), zp'(z); z) \prec \varphi_q(z) = z + \sqrt{1 + z^2}$. First, a condition on β is computed so that the subordination $p(z) + \beta zp'(z) \prec \varphi_q(z)$ implies $p \in \mathcal{P}[A, B]$. Similar results are obtained for expressions $p(z) + \beta zp'(z)/p(z)$ and $(1/p(z)) + \beta zp'(z)/p^k(z)$; $(k = 1, 2)$.

Consider the function q given by

$$q(z) = \frac{1 + Az}{1 + Bz}, \quad z \in \mathbb{D} \text{ and } -1 \leq B < A \leq 1. \tag{2.5}$$

Note that $q(0) = 1$, $\mathbf{E}(q) \subseteq \{1\}$ and q is univalent in $\overline{\mathbb{D}} \setminus \mathbf{E}(q)$. Thus $q \in \mathcal{Q}$ and the domain $q(\mathbb{D})$ is

$$\Delta = q(\mathbb{D}) = \left\{ w \in \mathbb{C} : \left| \frac{w - 1}{A - Bw} \right| < 1 \right\}.$$

We now define the condition of admissibility for the function q. The class of admissible functions is denoted by $\Psi_n[\Omega; A, B]$. It is easy to compute

$$q(\varsigma) = \frac{1 + Ae^{i\theta}}{1 + Be^{i\theta}}, \ q'(\varsigma) = \frac{A - B}{(1 + Be^{i\theta})^2} \text{ and } q''(\varsigma) = \frac{2B(A - B)}{-(1 + Be^{i\theta})^3}$$

for $\varsigma = e^{i\theta}$ and $0 < \theta < 2\pi$.

Hence, a simple calculation yields

$$\text{Re}\left(\frac{\varsigma q''(\varsigma)}{q'(\varsigma)} + 1\right) = \frac{1 - B^2}{1 + B^2 + 2B\cos\theta}$$

where $0 < \theta < 2\pi$ and $m \geq n \geq 1$. Thus the admissibility condition reduces to

$$\psi(r, s, t; z) \notin \Omega \quad \text{whenever} \quad (r, s, t; z) \in \text{dom}\,\psi$$

and

$$r = \frac{1 + Ae^{i\theta}}{1 + Be^{i\theta}}, \ s = \frac{m(A - B)e^{i\theta}}{(1 + Be^{i\theta})^2} \text{ and } \text{Re}\left(\frac{t}{s} + 1\right) \geq \frac{m(1 - B^2)}{1 + B^2 + 2B\cos\theta} \tag{2.6}$$

where $0 < \theta < 2\pi$ and $m \geq n \geq 1$.

To discuss our problems, we need the following result in the context of first-order differential subordination due to Theorem 2.1.

Theorem 2.2 *Let $p \in \mathcal{H}[1, n]$ with $n \in \mathbb{N}$. Let Ω be a subset of $\mathbb{C}$ and $\psi : \mathbb{C}^2 \times \mathbb{D} \to \mathbb{C}$ with dom ψ satisfy $\psi(r, s; z) \notin \Omega$ for all $z \in \mathbb{D}$, where*

$$r = \frac{1 + Ae^{i\theta}}{1 + Be^{i\theta}}, \ s = \frac{m(A - B)e^{i\theta}}{(1 + Be^{i\theta})^2}. \tag{2.7}$$

If $(p(z), zp'(z); z) \in \text{dom}\,\psi$ and $\psi(p(z), zp'(z); z) \in \Omega$ for $z \in \mathbb{D}$, then $p(z) \prec (1 + Az)/(1 + Bz)$.

Theorem 2.3 *Let p be an analytic function defined on $\mathbb{D}$ with $p(0) = 1$ and $\beta \neq 0$ and $-1 \leq B < A \leq 1$. We assume the following inequality holds:*

$$\begin{aligned}\big(|\beta|(A - B) - (1 + |A|)(1 + |B|)\big)^2 \geq 2\big((1 + |A|)^2(1 + |B|)^2 + (1 + |A|)(1 + |B|)^3 \\ + |\beta|(A - B)(1 + |B|)^2\big) + (1 + |B|)^4.\end{aligned} \tag{2.8}$$

If $p(z) + \beta zp'(z) \prec \varphi_q(z) = z + \sqrt{1 + z^2}$, then $p \in \mathcal{P}[A, B]$.

Proof Let $\Omega = \varphi_q(\mathbb{D}) = \{w \in \mathbb{C} : |w^2 - 1| < 2|w|\}$ be the domain. Define the analytic function $\psi : \mathbb{C}^2 \times \mathbb{D} \to \mathbb{C}$ as $\psi(r, s; z) = r + \beta s$. The required subordination is established if $\psi \in \Psi[\Omega, q]$. Thus by Theorem 2.2, it is enough to show that $|(\psi^2(r,s;z)-1)/\psi(r,s;z)| \geq 2$. Using (2.6), we have

$$\psi(r,s;z) = \frac{(1+Ae^{i\theta})(1+Be^{i\theta}) + \beta m(A-B)e^{i\theta}}{(1+Be^{i\theta})^2}$$

so that

$$\begin{aligned}
\left|\frac{\psi^2(r,s;z)-1}{\psi(r,s;z)}\right| &= \left|\frac{\begin{array}{c}(1+Ae^{i\theta})^2(1+Be^{i\theta})^2 + \beta^2 m^2 e^{2i\theta}(A-B)^2 \\ +\,2(1+Ae^{i\theta})(1+Be^{i\theta})\beta m e^{i\theta}(A-B) - (1+Be^{i\theta})^4\end{array}}{(1+Ae^{i\theta})(1+Be^{i\theta})^3 + \beta m e^{i\theta}(A-B)(1+Be^{i\theta})^2}\right| \\
&\geq \frac{\begin{array}{c}|\beta^2|m^2(A-B)^2 - (1+|A|)^2(1+|B|)^2 \\ -\,2(1+|A|)(1+|B|)\beta m(A-B) - (1+|B|)^4\end{array}}{|(1+Ae^{i\theta})(1+Be^{i\theta})^3| + |\beta|m(A-B)|(1+Be^{i\theta})^2|} \\
&\geq \frac{\begin{array}{c}|\beta^2|m^2(A-B)^2 - (1+|A|)^2(1+|B|)^2 \\ -\,2(1+|A|)(1+|B|)\beta m(A-B) - (1+|B|)^4\end{array}}{(1+|A|)(1+|B|)^3 + |\beta|m(A-B)(1+|B|)^2} \quad (\text{using } m \geq 1) \\
&\geq \frac{\begin{array}{c}(|\beta|m(A-B) - (1+|A|)(1+|B|))^2 \\ -\,(1+|B|)^4 - 2(1+|A|)^2(1+|B|)^2\end{array}}{(1+|A|)(1+|B|)^3 + m|\beta|(A-B)(1+|B|)^2}.
\end{aligned}$$

Since $m \geq 1$, it is possible to deduce

$$\left|\frac{\psi^2(r,s;z)-1}{\psi(r,s;z)}\right| \geq 2$$

whenever inequality (2.8) holds. Thus $\psi \in \Psi[\Omega, q]$. Using Theorem 2.2, we get the desired subordination $p(z) \prec (1+Az)/(1+Bz)$. ■

Theorem 2.4 *Let p be an analytic function defined on $\mathbb{D}$ with $p(0)=1$ and $-1 \leq B < A \leq 1$ and $\beta \neq 0$. Assume that*

$$\begin{aligned}\big(|\beta|(A-B) - (1+|A|)^2\big)^2 \geq 2\big((1+|A|)^4 + (1+|A|)^3(1+|B|) \\ +\,|\beta|(A-B)(1+|A|)(1+|B|)\big) + (1+|A|)^2(1+|B|)^2.\end{aligned} \tag{2.9}$$

If $p(z) + \beta z p'(z)/p(z) \prec \varphi_q(z)$, then $p \in \mathcal{P}[A,B]$.

Proof Consider Ω as in Theorem 2.3. Define the analytic function $\psi : \mathbb{C}\setminus\{0\} \times \mathbb{C} \times \mathbb{D} \to \mathbb{C}$ as $\psi(r,s;z) = r + \beta s/r$. The function $\psi(r,s;z)$ is given by

$$\psi(r,s;z) = \frac{(1+Ae^{i\theta})^2 + \beta m(A-B)e^{i\theta}}{(1+Ae^{i\theta})(1+Be^{i\theta})}.$$

A simple calculation yields

$$\frac{\psi^2(r,s;z)-1}{\psi(r,s;z)}=\frac{(1+Ae^{i\theta})^4+\beta^2m^2(A-B)^2e^{2i\theta}+2(1+Ae^{i\theta})^2\beta m(A-B)e^{i\theta}-(1+Ae^{i\theta})^2(1+Be^{i\theta})^2}{(1+Ae^{i\theta})^3(1+Be^{i\theta})+\beta m(A-B)e^{i\theta}(1+Ae^{i\theta})(1+Be^{i\theta})}$$

so that

$$\begin{aligned}\left|\frac{\psi^2(r,s;z)-1}{\psi(r,s;z)}\right| &= \left|\frac{(1+Ae^{i\theta})^4+\beta^2m^2(A-B)^2e^{2i\theta}+2(1+Ae^{i\theta})^2\beta m(A-B)e^{i\theta}-(1+Ae^{i\theta})^2(1+Be^{i\theta})^2}{(1+Ae^{i\theta})^3(1+Be^{i\theta})+\beta m(A-B)e^{i\theta}(1+Ae^{i\theta})(1+Be^{i\theta})}\right|\\
&\geq \frac{|\beta^2|m^2(A-B)^2-|(1+Ae^{i\theta})^2(1+Be^{i\theta})^2|-2|(1+Ae^{i\theta})|^2\beta m(A-B)-|(1+Ae^{i\theta})^4|}{|(1+Ae^{i\theta})^3(1+Be^{i\theta})|+|\beta|m(A-B)|1+Ae^{i\theta}||1+Be^{i\theta}|}\\
&\geq \frac{|\beta^2|m^2(A-B)^2-(1+|A|)^2(1+|B|)^2-2(1+|A|)^2\beta m(A-B)-(1+|A|)^4}{(1+|A|)^3(1+|B|)+|\beta|m(A-B)(1+|A|)(1+|B|)}\\
&= \frac{(|\beta|m(A-B)-(1+|A|)^2)^2-(1+|A|)^2(1+|B|)^2-2(1+|A|)^4}{(1+|A|)^3(1+|B|)+|\beta|m(A-B)(1+|A|)(1+|B|)}\\
&\geq \frac{(|\beta|(A-B)-(1+|A|)^2)^2-(1+|A|)^2(1+|B|)^2-2(1+|A|)^4}{(1+|A|)^3(1+|B|)+|\beta|m(A-B)(1+|A|)(1+|B|)} \quad (\because m\geq 1).\end{aligned}$$

Since $m \geq 1$, a computation gives $\left|(\psi^2(r,s;z)-1)/(\psi(r,s;z))\right| \geq 2$ if the inequality (2.9) holds. By Theorem 2.2 the desired result is proved. ■

Theorem 2.5 *Let p be an analytic function defined on $\mathbb{D}$ with $p(0)=1$. Suppose $-1 \leq B < A \leq 1$ and $\beta \neq 0$ and the following inequality holds:*

$$\big(|\beta|(A-B)-(1+|B|)^2\big)^2 \geq 2\big((1+|B|)^4+(1+|A|)(1+|B|)^3+|\beta|(A-B)(1+|A|)(1+|B|)\big)+(1+|A|)^2(1+|B|)^2. \tag{2.10}$$

If $(1/p(z))+\beta(zp'(z)/p(z)) \prec \varphi_q(z)$, then $p \in \mathcal{P}[A,B]$.

Proof Consider the analytic function ψ given as $\psi(r,s;z)=(1/r)+\beta(s/r)$. We have

$$\begin{aligned}\psi(r,s;z) &= \frac{1+Be^{i\theta}}{1+Ae^{i\theta}}+\beta\frac{me^{i\theta}(A-B)(1+Be^{i\theta})}{(1+Be^{i\theta})^2(1+Ae^{i\theta})}\\
&= \frac{(1+Be^{i\theta})^2+\beta me^{i\theta}(A-B)}{(1+Ae^{i\theta})(1+Be^{i\theta})}\end{aligned}$$

By Theorem 2.2 for $\psi \in \Psi[\Omega, q]$, it is enough to show that $|(\psi^2-1)/\psi| \geq 2$. Consider

$$\left|\frac{\psi^2(r,s;z)-1}{\psi(r,s;z)}\right| = \left|\frac{(1+Be^{i\theta})^4+\beta^2m^2e^{2i\theta}(A-B)^2+2(1+Be^{i\theta})^2\beta me^{i\theta}(A-B) - (1+Ae^{i\theta})^2(1+Be^{i\theta})^2}{(1+Ae^{i\theta})(1+Be^{i\theta})^3+\beta me^{i\theta}(A-B)(1+Ae^{i\theta})(1+Be^{i\theta})}\right|$$

$$\geq \frac{|\beta^2|m^2(A-B)^2-|(1+Ae^{i\theta})^2(1+Be^{i\theta})^2|-2|(1+Be^{i\theta})|^2 \beta m(A-B)-|(1+Be^{i\theta})^4|}{|(1+Ae^{i\theta})(1+Be^{i\theta})^3|+|\beta|m(A-B)|1+Ae^{i\theta}||1+Be^{i\theta}|}$$

$$= \frac{(|\beta|m(A-B)-(1+|B|)^2)^2-(1+|A|)^2(1+|B|)^2-2(1+|B|)^4}{|(1+|A|)(1+|B|)^3|+|\beta|m(A-B)(1+|A|)(1+|B|)}$$

$$\geq \frac{(|\beta|(A-B)-(1+|B|)^2)^2-(1+|A|)^2(1+|B|)^2-2(1+|B|)^4}{|(1+|A|)(1+|B|)^3|+|\beta|m(A-B)(1+|A|)(1+|B|)}$$

As analysis done in previous theorem, we conclude that $\left|(\psi^2(r,s;z)-1)/(\psi(r,s;z))\right| \geq 2$ if the inequality (2.10) holds. ■

Theorem 2.6 *Let $-1 \leq B < A \leq 1$ and $\beta \neq 0$. We assume the following inequality:*

$$\big(|\beta|(A-B)-(1+|A|)(1+|B|)\big)^2 \geq 2\big((1+|A|)^2(1+|B|)^2+(1+|A|)^3(1+|B|) + |\beta|(A-B)(1+|A|)^2\big)+(1+|A|)^4. \quad (2.11)$$

If the function $p \in \mathcal{P}$ satisfies $(1/p(z))+\beta(zp'(z)/p^2(z)) \prec \varphi_q(z)$, then $p \in \mathcal{P}[A,B]$.

Proof Consider Ω as in Theorem 2.3. Let $\psi : \mathbb{C}\setminus\{0\}\times\mathbb{C}\times\mathbb{D} \to \mathbb{C}$ be an analytic function given as

$$\psi(r,s;z) = \frac{1}{r}+\beta\frac{s}{r^2}.$$

By making use of Theorem 2.2, for $\psi \in \Psi[\Omega,q]$ it is enough to show that $|(\psi^2-1)/\psi| \geq 2$. We have

$$\psi(r,s;z) = \frac{(1+Be^{i\theta})(1+Ae^{i\theta})+\beta me^{i\theta}(A-B)}{(1+Ae^{i\theta})^2}$$

so that

$$\left|\frac{\psi^2(r,s;z)-1}{\psi(r,s;z)}\right| = \left|\frac{\left(\dfrac{(1+Be^{i\theta})(1+Ae^{i\theta})+\beta me^{i\theta}(A-B)}{(1+Ae^{i\theta})^2}\right)^2-1}{\dfrac{(1+Be^{i\theta})(1+Ae^{i\theta})+\beta me^{i\theta}(A-B)}{(1+Ae^{i\theta})^2}}\right|$$

$$= \left|\frac{(1+Ae^{i\theta})^2(1+Be^{i\theta})^2+\beta^2m^2e^{2i\theta}(A-B)^2 + 2(1+Ae^{i\theta})(1+Be^{i\theta})\beta me^{i\theta}(A-B)-(1+Ae^{i\theta})^4}{(1+Ae^{i\theta})^3(1+Be^{i\theta})+\beta me^{i\theta}(A-B)(1+Ae^{i\theta})^2}\right|$$

$$\geq \frac{|\beta^2|m^2(A-B)^2 - (1+|A|)^2(1+|B|)^2 - 2(1+|A|)(1+|B|)\beta m(A-B) - (1+|A|)^4}{|(1+Ae^{i\theta})^3(1+Be^{i\theta})| + |\beta|m(A-B)|(1+Ae^{i\theta})^2|}$$
$$\geq \frac{|\beta^2|m^2(A-B)^2 - (1+|A|)^2(1+|B|)^2 - 2(1+|A|)(1+|B|)\beta m(A-B) - (1+|A|)^4}{(1+|A|)^3(1+|B|) + |\beta|m(A-B)(1+|A|)^2}$$
$$\geq \frac{(|\beta|(A-B) - (1+|A|)(1+|B|))^2 - (1+|A|)^4 - 2(1+|A|)^2(1+|B|)^2}{(1+|A|)(1+|B|)^3 + |\beta|m(A-B)(1+|B|)^2}$$

A calculation shows that

$$\left|\frac{\psi^2(r,s;z) - 1}{\psi(r,s;z)}\right| \geq 2$$

whenever inequality (2.11) holds. By Theorem 2.2 the result is evident. ■

As an application of Theorems 2.3–2.6, we see that the following subordinations are sufficient for $f \in S^*[A, B]$:

(1)
$$\frac{zf'(z)}{f(z)}\left(1 + \beta\left(1 + \frac{zf''(z)}{f'(z)} - \frac{zf'(z)}{f(z)}\right)\right) \prec \varphi_q(z),$$

for some $\beta \neq 0$, $-1 \leq B < A \leq 1$ satisfying the inequality (2.8);

(2)
$$\frac{zf'(z)}{f(z)}\left(1 + \beta\left(\frac{zf'(z)}{f(z)}\right)^{-1}\left(1 + \frac{zf''(z)}{f'(z)} - \frac{zf'(z)}{f(z)}\right)\right) \prec \varphi_q(z),$$

whenever $-1 \leq B < A \leq 1$ and $\beta \neq 0$ and inequality (2.9) holds;

(3)
$$\frac{f(z)}{zf'(z)} + \beta\left(1 + \frac{zf''(z)}{f'(z)} - \frac{zf'(z)}{f(z)}\right) \prec \varphi_q(z),$$

where $-1 \leq B < A \leq 1$, $\beta \neq 0$ and the inequality (2.10) holds;

(4)
$$\left(\frac{zf'(z)}{f(z)}\right)^{-1}\left(1 + \beta\left(1 + \frac{zf''(z)}{f'(z)} - \frac{zf'(z)}{f(z)}\right)\right) \prec \varphi_q(z),$$

whenever $-1 \leq B < A \leq 1$ and $\beta \neq 0$ and the inequality (2.11) holds.

3 Further Results

This section deals with an alternative proof of the first-order differential subordination results by using admissibility condition, as demonstrated by authors [7], in which certain conditions are determined so that $p(z) \prec (1 + Az)/(1 + Bz)$ whenever $1 + \beta zp'(z)/p^k(z) \prec \varphi_q(z)$ $(k = 0, 1, 2)$ hold.

Theorem 3.1 *Let p be an analytic function defined on $\mathbb{D}$ with $p(0) = 1$. For $\beta \neq 0$ and $-1 \leq B < A \leq 1$, assume that*

$$|\beta|^2(A - B)^2 \geq 2\big((1 + |B|)^4 + 2|\beta|(A - B)(1 + |B|^2)\big) \tag{3.1}$$

If $1 + \beta z p'(z) \prec \varphi_q(z)$, then $p \in \mathcal{P}[A, B]$.

Proof Let Ω as in Theorem 2.3. Consider the analytic function $\psi(r, s; z) = 1 + \beta s$. By Theorem 2.2 for $\psi \in \Psi[\Omega; A, B]$, we have to show $\psi(r, s; z) \notin \Omega$. The function $\psi(r, s; z)$ is given by

$$\psi(r, s; z) = 1 + \beta \frac{m(A - B)e^{i\theta}}{(1 + Be^{i\theta})^2}$$

so that

$$\begin{aligned}\left|\frac{\psi^2(r, s; z) - 1}{\psi(r, s; z)}\right| &= \left|\frac{\beta^2 m^2(A - B)^2 e^{2i\theta} + 2\beta m e^{i\theta}(A - B)(1 + Be^{i\theta})^2}{(1 + Be^{i\theta})^2[(1 + Be^{i\theta})^2 + \beta m e^{i\theta}(A - B)]}\right| \\ &\geq \frac{\beta^2 m^2(A - B)^2 - 2|\beta| m(A - B)(1 + |B|)^2}{(1 + |B|)^4 + |\beta| m(A - B)(1 + |B|)^2} \\ &= \frac{\big(|\beta| m(A - B) - (1 + |B|)^2\big)^2 - (1 + |B|)^4}{(1 + |B|)^4 + |\beta| m(A - B)(1 + |B|)^2} \\ &\geq \frac{\big(|\beta|(A - B) - (1 + |B|)^2\big)^2 - (1 + |B|)^4}{(1 + |B|)^4 + |\beta| m(A - B)(1 + |B|)^2}\end{aligned}$$

A simple calculation shows that

$$\left|\frac{\psi^2(r, s; z) - 1}{\psi(r, s; z)}\right| \geq 2$$

whenever the inequality (3.1) holds. Hence Theorem 2.2 concludes the desired proof. ∎

Theorem 3.2 *Let p be an analytic function defined on $\mathbb{D}$ with $p(0) = 1$ and $\beta \neq 0$ and $-1 \leq B < A \leq 1$. We assume that*

$$|\beta|^2(A - B)^2 \geq 2\big((1 + |B|)^2(1 + |A|)^2 + 2|\beta|(A - B)(1 + |A|)(1 + |B|)\big). \tag{3.2}$$

If $1 + \beta z p'(z)/p(z) \prec \varphi_q(z)$, then $p \in \mathcal{P}[A, B]$.

Proof Let q and Ω be as in Theorem 3.1. Consider the analytic function $\psi : \mathbb{C} \setminus \{0\} \times \mathbb{C} \times \mathbb{D} \to \mathbb{C}$ defined as $\psi(r, s, ; z) = 1 + \beta \frac{s}{r}$. The required subordination is established if $\psi \in \Psi[\Omega; A, B]$. Thus by Theorem 2.2 it is enough to show that $|(\psi^2 - 1)/\psi| \geq 2$. We have

$$\psi(r,s,;z) = 1 + \beta \frac{m(A-B)e^{i\theta}}{(1+Be^{i\theta})(1+Ae^{i\theta})}$$

so that

$$\begin{aligned}\left|\frac{\psi^2(r,s;z)-1}{\psi(r,s;z)}\right| &= \left|\frac{\beta^2 m^2 (A-B)^2 e^{i\theta} + 2\beta m(A-B)(1+Ae^{i\theta})(1+Be^{i\theta})}{(1+Ae^{i\theta})^2(1+Be^{i\theta})^2 + \beta m e^{i\theta}(1+Ae^{i\theta})(1+Be^{i\theta})}\right| \\ &\geq \frac{|\beta|^2 m^2 (A-B)^2 - 2|\beta| m(A-B)(1+|A|)(1+|B|)}{(1+|A|)^2(1+|B|)^2 + |\beta| m(A-B)(1+|A|)(1+|B|)} \\ &= \frac{(|\beta| m(A-B) - (1+|A|)(1+|B|))^2 - (1+|A|)^2(1+|B|)^2}{(1+|A|)^2(1+|B|)^2 + |\beta| m(A-B)(1+|A|)(1+|B|)} \\ &\geq \frac{(|\beta|(A-B) - (1+|A|)(1+|B|))^2 - (1+|A|)^2(1+|B|)^2}{(1+|A|)^2(1+|B|)^2 + |\beta| m(A-B)(1+|A|)(1+|B|)} \geq 2\end{aligned}$$

whenever the inequality (3.2) holds. Thus $\psi \notin \Omega$ and Theorem 2.2 yields the desired subordination. ■

Theorem 3.3 *Let p be an analytic function defined on $\mathbb{D}$ with $p(0) = 1$ and $\beta \neq 0$ and $-1 \leq B < A \leq 1$. We assume that*

$$|\beta|^2(A-B)^2 \geq 2\big((1+|A|)^4 + 2|\beta|(A-B)(1+|A|^2)\big) \tag{3.3}$$

If $1 + \beta z p'(z)/p^2(z) \prec \varphi_q(z)$ then $p(z) \prec (1+Az)/(1+Bz)$.

Proof Consider the function q and the domain Ω as in Theorem 3.1. Let $\psi : \mathbb{C} \setminus \{0\} \times \mathbb{C} \times \mathbb{D} \to \mathbb{C}$ be the analytic function given by $\psi(r,s;z) = 1 + \beta s/r^2$. By making use of Theorem 2.2, $\psi \in \Psi[\Omega, q]$ whenever $|(\psi^2 - 1)/\psi| \geq 2$. We have

$$\psi(r,s;z) = 1 + \beta \frac{m(A-B)e^{i\theta}}{(1+Ae^{i\theta})^2}$$

so that

$$\begin{aligned}\left|\frac{\psi^2(r,s;z)-1}{\psi(r,s;z)}\right| &= \frac{\begin{array}{r}|2\beta m(A-B) + (\beta^2 m^2 (A-B)^2 + 2\beta m(A-B)2A)e^{i\theta} \\ + 2\beta m(A-B)A^2 e^{2i\theta}|\end{array}}{|(1+Ae^{i\theta})^4 + \beta m e^{i\theta}(A-B)(1+Ae^{i\theta})^2|} \\ &\geq \frac{\beta^2 m^2 (A-B)^2 - 2|\beta| m(A-B)(1+|A|)^2}{(1+|A|)^4 + |\beta| m(A-B)(1+|A|)^2} \\ &= \frac{(|\beta| m(A-B) - (1+|A|)^2)^2 - (1+|A|)^4}{(1+|A|)^4 + |\beta| m(A-B)(1+|A|)^2} \\ &\geq \frac{(|\beta|(A-B) - (1+|A|)^2)^2 - (1+|A|)^4}{(1+|A|)^4 + |\beta| m(A-B)(1+|A|)^2} \\ &\geq 2\end{aligned}$$

whenever the inequality (3.3) holds. Hence by Theorem 2.2 we get the desired subordination. ■

As an application of Theorems 3.1–3.3, we see that the following subordinations are sufficient for $f \in S^*[A, B]$:

(1)
$$1 + \beta \frac{zf'(z)}{f(z)} \left(1 + \frac{zf''(z)}{f'(z)} - \frac{zf'(z)}{f(z)} \right) \prec \varphi_q(z)$$

for some $\beta \neq 0$ and $-1 \leq B < A \leq 1$ and the inequality (3.1) holds;

(2)
$$1 + \beta \left(1 + \frac{zf''(z)}{f'(z)} - \frac{zf'(z)}{f(z)} \right) \prec \varphi_q(z)$$

for some $\beta \neq 0$, if the inequality (3.2) holds;

(3)
$$1 + \beta \left(\frac{zf'(z)}{f(z)} \right)^{-1} \left(1 + \frac{zf''(z)}{f'(z)} - \frac{zf'(z)}{f(z)} \right) \prec \varphi_q(z)$$

for some $\beta \neq 0$, if the inequality (3.3) holds.

References

1. O.P. Ahuja, S. Kumar, V. Ravichandran, Applications of first order differential subordination for functions with positive real part. Stud. Univ. Babeş-Bolyai Math. **63**(3), 303–311 (2018)
2. R.M. Ali, V. Ravichandran, N. Seenivasagan, Sufficient conditions for Janowski starlikeness. Int. J. Math. Math. Sci. **2007**, 7 (2007)
3. R.M. Ali, N.E. Cho, V. Ravichandran, S.S. Kumar, Differential subordination for functions associated with the lemniscate of Bernoulli. Taiwan. J. Math. **16**(3), 1017–1026 (2012)
4. N. Bohra, S. Kumar, V. Ravichandran, Some special differential subordinations, Hacet. J. Math. Stat. (2018)
5. N.E. Cho, H.J. Lee, J.H. Park, R. Srivastava, Some applications of the first-order differential subordinations. Filomat **30**(6), 1465–1474 (2016)
6. N.E. Cho, S. Kumar, V. Kumar, V. Ravichandran, Differential subordination and radius estimates for starlike functions associated with the Booth lemniscate. Turk. J. Math. **42**(3), 1380–1399 (2018)
7. S. Gandhi, S. Kumar, V. Ravichandran, First order differential subordinations for caratheodory functions. Kyungpook Math. J. **58**, 257–270 (2018)
8. A.W. Goodman, *Univalent Functions*, vol. 2 (Mariner Publishing Co., Inc, Tampa, FL, 1983)
9. W. Janowski, Extremal problems for a family of functions with positive real part and for some related families, Ann. Polon. Math. **23** (1970/1971), 159–177
10. S. Kanas, Techniques of the differential subordination for domains bounded by conic sections. Int. J. Math. Math. Sci. **38**, 2389–2400 (2003)
11. S. Kanas, A. Lecko, Differential subordination for domains bounded by hyperbolas. Zeszyty Nauk. Politech. Rzeszowskiej Mat. **175**(23), 61–70 (1999)
12. S. Kumar, V. Ravichandran, Subordinations for functions with positive real part. Complex Anal. Oper. Theory **12**(5), 1179–1191 (2018)
13. S.S. Kumar, V. Kumar, V. Ravichandran, N.E. Cho, Sufficient conditions for starlike functions associated with the lemniscate of Bernoulli, J. Inequal. Appl. **176**, 13 (2013)

14. S. Kumar, S. Nagpal, V. Ravichandran, Coefficient inequalities for Janowski starlikeness. Proc. Jangjeon Math. Soc. **19**(1), 83–100 (2016)
15. S.S. Miller, P.T. Mocanu, *Differential Subordinations*, Monographs and Textbooks in Pure and Applied Mathematics, vol. 225 (Marcel Dekker Inc, New York, 2000)
16. M. Nunokawa, M. Obradović, S. Owa, One criterion for univalency. Proc. Am. Math. Soc. **106**(4), 1035–1037 (1989)
17. R.K. Raina, J. Sokół, On coefficient estimates for a certain class of starlike functions. Hacet. J. Math. Stat. **44**(6), 1427–1433 (2015)
18. V. Ravichandran, K. Sharma, Sufficient conditions for starlikeness. J. Korean Math. Soc. **52**(4), 727–749 (2015)

Coefficient Bounds for a Unified Class of Holomorphic Functions

Mridula Mundalia and Sivaprasad Kumar Shanmugam

Abstract In the present paper, sharp initial coefficient bounds have been estimated for functions in the newly defined classes $\mathcal{S}^k_{\gamma,\delta}(\Phi)$ and $\mathcal{S}^k_{\gamma,\delta,h}(\Phi)$, which in fact, unifies many earlier known classes. Further, sharp bounds of the Fekete–Szegö coefficient functional for functions in the classes introduced here are obtained and special cases of our results are also pointed out.

Keywords Univalent functions · Starlike functions · Convex functions · Fekete–Szegö coefficient functional · Subordination

2010 Mathematics Subject Classification 30C45 · 30C80

1 Introduction and Preliminaries

Let $\mathcal{A}$ be the class all of functions f that are holomorphic in the open unit disk $\mathbb{D} = \{z \in \mathbb{C} : |z| < 1\}$, possessing the series expansion of the form

$$f(z) = z + \sum_{n=2}^{\infty} a_n z^n. \tag{1.1}$$

Let $\mathcal{S}$ be the subclass of $\mathcal{A}$ consisting of univalent functions. Let h and g be holomorphic functions defined in $\mathbb{D}$, h is said to be *subordinate* to g, denoted by $h \prec g$, if there exists a Schwarz function $v : \mathbb{D} \to \mathbb{D}$ with $v(0) = 0$ such that $h(z) =$

The work presented here by first author was supported by a Research Fellowship from Delhi Technological University, New Delhi.

M. Mundalia (✉) · S. K. Shanmugam
Department of Applied Mathematics, Delhi Technological University, Delhi 110042, India
e-mail: mridulamundalia@yahoo.co.in

S. K. Shanmugam
e-mail: spkumar@dce.ac.in

N. Deo et al. (eds.), *Mathematical Analysis I: Approximation Theory*, Springer Proceedings in Mathematics & Statistics 306,
https://doi.org/10.1007/978-981-15-1153-0_17

$g(v(z))$. Equivalently, if g is univalent in $\mathbb{D}$, then $h(z) \prec g(z)$ $(z \in \mathbb{D})$ if and only if $h(0) = g(0)$ and $h(\mathbb{D}) \subset g(\mathbb{D})$.

In the past, many authors found coefficient bounds for the class of functions defined through subordination involving $zf'(z)/f(z)$ or $1 + zf''(z)/f'(z)$ or $f(z)/z$ or $f'(z)$ or their ratios or product of powers of these expressions or in terms of their weighted sum or product (See [6, 8, 11, 13, 16, 18–21, 27–31]). In the present paper, an attempt has been made to unify all these analytic characterizing expressions into one, which is given below in (1.2) and thereby many well known classes have been clubbed together, which in fact, allows for various cross combinations of the abovementioned expressions.

$$\left(\frac{zf'(z) + \alpha z^2 f''(z)}{\alpha z f'(z) + (1-\alpha) f(z)}\right)^k \left(\beta f'(z) + (1-\beta)\frac{f(z)}{z}\right)^{1-k}. \tag{1.2}$$

For brevity we shall assume $F_m(z) := mzf'(z) + (1-m)f(z)$, so that the expression in (1.2) becomes: $\left(zF'_\alpha(z)/F_\alpha(z)\right)^k \left(F_\beta(z)/z\right)^{1-k}$. We further choose $\delta = \alpha k + \beta(1-k)$, since α and β vanish along with k and $1-k$, when they reduce to zero, respectively. Throughout the paper we shall assume that Φ is a holomorphic univalent function in $\mathbb{D}$ such that Re $\Phi(z) > 0$ $(z \in \mathbb{D})$. Also let $\Phi(\mathbb{D})$ be symmetric with respect to the real axis and starlike with respect to 1 satisfying $\Phi(0) = 1$ and $\Phi'(0) > 0$. We assume that Φ is of the form

$$\Phi(z) = 1 + \sum_{n=1}^{\infty} B_n z^n.$$

Clearly, $\Phi'(0) = B_1 > 0$.

Definition 1.1 A function f in $\mathcal{A}$ is said to be in the class $\mathcal{S}^k_{\gamma,\delta}(\Phi)$, if it satisfies:

$$1 + \frac{1}{\gamma}\left(\left(\frac{zF'_\alpha(z)}{F_\alpha(z)}\right)^k \left(\frac{F_\beta(z)}{z}\right)^{1-k} - 1\right) \prec \Phi(z), \tag{1.3}$$

where $F_m(z) := mzf'(z) + (1-m)f(z)$ with $m = \alpha$ or β and $\delta = \alpha k + \beta(1-k)$, with $\gamma \in \mathbb{C}\backslash\{0\}$, $0 \le \alpha, \beta, k \le 1$.

The Hadamard product (or convolution) of f and h in $\mathcal{A}$, of the form $f(z) = z + a_2z^2 + a_3z^3 + \cdots$ and $h(z) = z + h_2z^2 + h_3z^3 + \cdots$, respectively, is defined by

$$f * h = z + \sum_{n=2}^{\infty} a_n h_n z^n.$$

We assume that the coefficients of h are positive. We define the class $\mathcal{S}^k_{\gamma,\delta,h}(\Phi)$ consisting of functions f in $\mathcal{A}$ satisfying $f * h \in \mathcal{S}^k_{\gamma,\delta}(\Phi)$. If $h(z) = z(1-z)^{-1}$, the class $\mathcal{S}^k_{\gamma,\delta,h}(\Phi)$ reduces to $\mathcal{S}^k_{\gamma,\delta}(\Phi)$. Clearly the class $\mathcal{S}^k_{\gamma,\delta}(\Phi)$ reduces to numerous well known classes for some appropriate choice of parameter. We illustrate some of the important subclasses studied in the past. For $k = 1$ and $\Phi(z) = (1 + Az)/(1 + Bz)$, $-1 \leq B < A \leq 1$, we obtain the class $\mathcal{S}^1_{\gamma,\alpha}(A, B) = \mathcal{S}^1_{\gamma,\alpha}((1 + Az)/(1 + Bz))$. Additionally, for $\alpha = 0$, $A = 1$ and $B = -1$, the class $\mathcal{S}^1_{\gamma,0}(1, -1)$ coincides with $\mathcal{S}^*(\gamma)$, the class of starlike functions of complex order γ ($\gamma \in \mathbb{C}\backslash\{0\}$), introduced and studied by Nasr and Aouf [21], consisting of functions f in $\mathcal{A}$ satisfying $f(z)/z \neq 0$ and

$$\operatorname{Re}\left(1 + \frac{1}{\gamma}\left(\frac{zf'(z)}{f(z)} - 1\right)\right) > 0.$$

The class of convex functions of complex order γ ($\gamma \in \mathbb{C}\backslash\{0\}$), introduced by Wiatrowski [31], consists of functions f in $\mathcal{A}$ satisfying the conditions $f(z)/z \neq 0$ and

$$\operatorname{Re}\left(1 + \frac{zf''(z)}{\gamma f'(z)}\right) > 0.$$

This class can be obtained by taking $k = \alpha = 1$, $A = 1$ and $B = -1$, $\mathcal{S}^1_{\gamma,1}(1, -1) = \mathcal{C}(\gamma)$. When $k = 0$ and $\beta = 1$, we obtain the class $\mathcal{S}^0_{\gamma,1}(\Phi) \equiv \mathcal{R}_\gamma(\Phi)$ consisting of functions which are closely related to the class of functions with positive real part. Dixit et al. [5] introduced the class $\mathcal{R}_\gamma(A, B)$, $-1 \leq B < A \leq 1$ consisting of functions f in $\mathcal{A}$ satisfying:

$$\left|\frac{f'(z) - 1}{\gamma(A - B) - B(f'(z) - 1)}\right| < 1.$$

Note that $\mathcal{S}^0_{\gamma,1}(A, B) = \mathcal{R}_\gamma(A, B)$, $-1 \leq B < A \leq 1$.

Here below, we discuss some special cases of our class when $\gamma = 1$. For $\Phi(z) = (1 + Az)/(1 + Bz)$, we obtain the following new class

$$\begin{aligned}\mathcal{S}^k_\delta(A, B) &= \mathcal{S}^k_{1,\delta}((1 + Az)/(1 + Bz))\\ &= \left\{f \in \mathcal{A} : \left(\frac{zf'(z) + \alpha z^2 f''(z)}{\alpha zf'(z) + (1 - \alpha)f(z)}\right)^k \left(\beta f'(z) + (1 - \beta)\frac{f(z)}{z}\right)^{1-k} \prec \frac{1 + Az}{1 + Bz}\right\}.\end{aligned}$$

Particularly for $\Phi(z) = (1 + (1 - 2\tau)z)/(1 - z)$, $0 \leq \tau < 1$, the class $\mathcal{S}^k_{\gamma,\delta}(\Phi)$ reduces to

$$\begin{aligned}\mathcal{S}^k_\delta(\tau) &= \mathcal{S}^k_{1,\delta}((1 + (1 - 2\tau)z)/(1 - z))\\ &= \left\{f \in \mathcal{A} : \operatorname{Re}\left(\left(\frac{zf'(z) + \alpha z^2 f''(z)}{\alpha zf'(z) + (1 - \alpha)f(z)}\right)^k \left(\beta f'(z) + (1 - \beta)\frac{f(z)}{z}\right)^{1-k}\right) > \tau\right\}.\end{aligned}$$

By taking $k = 1$ and $\alpha = 0$, we obtain Ma and Minda class of starlike functions [16] $\mathcal{S}^*(\Phi) = \mathcal{S}^1_{1,0}(\Phi)$. Further, if $\Phi(z) = (1 + Az)/(1 + Bz)$, $-1 \leq B < A \leq 1$, we obtain the class of Janowski starlike functions [10]. Note that for $A = 1 - 2\tau$ and $B = -1$, we obtain the class of starlike functions of order τ $(0 \leq \tau < 1)$, $\mathcal{S}^1_{1,0}(1-2\tau, -1) = \mathcal{S}^*(\tau)$ introduced by Robertson [26]. Further if $\tau = 0$, we obtain $\mathcal{S}^1_{1,0}(1, -1) = \mathcal{S}^*$ [6, 8, 23], the class of starlike functions. By taking k=1 and $\alpha = 1$, we obtain the Ma and Minda class of convex functions $\mathcal{S}^1_{1,1}(\Phi) = \mathcal{C}(\Phi)$[16]. Particularly, if $\Phi(z) = (1 + Az)/(1 + Bz)$, $-1 \leq B < A \leq 1$, we obtain $\mathcal{S}^1_{1,1}(A, B) = \mathcal{K}(A, B)$, the class of Janowski convex functions [9]. Further, if $A = -B = 1$, the class $\mathcal{S}^1_{1,1}(A, B)$ reduces to give the well known class of convex functions $\mathcal{S}^1_{1,1}(1, -1) = \mathcal{C}$ (See [6, 8, 23]). Observe that in the case when $k = 0$ and $\beta = 1$, the class $\mathcal{S}^0_{1,1}(\Phi)$ coincides with $\mathcal{R}(\Phi)$, a subclass of close-to-convex functions. Particularly, for $\Phi(z) = (1 + Az)/(1 + Bz)$, the class $\mathcal{S}^0_{1,1}(A, B) = \mathcal{R}(A, B)$, $-1 \leq B < A \leq 1$, studied by Goel and Mehrok [7], consists of functions $f \in \mathcal{A}$ for which $\left|f'(z) - 1\right| < |Bf'(z) - A|$. MacGregor [17] systematically studied the class $\mathcal{R}$ consisting of functions f in $\mathcal{A}$, whose derivative has a positve real part. Note that the class $\mathcal{S}^0_{1,1}(1, -1)$ coincides with $\mathcal{R}$. With $A = 1$ and $B = 0$, the class $\mathcal{S}^0_{1,1}(1, 0)$ coincides with a subclass $\mathcal{R}(1)$ of $\mathcal{R}$, studied by MacGregor [17], consisting of functions f in $\mathcal{A}$ satisfying the inequality $|f'(z) - 1| < 1$.

The famous Beiberbach's conjecture, proved by de Branges in 1985, states that $|a_n| \leq n$ for any $f \in \mathcal{S}$. This is a sharp bound and koebe function works as the extremal function for this estimate. Fekete and Szegö established a sharp bound for the following functional for functions in class $\mathcal{S}$,

$$|a_3 - \mu a_2^2| \leq \begin{cases} 4\mu - 3, & \mu \geq 1; \\ 1 + \exp\left(\frac{-2\mu}{1-\mu}\right), & 0 \leq \mu \leq 1; \\ 3 - 4\mu, & \mu \leq 0. \end{cases}$$

This functional is commonly known as the Fekete–Szegö functional which was studied by Fekete and Szegö in 1933. Ma and Minda [16] investigated this problem for some special classes of convex and starlike functions, $\mathcal{C}(\Phi)$ and $\mathcal{S}^*(\Phi)$ respectively. The bounds for the quantity $|a_3 - \mu a_2^2|$, for close-to-convex functions was studied by Kim et al. [14] and Cho et al. [4]. Ali et al. in [2] established a sharp result for this functional for some subclasses of p-valent functions. Many authors [1, 3, 15, 22, 25] have obtained this bound for many subclasses of $\mathcal{S}$.

Let Υ be the class of all Schwarz functions v of the form

$$v(z) = \sum_{n=1}^{\infty} v_n z^n \tag{1.4}$$

defined in $\mathbb{D}$, satisfying $v(0) = 0$ and $|v(z)| < 1$.

Lemmas stated below are required in sequel to derive our main results.

Lemma 1.2 ([2]) *If* $v \in \Upsilon$, *then*

$$|v_2 - tv_1^2| \leq \begin{cases} -t \text{ } if t \leq -1; \\ 1 \text{ } if -1 \leq t \leq 1; \\ t \text{ } if t \geq 1. \end{cases}$$

When $t < -1$ *or* $t > 1$*, equality holds if and only if* $v(z) = z$ *or one of its rotations. If* $-1 < t < 1$*, then equality holds if and only if* $v(z) = z^2$ *or one of its rotations. When* $t = -1$ *then equality holds if and only if* $v(z) = z(\lambda + z)/(1 + \lambda z)$*,* $(0 \leq \lambda \leq 1)$ *or one of its rotations. For* $t = 1$*, equality holds if and only if* $v(z) = -z(\lambda + z)/(1 + \lambda z)$ $(0 \leq \lambda \leq 1)$ *or one of its rotations. Also the sharp upper bound above can be improved as follows when* $-1 < t < 1$*:*

$$|v_2 - tv_1^2| + (1+t)|v_1|^2 \leq 1 \, (-1 < t \leq 0) \tag{1.5}$$

and

$$|v_2 - tv_1^2| + (1-t)|v_1|^2 \leq 1 \, (0 < t < 1). \tag{1.6}$$

Lemma 1.3 ([12]) *Let* $v \in \Upsilon$*, then for any* $t \in \mathbb{C}$*,*

$$|v_2 - tv_1^2| \leq \max\{1; |t|\}.$$

Extremal functions are $v(z) = z^2$ *or* $v(z) = z$*.*

In this paper, sharp bound for the Fekete-Szegö functional $|a_3 - \mu a_2^2|$, where $\mu \in \mathbb{C}$, is obtained for functions belonging to $\mathcal{S}^k_{\gamma,\delta}(\Phi)$. Apart from that sharp bounds for initial coefficients a_2 and a_3 have been found and their extremal functions have been obtained. A sharp bound for a_4 is obtained for $\gamma = 1$.

2 Main Results

In the following sections Lemmas 1.2 and 1.3 have been used to derive our main results.

Theorem 2.1 *Let* f *be in the class* $S^k_{\gamma,\delta}(\Phi)$ *and* $\Phi(z) = 1 + B_1 z + B_2 z^2 + \cdots$*. Then, for any* $\mu \in \mathbb{C}$*, we have*

$$|a_3 - \mu a_2^2| \leq \frac{|\gamma| B_1}{M_1} \max\left\{1; \left|\frac{\gamma B_1(2\mu M_1 - M_2)}{2(1+\delta)^2} - \frac{B_2}{B_1}\right|\right\}, \tag{2.1}$$

where

$$M_1 = 1 + k(1 + 4\alpha) + 2\beta(1 - k) \tag{2.2}$$

and

$$M_2 = k((3-k)(1+\alpha)^2 - (1-k)(1+\beta)(1+2\alpha-\beta)). \tag{2.3}$$

Further,

$$|a_2| \leq \frac{|\gamma| B_1}{1+\delta} \quad \textit{and} \quad |a_3| \leq \frac{|\gamma| B_1}{M_1} \max\left\{1; \left|\frac{B_2}{B_1} + \frac{\gamma B_1 M_2}{2(1+\delta)^2}\right|\right\}. \tag{2.4}$$

These estimates are sharp.

Proof Since f is in $S^k_{\gamma,\delta}(\Phi)$, then there exists a holomorphic function v in Υ such that

$$1 + \frac{1}{\gamma}\left(\left(\frac{zF'_\alpha(z)}{F_\alpha(z)}\right)^k \left(\frac{F_\beta(z)}{z}\right)^{1-k} - 1\right) = \Phi(v(z)). \tag{2.5}$$

Upon expanding F_α, F_β in terms of f and further using power series expansion of f, we obtain

$$\begin{aligned}\left(\frac{zF'_\alpha(z)}{F_\alpha(z)}\right)^k \left(\frac{F_\beta(z)}{z}\right)^{1-k} &= \left(\frac{zf'(z) + \alpha z^2 f''(z)}{\alpha z f'(z) + (1-\alpha) f(z)}\right)^k \left(\beta f'(z) + (1-\beta)\frac{f(z)}{z}\right)^{1-k} \\ &= 1 + (1+\delta) a_2 z + \frac{1}{2}\left(a_2^2 M_2 + 2a_3 M_1\right) z^2 + \cdots.\end{aligned} \tag{2.6}$$

Also,

$$\Phi(v(z)) = 1 + B_1 v_1 z + \left(B_2 v_1^2 + B_1 v_2\right) z^2 + \cdots.$$

Therefore, using (2.5), we obtain the coefficients a_2 and a_3 as follows:

$$a_2 = \frac{\gamma B_1 v_1}{1+\delta} \tag{2.7}$$

and

$$a_3 = \gamma\left(\frac{B_2}{M_1} + \frac{\gamma B_1^2 M_2}{2(1+\delta)^2 M_1}\right) v_1^2 + \frac{\gamma B_1}{M_1} v_2. \tag{2.8}$$

On substituting these values in the Fekete–Szegö coefficient functional, it reduces to

$$|a_3 - \mu a_2^2| \leq \frac{|\gamma| B_1}{M_1} \left| v_2 - \left(\frac{\gamma B_1 (2\mu M_1 - M_2)}{2(1+\delta)^2} - \frac{B_2}{B_1}\right) v_1^2\right|. \tag{2.9}$$

The result now follows by applying Lemma 1.3. Bounds for the first two coefficients a_2 and a_3, can be obtained directly from inequality (2.1). Following functions play the role of extremal functions:

$$f_1(z) = z + \frac{\gamma B_1}{(1+\delta)} z^2 + \frac{\gamma B_1}{M_1}\left(\frac{B_2}{B_1} + \frac{\gamma M_2 B_1}{2(1+\delta)^2}\right) z^3$$

$$f_2(z) = z + \frac{\gamma B_1}{M_1} z^3,$$

where $f_1(z)$ is the extremal function for the second coefficient and Fekete–Szegö functional. Extremal function for the third coefficient is given by

$$\begin{cases} f_1(z) \text{ when } |2(1+\delta)^2 B_2 + \gamma B_1^2 M_2| > 2B_1(1+\delta)^2, \\ f_2(z) \text{ when } |2(1+\delta)^2 B_2 + \gamma B_1^2 M_2| \le 2B_1(1+\delta)^2. \end{cases}$$

By choosing $v(z) = z$ and z^2 respectively, in Eq. (2.5) the above extremal functions can be obtained. ■

Remark 2.2 By taking $k = 0$ and $\beta = 1$, in the above theorem, inequality in (2.1) reduces to an inequality given in [2, Theorem 3 (for $p = 1$)]. Further, with $\Phi(z) = (1 + Az)/(1 + Bz)$, $(-1 \le B < A \le 1)$, Theorem 2.1 reduces to [5, Theorem 4]. Also note that with $\gamma = k = 1$ and $\alpha = 0$, inequality (2.1) reduces to give the inequality in [2, Theorem 1 (for $p = 1$)].

It is presumed that, M_1 and M_2 carry their expressions as stated in Eqs. (2.2) and (2.3), respectively.

By choosing suitable values of α, β and k, in Theorem 2.1, we obtain the following corollary:

Corollary 2.3 *Let f be in the class $\mathcal{S}^k_{\gamma,\delta}(\Phi)$, then for $\mu \in \mathbb{C}$,*

(i) If $k = 1$, then

$$|a_3 - \mu a_2^2| \le \frac{|\gamma| B_1}{2(1+2\alpha)} \max\left\{1; \left|\gamma B_1 \left(\frac{2\mu(1+2\alpha)}{(1+\alpha)^2} - 1\right) - \frac{B_2}{B_1}\right|\right\}.$$

(ii) If $k = 0$, then

$$|a_3 - \mu a_2^2| \le \frac{|\gamma| B_1}{(1+2\beta)} \max\left\{1; \left|\frac{\mu\gamma B_1(1+2\beta)}{(1+\beta)^2} - \frac{B_2}{B_1}\right|\right\}.$$

For $\gamma = 1$ and $\alpha = 1$ Theorem 2.1 reduces to give the following bound for the Ma and Minda class of convex functions $\mathcal{S}^1_{1,1}(\Phi)$,

$$|a_3 - \mu a_2^2| \le \frac{B_1}{6} \max\left\{1; \left|B_1\left(\frac{3\mu}{2} - 1\right) - \frac{B_2}{B_1}\right|\right\}.$$

This result is sharp.

Remark 2.4 For any real μ, second part of Corollary 2.3 for $\gamma = 1$ and $\alpha = 1$ coincides to give a result derived by Ma and Minda in [16, Theorem 3]. Further by taking $\Phi(z) = (1+z)/(1-z)$, a sharp estimate, $|a_3 - \mu a_2^2| \leq \max\{1/3, |\mu - 1|\}$ is obtained for any $\mu \in \mathbb{C}$. This result was obtained by Keogh et al. [12, Corollary 1].

Theorem 2.5 *Let f be in the class $S^k_{\gamma,\delta}(\Phi)$ and $\Phi(z) = 1 + B_1 z + B_2 z^2 + \cdots$. Then, we have*

$$|a_4| \leq \frac{|\gamma| B_1}{J_1}\left(H(q_1, q_2) + \frac{|\gamma J_2| B_1}{(1+\delta)M_1}\max\{1; |t|\}\right), \tag{2.10}$$

where $H(q_1, q_2)$ is as defined in [24, Lemma 2], with

$$q_1 = \frac{2B_2}{B_1}, \quad q_2 = \frac{B_3}{B_1} \quad \text{and} \quad t = \frac{\gamma B_1 J_3}{6J_2(1+\delta)^2} - \frac{B_2}{B_1},$$

where

$$\begin{aligned} J_1 &= 1 + k(2 + 9\alpha) + 3\beta(1-k), \\ J_2 &= k\left(2\alpha^2(2k-5) + \alpha(6\beta - 6\beta k + k - 10) + \beta(2\beta - 1)(k-1) - 3\right) \end{aligned}$$

and

$$\begin{aligned} J_3 = &-k(\alpha+1)((\alpha+1)^2(k-7)(k-2)M_1 + 3(2\alpha+1)(2k-5)M_2) \\ &+ (k-1)k(k(-3\alpha\beta + 3\alpha(\alpha+1) + \beta^2 - \beta + 1) + (\beta+1)M_1((3\alpha+\beta+4) \\ &\times (-3\alpha + \beta - 2)) + 3M_2(\alpha(6\beta+5) - 2\beta^2 + \beta + 2)). \end{aligned}$$

Proof Upon using Eq. (2.6), the fourth coefficient is given by

$$a_4 = \frac{\gamma B_1}{J_1}\left(\left(v_3 + \frac{2B_2}{B_1}v_1 v_2 + \frac{B_3}{B_1}v_1^3\right) - \frac{\gamma B_1 J_2 v_1}{(1+\delta)M_1}(v_2 - t v_1^2)\right).$$

Now by applying Lemma 1.3 to the above expression together with [24, Lemma 2], bound for the fourth coefficient can be established. ■

Remark 2.6 For $k = 0$ and $\beta = 1$, inequality in Theorem 2.5 reduces to give the inequality in [2, Theorem 3 (for $p = 1$)].

We now derive the following result for functions in the class $\mathcal{S}^k_{1,\delta}(\Phi) = \mathcal{S}^k_\delta(\Phi)$.

Theorem 2.7 *Let f be in the class $S^k_\delta(\Phi)$ and $\Phi(z) = 1 + B_1 z + B_2 z^2 + \cdots$. Then*

$$|a_3 - \mu a_2^2| \leq \begin{cases} \dfrac{-B_1 t}{M_1} & \text{when } \mu \leq \sigma_1, \\[2ex] \dfrac{B_1}{M_1} & \text{when } \sigma_1 \leq \mu \leq \sigma_2, \\[2ex] \dfrac{B_1 t}{M_1} & \text{when } \mu \geq \sigma_2, \end{cases} \tag{2.11}$$

where

$$t = \frac{B_1(2\mu M_1 - M_2)}{2(1+\delta)^2} - \frac{B_2}{B_1}.$$

Further, if $\sigma_1 \leq \mu \leq \sigma_3$, *then*

$$|a_3 - \mu a_2^2| + \frac{(1+\delta)^2}{B_1 M_1}\left\{1 - \frac{B_2}{B_1} + \frac{B_1(2\mu M_1 - M_2)}{2(1+\delta)^2}\right\}|a_2|^2 \leq \frac{B_1}{M_1}. \tag{2.12}$$

If $\sigma_3 \leq \mu \leq \sigma_2$, *then*

$$|a_3 - \mu a_2^2| + \frac{(1+\delta)^2}{B_1 M_1}\left\{1 + \frac{B_2}{B_1} - \frac{B_1(2\mu M_1 - M_2)}{2(1+\delta)^2}\right\}|a_2|^2 \leq \frac{B_1}{M_1}. \tag{2.13}$$

where

$$\sigma_1 = \frac{(1+\delta)^2}{B_1 M_1}\left(\frac{B_2}{B_1} - 1\right) + \frac{M_2}{2M_1}, \quad \sigma_2 = \frac{(1+\delta)^2}{B_1 M_1}\left(\frac{B_2}{B_1} + 1\right) + \frac{M_2}{2M_1} \quad \text{and} \quad \sigma_3 = \frac{M_2}{2M_1} + \frac{(1+\delta)^2 B_2}{B_1^2 M_1}. \tag{2.14}$$

Further,

$$|a_2| \leq \frac{B_1}{1+\delta}$$

and

$$|a_3| \leq \begin{cases} \dfrac{B_2}{M_1} + \dfrac{B_1^2 M_2}{2(1+\delta)^2 M_1} & \text{when } 2(B_1 - B_2)(1+\delta)^2 \leq B_1^2 M_2, \\[2ex] \dfrac{B_1}{M_1} & \text{when } 2(B_1 - B_2)(1+\delta)^2 \geq B_1^2 M_2 \text{ or} \\ & \quad -2(B_1 + B_2)(1+\delta)^2 \leq B_1^2 M_2, \\[2ex] -\dfrac{B_2}{M_1} - \dfrac{B_1^2 M_2}{2(1+\delta)^2 M_1} & \text{when } -2(B_1 + B_2)(1+\delta)^2 \geq B_1^2 M_2. \end{cases}$$

These estimates are sharp.

Proof Proceeding as in Theorem 2.1, the bounds in inequalities (2.11)–(2.13) can be established by applying Lemma 1.2. For sharpness we define the functions $K_{\Phi n}$: $\mathbb{D} \to \mathbb{C}$ ($n = 2, 3, \dots$), satisfying:

$$1+\frac{1}{\gamma}\left(\left(\frac{zK'_{\Phi n}(z)+\alpha z^2K''_{\Phi n}(z)}{\alpha zK'_{\Phi n}(z)+(1-\alpha)K_{\Phi n}(z)}\right)^k\left(\beta K'_{\Phi n}(z)+(1-\beta)\frac{K_{\Phi n}(z)}{z}\right)^{1-k}-1\right)=\Phi(z^{n-1}),$$

with $K_{\Phi n}(0) = 0$, $K'_{\Phi n}(0) = 1$, H_λ and G_λ ($0 \le \lambda \le 1$) with $H_\lambda(0) = 0$, $H'_\lambda(0) = 1$ and $G_\lambda(0) = 0$, $G'_\lambda(0) = 1$,, respectively, satisfying the following:

$$1+\frac{1}{\gamma}\left(\left(\frac{zH'_\lambda(z)+\alpha z^2H''_\lambda(z)}{\alpha zH'_\lambda(z)+(1-\alpha)H_\lambda(z)}\right)^k\left(\beta H'_\lambda(z)+(1-\beta)\frac{H_\lambda(z)}{z}\right)^{1-k}-1\right)=\Phi\left(\frac{z(\lambda+z)}{1+\lambda z}\right)$$

and

$$1+\frac{1}{\gamma}\left(\left(\frac{zG'_\lambda(z)+\alpha z^2G''_\lambda(z)}{\alpha zG'_\lambda(z)+(1-\alpha)G_\lambda(z)}\right)^k\left(\beta G'_\lambda(z)+(1-\beta)\frac{G_\lambda(z)}{z}\right)^{1-k}-1\right)=\Phi\left(\frac{-z(\lambda+z)}{1+\lambda z}\right).$$

Clearly functions $K_{\Phi n}$, H_λ, $G_\lambda \in \mathcal{S}^k_\delta(\Phi)$. For $\mu < \sigma_1$ or $\mu > \sigma_2$ extremal function for inequality (2.11) is $K_\Phi = K_{\Phi 2}$ or one of its rotations. Extremal function for $\sigma_1 < \mu < \sigma_2$ is K_{Φ_3} or any of its rotations. When $\mu = \sigma_1$, H_λ or any of its rotations works as the extremal function. For $\mu = \sigma_2$ extremal function is G_λ or any of its rotations. Bounds for a_2 and a_3 can be directly obtained from inequality (2.11). ■

Theorem 2.8 *Let f be in the class $S^k_\delta(\Phi)$ and $\Phi(z) = 1 + B_1z + B_2z^2 + \cdots$. Then, for any $\mu \in \mathbb{C}$, we have*

$$|a_3-\mu a_2^2|\le\frac{B_1}{M_1}\max\left\{1;\left|\frac{B_1(2\mu M_1-M_2)}{2(1+\delta)^2}-\frac{B_2}{B_1}\right|\right\}. \tag{2.15}$$

This result is sharp.

Proof Inequality (2.15) can be derived by applying Lemma 1.3. ■

Remark 2.9 Let μ be a real number. When $k = \alpha = 1$, Theorem 2.7 reduces to a result proved in [16, Theorem 3]. When $\beta = 1$ and $\alpha = 0$, Theorem 2.7 coincides with a known result proved in [13, Theorem 2.11]. For $\alpha = \beta = 1$, Theorem 2.7 reduces to give a result derived in [13, Theorem 2.15]. For $\alpha = \beta = 0$, and $\alpha = 1$, $\beta = 0$, Theorem 2.7 reduces to [13, Theorem 2.19] and [13, Theorem 2.23] respectively. Further, all the special cases referred therein also become particular cases of our result.

Theorem 2.10 *Let f be in the class $S^k_\delta(\Phi)$ and $\Phi(z) = 1 + B_1z + B_2z^2 + \cdots$. Then, we have*

$$|a_4|\le\frac{B_1}{J_1}H(q_1,q_2), \tag{2.16}$$

where $H(q_1, q_2)$ is as defined in [24, Lemma 2], with

$$q_1 = \frac{2B_2}{B_1} - \frac{B_1 J_2}{(1+\delta)M_1} \quad and \quad q_2 = \frac{B_3}{B_1} - \frac{B_2 J_2}{M_1(1+\delta)} + \frac{B_1^2 J_3}{6M_1(1+\delta)^3},$$

where J_1, J_2 and J_3 are as defined in Theorem 2.5. This result is sharp.

Proof Using (2.6) with suitable rearrangement of terms, we obtain the following expression for the fourth coefficient:

$$a_4 = \frac{B_1}{J_1}\left(v_3 + \left(\frac{2B_2}{B_1} - \frac{B_1 J_2}{(1+\delta)M_1}\right) v_1 v_2 + \left(\frac{B_3}{B_1} - \frac{B_2 J_2}{M_1(1+\delta)} + \frac{B_1^2 J_3}{6M_1(1+\delta)^3}\right) v_1^3\right).$$

Now by an application of Lemma 2 in [24], we arrive at the bound of fourth coefficient as stated above. ∎

Remark 2.11 When $k = 1$ and $\alpha = 0$, inequality (2.16) reduces to give the inequality in [2, Theorem 1 (for $p = 1$)]. For $k = 0$ and $\beta = 1$, inequality (2.16) of Theorem 2.10 reduces to give the inequality in [2, Theorem 3 (for $p = 1$ and $b = 1$)]. If $k = \alpha = 1$ in the above theorem, we obtain a sharp bound for a_4 for functions in the class $\mathcal{S}_{1,1}^1(\Phi)$, given by:

$$|a_4| \leq \frac{B_1}{12} H(q_1, q_2), \text{ where } q_1 = \frac{4B_2 + 3B_1^2}{2B_1}, q_2 = \frac{2B_3 + 3B_2 B_1 + B_1^3}{2B_1}.$$

Further, if $\Phi(z) = (1+z)/(1-z)$, then $|a_4| \leq 1$ [8], which is a sharp estimate for the class of convex functions.

Proceeding as in the previous results we now establish the coefficient bounds for functions in the class $\mathcal{S}_{\gamma,\delta,h}^k(\Phi)$.

Theorem 2.12 *Let f be in the class $S_{\gamma,\delta,h}^k(\Phi)$ and $\Phi(z) = 1 + B_1 z + B_2 z^2 + \cdots$. Then, for any $\mu \in \mathbb{C}$, we have*

$$|a_3 - \mu a_2^2| \leq \frac{|\gamma| B_1}{h_3 M_1} \max\left\{1; \left|\frac{\gamma B_1(2\mu h_3 M_1 - h_2^2 M_2)}{2h_2^2(1+\delta)^2} - \frac{B_2}{B_1}\right|\right\}. \tag{2.17}$$

Further,

$$|a_2| \leq \frac{|\gamma| B_1}{h_2(1+\delta)} \quad and \quad |a_3| \leq \frac{|\gamma| B_1}{h_3 M_1} \max\left\{1; \left|\frac{B_2}{B_1} + \frac{\gamma h_2^2 B_1 M_2}{2h_2^2(1+\delta)^2}\right|\right\}. \tag{2.18}$$

These estimates are sharp.

Theorem 2.13 *Let f be in the class $S_{\gamma,\delta,h}^k(\Phi)$ and $\Phi(z) = 1 + B_1 z + B_2 z^2 + \cdots$. Then, we have*

$$|a_4| \leq \frac{|\gamma|B_1}{h_4 J_1}\left(H(q_1, q_2) + \frac{|\gamma J_2|B_1}{M_1(1+\delta)}\max\{1; |t|\}\right), \tag{2.19}$$

where $q_1, q_2, H(q_1, q_2), J_1$ *and* J_2 *are as defined in Theorem 2.5.*

Theorem 2.14 *Let* f *be in the class* $S^k_{\delta,h}(\Phi)$ *and* $\Phi(z) = 1 + B_1 z + B_2 z^2 + \cdots$. *Then, we have*

$$|a_3 - \mu a_2^2| \leq \begin{cases} \dfrac{-B_1 t}{h_3 M_1} & \text{when } \mu \leq \sigma_1, \\[2ex] \dfrac{B_1}{h_3 M_1} & \text{when } \sigma_1 \leq \mu \leq \sigma_2, \\[2ex] \dfrac{B_1 t}{h_3 M_1} & \text{when } \mu \geq \sigma_2, \end{cases} \tag{2.20}$$

where

$$t = \frac{B_1(2\mu h_3 M_1 - h_2^2 M_2)}{2h_2^2(1+\delta)^2} - \frac{B_2}{B_1}.$$

Further, if $\sigma_1 \leq \mu \leq \sigma_3$, *then,*

$$|a_3 - \mu a_2^2| + \frac{(1+\delta)^2 h_2^2}{h_3 B_1 M_1}\left\{1 - \frac{B_2}{B_1} + \frac{B_1(2\mu h_3 M_1 - h_2^2 M_2)}{2h_2^2(1+\delta)^2}\right\}|a_2|^2 \leq \frac{B_1}{h_3 M_1}. \tag{2.21}$$

If $\sigma_3 \leq \mu \leq \sigma_2$, *then,*

$$|a_3 - \mu a_2^2| + \frac{(1+\delta)^2 h_2^2}{h_3 B_1 M_1}\left\{1 + \frac{B_2}{B_1} - \frac{B_1(2\mu h_3 M_1 - h_2^2 M_2)}{2h_2^2(1+\delta)^2}\right\}|a_2|^2 \leq \frac{B_1}{h_3 M_1}. \tag{2.22}$$

where

$$\sigma_1 = \frac{h_2^2(1+\delta)^2}{h_3 B_1 M_1}\left(\frac{B_2}{B_1} - 1\right) + \frac{h_2^2 M_2}{2h_3 M_1}, \quad \sigma_2 = \frac{h_2^2(1+\delta)^2}{h_3 B_1 M_1}\left(\frac{B_2}{B_1} + 1\right) + \frac{h_2^2 M_2}{2h_3 M_1} \tag{2.23}$$

$$\text{and} \quad \sigma_3 = \frac{h_2^2 M_2}{2h_3 M_1} + \frac{h_2^2(1+\delta)^2 B_2}{h_3 B_1^2 M_1}. \tag{2.24}$$

Further,

$$|a_2| \leq \frac{B_1}{h_2(1+\delta)}$$

and

$$|a_3| \leq \begin{cases} \dfrac{B_2}{h_3 M_1} + \dfrac{h_2^2 B_1^2 M_2}{2h_2^2 h_3 (1+\delta)^2 M_1} & \text{when } 2h_2^2(1+\delta)^2(B_1 - B_2) \leq h_2^2 B_1^2 M_2, \\ \dfrac{B_1}{h_3 M_1} & \text{when } 2h_2^2(1+\delta)^2(B_1 - B_2) \geq B_1^2 M_2 \text{ or} \\ & \quad -2h_2^2(B_1 + B_2)(1+\delta)^2 \leq B_1^2 M_2, \\ -\dfrac{B_2}{h_3 M_1} - \dfrac{h_2^2 B_1^2 M_2}{2h_2^2 h_3 (1+\delta)^2 M_1} & \text{when } -2h_2^2(B_1 + B_2)(1+\delta)^2 \geq h_2^2 B_1^2 M_2. \end{cases}$$

These estimates are sharp.

Theorem 2.15 *Let f be in the class $S^k_{\delta,h}(\Phi)$ and $\Phi(z) = 1 + B_1 z + B_2 z^2 + \cdots$. Then, for any $\mu \in \mathbb{C}$, we have*

$$|a_3 - \mu a_2^2| \leq \frac{B_1}{h_3 M_1} \max\left\{1; \left|\frac{B_1(2\mu h_3 M_1 - h_2^2 M_2)}{2h_2^2(1+\delta)^2} - \frac{B_2}{B_1}\right|\right\}. \tag{2.25}$$

This result is sharp.

Theorem 2.16 *Let f be in the class $S^k_{\delta,h}(\Phi)$ and $\Phi(z) = 1 + B_1 z + B_2 z^2 + \cdots$. Then, we have*

$$|a_4| \leq \frac{B_1}{h_4 J_1} H(q_1, q_2), \tag{2.26}$$

where q_1, q_2, J_1 and $H(q_1, q_2)$ are as defined in Theorem 2.10. This result is sharp.

References

1. S. Abdul Halim, On a class of functions of complex order. Tamkang J. Math. **30**(2), 147–153 (1999)
2. R.M. Ali, V. Ravichandran, N. Seenivasagan, Coefficient bounds for p-valent functions. Appl. Math. Comput. **187**(1), 35–46 (2007)
3. N.E. Cho, S. Owa, On the Fekete-Szegö problem for strongly α-quasiconvex functions. Tamkang J. Math. **34**(1), 21–28 (2003)
4. N.E. Cho, B. Kowalczyk, A. Lecko, Fekete-Szegö problem for close-to-convex functions with respect to a certain convex function dependent on a real parameter. Front. Math. China **11**(6), 1471–1500 (2016)
5. K.K. Dixit, S.K. Pal, On a class of univalent functions related to complex order. Indian J. Pure Appl. Math. **26**(9), 889–896 (1995)
6. P.L. Duren, *Univalent Functions, Grundlehren der Mathematischen Wissenschaften*, vol. 259 (Springer, New York, 1983)
7. R.M. Goel, B.S. Mehrok, A subclass of univalent functions. J. Austral. Math. Soc. Ser. A **35**(1), 1–17 (1983)
8. A.W. Goodman, *Univalent Functions*, vol. 2 (Mariner Publishing Co. Inc., Tampa, 1983)

9. W. Janowski, Extremal problems for a family of functions with positive real part and for some related families, Ann. Polon. Math. **23**, 159–177 (1970/1971)
10. W. Janowski, Some extremal problems for certain families of analytic functions I. Ann. Polon. Math. **28**, 297–326 (1973)
11. A. Janteng, S.A. Halim, M. Darus, Coefficient inequality for a function whose derivative has a positive real part, JIPAM. J. Inequal. Pure Appl. Math. **7**(2), Article 50, 5 (2006)
12. F.R. Keogh, E.P. Merkes, A coefficient inequality for certain classes of analytic functions. Proc. Am. Math. Soc. **20**, 8–12 (1969)
13. K. Khatter, S.K. Lee, S.S. Kumar, Coefficient bounds for certain analytic functions. Bull. Malays. Math. Sci. Soc. **41**(1), 455–490 (2018)
14. Y.C. Kim, J.H. Choi, T. Sugawa, Coefficient bounds and convolution properties for certain classes of close-to-convex functions. Proc. Japan Acad. Ser. A Math. Sci. **76**(6), 95–98 (2000)
15. O.S. Kwon, N.E. Cho, On the Fekete-Szegö problem for certain analytic functions. J. Korea Soc. Math. Educ. Ser. B Pure Appl. Math. **10**(4), 265–271 (2003)
16. W.C. Ma, D. Minda, A unified treatment of some special classes of univalent functions, in *Proceedings of the Conference on Complex Analysis, Tianjin,1992, Lecture Notes Analysis. I* (International Press, Cambridge, 1992), pp. 157–169
17. T.H. MacGregor, Functions whose derivative has a positive real part. Trans. Am. Math. Soc. **104**, 532–537 (1962)
18. G. Murugusundaramoorthy, H.M. Srivastava, Neighborhoods of certain classes of analytic functions of complex order, JIPAM. J. Inequal. Pure Appl. Math. **5**(2), Article 24, 8 (2004)
19. M.A. Nasr, M.K. Aouf, Bounded starlike functions of complex order. Proc. Indian Acad. Sci. Math. Sci. **92**(2), 97–102 (1983)
20. M.A. Nasr, M.K. Aouf, Radius of convexity for the class of starlike functions of complex order. Bull. Fac. Sci. Assiut Univ. A **12**(1), 153–159 (1983)
21. M.A. Nasr, M.K. Aouf, Starlike function of complex order. J. Nat. Sci. Math. **25**(1), 1–12 (1985)
22. H. Orhan, The Fekete-Szegö inequality for a subclass of analytic functions involving Hadamard product, Mathematica **49**(72), 55–61 (2007)
23. C. Pommerenke, *Univalent Functions* (Vandenhoeck & Ruprecht, Göttingen, 1975)
24. D.V. Prokhorov, J. Szynal, Inverse coefficients for (α, β)-convex functions. Ann. Univ. Mariae Curie-Skłodowska Sect. A **35**(1981), 125–143 (1984)
25. V. Ravichandran, M Darus, M.H. Khan, K.G. Subramanian, Fekete-Szegö inequality for certain class of analytic functions, Aust. J. Math. Anal. Appl. **1**(2), Article 4, 7 (2004)
26. M.I.S. Robertson, On the theory of univalent functions. Ann. Math. **37**(2), 374–408 (1936)
27. H.M. Srivastava, A.Y. Lashin, Some applications of the Briot-Bouquet differential subordination, JIPAM. J. Inequal. Pure Appl. Math. **6**(2), Article 41, 7 (2005)
28. H.M. Srivastava, S. Owa, *Current Topics in Analytic Function Theory*, (World Scientific Publishing Co., Inc., River Edge, 1992)
29. H.M. Srivastava, O. Altıntaş, S.K. Serenbay, Coefficient bounds for certain subclasses of starlike functions of complex order. Appl. Math. Lett. **24**(8), 1359–1363 (2011)
30. H.M. Srivastava, A.K. Mishra, P. Gochhayat, Certain subclasses of analytic and bi-univalent functions. Appl. Math. Lett. **23**(10), 1188–1192 (2010)
31. P. Wiatrowski, The coefficients of a certain family of holomorphic functions. Zeszyty Nauk. Uniw. Łódz. Nauki Mat. Przyrod. Ser. II. **39**, 75–85 (1971)

Bohr Radius for Certain Analytic Functions

Naveen Kumar Jain and Shalu Yadav

Abstract For an analytic self-mapping $f(z) = \sum_{n=0}^{\infty} a_n z^n$ of the unit disk $\mathbb{D}$, it is well-known that $\sum_{n=0}^{\infty} |a_n|\,|z|^n \leq 1$ for $|z| \leq 1/3$ and the number 1/3, known as the Bohr radius for the class of analytic self-mappings of $\mathbb{D}$, is sharp. We have obtained the Bohr radius for the class of α-spiral functions of order ρ and the Bohr radius for the class of analytic functions f defined on the unit disk satisfying the differential subordination $f(z) + \beta z f'(z) + \gamma z^2 f''(z) \prec h(z)$.

Keywords Bohr radius · Convex function · Starlike function · Spiral function · Convolution

2010 Mathematics Subject Classification 30A10 · 32A05 · 30C80 · 30C45

1 Introduction

Let $\mathbb{D} := \{z \in \mathbb{C} : |z| < 1\}$ denote the unit disk in $\mathbb{C}$ and $\mathcal{H}$ be the class of analytic functions defined in $\mathbb{D}$. Bohr [8] in 1914 proved that if $f(z) = \sum_{n=0}^{\infty} a_n z^n$ is analytic in $\mathbb{D}$ and $|f(z)| \leq 1$ for all $z \in \mathbb{D}$, then $\sum_{n=0}^{\infty} |a_n z^n| \leq 1$ in the disk $|z| \leq k$ where $k \geq 1/6$. Bohr [8] pointed out in his paper that the exact value of k was determined by Wiener, Riesz, and Schur independently. He has also reproduced Wiener's proof that $k = 1/3$. The number $1/3$ is known as the Bohr radius for the class of analytic functions f defined on $\mathbb{D}$ and satisfying $|f(z)| \leq 1$. Other proofs on the Bohr radius can be found in [15–17, 20, 22]. The Bohr's theorem received much attention after

N. K. Jain (✉)
Department of Mathematics, Aryabhatta College, Delhi 110021, India
e-mail: naveenjain05@gmail.com

S. Yadav
Department of Mathematics, University of Delhi, Delhi 110007, India
e-mail: shaluyadav903@gmail.com

N. Deo et al. (eds.), *Mathematical Analysis I: Approximation Theory*,
Springer Proceedings in Mathematics & Statistics 306,
https://doi.org/10.1007/978-981-15-1153-0_18

the work by Dixon [9]. He showed a connection between the inequality and characterization of Banach algebras that satisfy von Neumann's inequality. Many authors have generalized Bohr's theorem; Aizenberg, Aytuna and Djakov [2, 5] studied the Bohr's property of bases for holomorphic function and Ali, Abdulhadi and Ng [4] found the Bohr radius for the class of starlike logharmonic mappings. Paulsen and Singh [15] extended Bohr's inequality to Banach algebras. The relation between Banach theory and Bohr's theorem was explored in [7, 10, 11].

The inequality $\sum_{n=0}^{\infty}|a_n z^n| \leq 1$ is known as the Bohr's inequality. The Bohr's inequality can be written in the distance formulation as

$$d\left(\sum_{n=0}^{\infty}|a_n z^n|, |a_0|\right) = \sum_{n=1}^{\infty}|a_n z^n| \leq 1 - |a_0| = 1 - |f(0)| = d(f(0), \partial\mathbb{D}),$$

where d is the Euclidean distance, $\partial\mathbb{D}$ is the boundary of the unit disc. In this form, the notion of Bohr's inequality can be generalized to the class of analytic functions which maps the unit disk onto a domain Ω as follows:

$$d\left(\sum_{n=0}^{\infty}|a_n z^n|, |a_0|\right) \leq d(f(0), \partial\Omega). \tag{1.1}$$

The Bohr radius for a class $\mathcal{B}$ of analytic functions $f(z) = \sum_{n=0}^{\infty} a_n z^n$ mapping the unit disk into a domain Ω is the largest radius $r^* \in (0, 1]$ such that every function $f \in \mathcal{B}$ satisfies the inequality (1.1) for all $z \in \mathbb{D}_r := \{z \in \mathbb{C} : |z| < r\}$ for every $0 < r \leq r^*$.

This paper studies the Bohr radius for the well-known classes of analytic functions which includes the familiar classes consisting of starlike, convex, and close-to-convex functions. Let $\mathcal{A}$ denote the subclass of $\mathcal{H}$ consisting of functions normalized by $f(0) = 0 = f'(0) - 1$ and let $\mathcal{S}$ denotes the subclass of $\mathcal{A}$ consisting of univalent functions. For $0 \leq \alpha < 1$, let $\mathcal{ST}(\alpha)$ and $\mathcal{CV}(\alpha)$ be the subclasses of $\mathcal{A}$ consisting of functions starlike and convex of order α, respectively. The classes $\mathcal{ST} = \mathcal{ST}(0)$ and $\mathcal{CV} = \mathcal{CV}(0)$ are, respectively, the classes of starlike and convex functions. In Sect. 2, we study the Bohr radius for the class $\mathcal{SP}(\alpha, \rho)$ of α-spiral-like functions of order ρ ($|\alpha| < \pi/2, 0 \leq \rho < 1$), introduced by Libera in 1967, and defined by

$$\mathcal{SP}(\alpha, \rho) = \left\{ f \in \mathcal{A} : \operatorname{Re}\left(e^{i\alpha}\frac{zf'(z)}{f(z)}\right) > \rho\cos\alpha \right\}.$$

In Sect. 3, the Bohr radius is found for the class $\mathcal{R}(\beta, \gamma, h)$ defined as

$$\mathcal{R}(\beta, \gamma, h) := \{f \in \mathcal{A} : f(z) + \beta z f'(z) + \gamma z^2 f''(z) \prec h(z), z \in \mathbb{D}\}$$

where $\beta \geq \gamma \geq 0$, and h is an analytic convex (starlike) function of order α ($\alpha < 1$).

2 Bohr Radius for the Class of α-Spiral Functions of Order ρ

For $|\alpha| < \frac{\pi}{2}, 0 \leq \rho < 1$, let

$$\mathcal{SP}(\alpha, \rho) = \left\{ f \in \mathcal{S} : \operatorname{Re}\left(e^{i\alpha} \frac{zf'(z)}{f(z)} \right) > \rho \cos\alpha \right\}.$$

In the present section, we find the Bohr radius for the class $\mathcal{SP}(\alpha, \rho)$.

Lemma 2.1 *If $f \in \mathcal{SP}(\alpha, \rho)$, then*

$$d(0, \partial f(\mathbb{D})) \geq (4\cos^2\alpha)^{-(1-\rho)\cos^2\alpha} \exp(-\alpha(1-\rho)\sin 2\alpha).$$

The result is sharp.

Proof By [12, Theorem 1, p. 3], for $|z| \leq r$,

$$\log\left|\frac{f(z)}{z}\right| \geq \phi(r),$$

where

$$\begin{aligned}\phi(r) =& -(1-\rho)\log(1 + 2r\sqrt{1 - r^2\sin^2\alpha}\cos\alpha + r^2\cos 2\alpha)\cos^2\alpha \\ & - 2(1-\rho)\arctan\left[\frac{r(\sqrt{1-r^2\sin^2\alpha} + r\cos\alpha)\sin\alpha}{1 + r(\sqrt{1-r^2\sin^2\alpha})\cos\alpha - r^2\sin^2\alpha}\right]\cos\alpha\sin\alpha.\end{aligned}$$

It gives

$$\left|\frac{f(z)}{z}\right| \geq \exp(\phi(r))$$

so that

$$\lim_{|z|\to 1^-} |f(z)| \geq \lim_{r\to 1^-} \exp(\phi(r)) = \exp(\lim_{r\to 1^-} \phi(r)). \tag{2.1}$$

Now

$$\begin{aligned}\lim_{r\to 1^-} \phi(r) =& -(1-\rho)\log(1 + 2\cos^2\alpha + \cos 2\alpha)\cos^2\alpha \\ & - 2(1-\rho)\arctan\left(\frac{2\sin\alpha\cos\alpha}{1+\cos^2\alpha - \sin^2\alpha}\right)\cos\alpha\sin\alpha \\ =& -(1-\rho)\log(4\cos^2\alpha)\cos^2\alpha - 2(1-\rho)\arctan(\tan\alpha)\cos\alpha\sin\alpha \\ =& -(1-\rho)\log(4\cos^2\alpha)\cos^2\alpha - 2(1-\rho)\alpha\cos\alpha\sin\alpha \\ =& -(1-\rho)\log(4\cos^2\alpha)\cos^2\alpha - (1-\rho)\alpha\sin 2\alpha. \end{aligned} \tag{2.2}$$

Equations (2.1) and (2.2) yield

$$\begin{aligned} d(0, \partial f(\mathbb{D})) &\geq \exp(-(1-\rho)\log(4\cos^2\alpha)\cos^2\alpha - (1-\rho)\alpha\sin 2\alpha) \\ &= (4\cos^2\alpha)^{-(1-\rho)\cos^2\alpha}\exp(-\alpha(1-\rho)\sin 2\alpha). \end{aligned}$$

The result is sharp for the function

$$f(z) = \frac{z}{(1-z)^{2s(1-\rho)}}, \quad s = \exp(-i\alpha\cos\alpha).$$

□

Theorem 2.2 *Let* $f \in \mathcal{SP}(\alpha, \rho)$, $0 \leq \rho < 1$ *and* $f(z) = z + \sum_{n=2}^{\infty} a_n z^n$ *then*

$$|z| + \sum_{n=2}^{\infty} |a_n z^n| \leq d(0, \partial f(\mathbb{D}))$$

for $|z| < r^*$, *where* $r^* \in (0, 1]$ *is given by the root of the equation*

$$r + \sum_{n=2}^{\infty}\left(\prod_{k=1}^{n-1}\frac{((k-1)^2 + 4(1-\rho)(k-\rho)\cos^2\alpha)^{1/2}}{k}\right) r^n = (4\cos^2\alpha)^{-(1-\rho)\cos^2\alpha}\exp(-\alpha(1-\rho)\sin 2\alpha).$$

The result is sharp.

Proof By [13], we have,

$$\begin{aligned} |z| + \sum_{n=2}^{\infty}|a_n z^n| &\leq r + \sum_{n=2}^{\infty}\left(\prod_{k=1}^{n-1}\frac{|k + 2s(1-\rho) - 1|}{k}\right) r^n, s = \exp(-i\alpha\cos\alpha) \\ &= r + \sum_{n=2}^{\infty}\left(\prod_{k=1}^{n-1}\frac{((k-1)^2 + 4(1-\rho)(k-\rho)\cos^2\alpha)^{1/2}}{k}\right) r^n. \end{aligned}$$

It follows from Lemma 2.1 that

$$(4\cos^2\alpha)^{-(1-\rho)\cos^2\alpha}\exp(-\alpha(1-\rho)\sin 2\alpha) \leq d(0, \partial f(\mathbb{D})). \tag{2.3}$$

Thus,

$$|z| + \sum_{n=2}^{\infty}|a_n z^n| \leq d(0, \partial f(\mathbb{D}))$$

if

$$r+\sum_{n=2}^{\infty}\left(\prod_{k=1}^{n-1}\frac{((k-1)^2+4(1-\rho)\cos^2\alpha(k-\rho))^{1/2}}{k}\right)r^n$$
$$\leq (4\cos^2\alpha)^{-(1-\rho)\cos^2\alpha}\exp(-(1-\rho)\alpha\sin 2\alpha),$$

which provides the required radius.

To prove the sharpness, consider the function $f(z)=z/(1-z)^{2s(1-\rho)}$, $s=\exp(-i\alpha)\cos\alpha$. For $|z|=r^*$,

$$\begin{aligned}|z|+\sum_{n=2}^{\infty}|a_nz^n| &= r^*+\sum_{n=2}^{\infty}\left(\prod_{k=1}^{n-1}\frac{|k+2s(1-\rho)-1|}{k}\right)(r^*)^n\\ &= r^*+\sum_{n=2}^{\infty}\left(\prod_{k=1}^{n-1}\frac{((k-1)^2+4(1-\rho)(k-\rho)\cos^2\alpha)^{1/2}}{k}\right)(r^*)^n\\ &= (4\cos^2\alpha)^{-(1-\rho)\cos^2\alpha}\exp(-\alpha(1-\rho)\sin 2\alpha)\\ &= d(0,\partial f(\mathbb{D})).\end{aligned}$$

□

For $|\alpha|<\pi/2$, let

$$\mathcal{SP}(\alpha)=\left\{f\in\mathcal{S}:\mathrm{Re}\left(\exp(i\alpha)\frac{zf'(z)}{f(z)}\right)>0\right\}.$$

It is easy to see that putting $\rho=0$ in Theorem 2.2, leads the Bohr radius for the class $\mathcal{SP}(\alpha)$.

Corollary 2.3 *Let $f\in\mathcal{SP}(\alpha)$, where $f(z)=z+\sum_{n=2}^{\infty}a_nz^n$. Then*

$$|z|+\sum_{n=2}^{\infty}|a_nz^n|<d(0,\partial f(\mathbb{D})),\ |z|<r^*$$

if $r^*\in(0,1]$ is the smallest root of the equation

$$r+\sum_{n=2}^{\infty}\left(\prod_{k=1}^{n-1}\frac{((k-1)^2+4k\cos^2\alpha)^{1/2}}{k}\right)r^n=(4\cos^2\alpha)^{-\cos^2\alpha}\exp-\alpha\sin 2\alpha).$$

The result is sharp.

Remark 2.1 Bhowmik et al. [6, Theorem 3, p. 1093] obtained the Bohr radius for the class of starlike functions of order $\alpha(0\leq\alpha\leq 1/2)$. Putting $\alpha=0$ in Theorem 2.2 and replacing ρ by α, we obtain the Bohr radius for the class $\mathcal{ST}(\alpha)(0\leq\alpha<1)$.

Corollary 2.4 *Let* $f \in \mathcal{ST}(\alpha)$ *and* $f(z) = z + \sum_{n=2}^{\infty} a_n z^n$. *Then*

$$|z| + \sum_{n=2}^{\infty} |a_n z^n| \leq d(0, \partial f(\mathbb{D})), \qquad for \ \ |z| \leq r^* \tag{2.4}$$

where $r^* \in (0, 1]$ *is the root of the equation*

$$(1-r)^{2(1-\alpha)} - 2^{2(1-\alpha)} r = 0.$$

The result is sharp.

For $\alpha = 0$, we obtain the Bohr radius for the class of starlike functions.

Corollary 2.5 *Let* $f \in \mathcal{ST}$ *and* $f(z) = z + \sum_{n=2}^{\infty} a_n z^n$. *Then the Bohr radius is* $3 - 2\sqrt{2}$. *The result is sharp.*

3 Bohr Radius for Second-Order Differential Subordination

For $\beta \geq 0$, define

$$\mathcal{R}(\beta, h) := \{f \in \mathcal{A} : f(z) + \beta z f'(z) \prec h(z), z \in \mathbb{D}\}.$$

Many authors have studied the class for some analytic function h. The results for the classes can be found in [21, 23].

In this section, we study an extension of the above class. Let

$$\mathcal{R}(\beta, \gamma, h) := \{f \in \mathcal{A} : f(z) + \beta z f'(z) + \gamma z^2 f''(z) \prec h(z), z \in \mathbb{D}\}$$

where $\beta \geq \gamma \geq 0$, and h is an analytic convex (starlike) function of order α, ($\alpha \leq 1$). Ali et al. [3] have shown that $f(z) \prec h(z)$ whenever $f \in \mathcal{R}(\beta, \gamma, h)$. Muhanna et al.[1] found the Bohr radius for the class $\mathcal{R}(\beta, \gamma, h)$ when h is convex or starlike, respectively.

For two analytic functions $f(z) = \sum_{n=0}^{\infty} a_n z^n$ and $g(z) = \sum_{n=0}^{\infty} b_n z^n$, their Hadamard product (or convolution) is the function $f * g$ defined by

$$(f * g)(z) = \sum_{n=0}^{\infty} a_n b_n z^n.$$

Define a function

$$\phi_\lambda(z) = \int_0^1 \frac{dt}{1 - zt^\lambda} = \sum_{n=0}^{\infty} \frac{z^n}{1 + \lambda n}.$$

Ruscheweyh [19] has shown that ϕ_λ is convex in $\mathbb{D}$ provided Re $\lambda \geq 0$.

For $\beta \geq \lambda \geq 0$, let

$$\nu + \mu = \beta - \gamma, \quad \nu\mu = \gamma,$$

and

$$q(z) = \int_0^1 \int_0^1 h(zt^\mu s^\mu) dt ds = (\phi_\nu * \phi_\mu) * h(z). \tag{3.1}$$

Since $\phi_\lambda * \phi_\mu$ is a convex function and h is a convex function of order α, from [14, Theorem 5, p.167], it follows that q is a convex function of order α. It is easy to see that $q \in \mathcal{R}(\beta, \gamma, h)$. It was shown by Ali et al. [3] that

$$f(z) \prec q(z) \prec h(z)$$

for all $f \in \mathcal{R}(\beta, \gamma, h)$. Thus, $\mathcal{R}(\beta, \gamma, h) \subset \mathcal{S}(h)$.

Theorem 3.1 *Let $f(z) = \sum_{n=0}^{\infty} a_n z^n \in \mathcal{R}(\beta, \gamma, h)$, and h be convex of order α. Then*

$$\sum_{n=1}^{\infty} |a_n z^n| \leq d(h(0), \partial h(\mathbb{D}))$$

for all $|z| \leq r^*$, where $r^* \in (0, 1]$ is the smallest positive root of the equation

$$\frac{r}{1 + (\mu + \nu)n + \mu\nu n^2} + \sum_{n=2}^{\infty} \frac{\frac{1}{n!}\prod_{k=2}^{n}(k - 2\alpha)}{1 + (\mu + \nu)n + \mu\nu n^2} r^n = -l_\alpha(-1)$$

The result is sharp.

Proof Let $F \in \mathcal{R}(\beta, \gamma, h)$. Then

$$F(z) = f(z) + \beta z f'(z) + \gamma z^2 f''(z) = \sum_{n=0}^{\infty} [1 + \beta n + \gamma n(n-1)] a_n z^n \prec h(z) \; (a_1 = 1)$$

and

$$\frac{1}{h'(0)} \sum_{n=1}^{\infty} [1 + \beta n + \gamma n(n-1)] a_n z^n = \frac{F(z) - F(0)}{h'(0)} \prec \frac{h(z) - h(0)}{h'(0)} = H(z).$$

Thus, by [18], we have

$$\left| \frac{1 + \beta n + \gamma n(n-1)}{h'(0)} \right| |a_n| \leq \frac{1}{n!} \prod_{k=2}^{n} (k - 2\alpha), n \geq 2,$$

which yields

$$\sum_{n=1}^{\infty}|a_n|r^n \leq \frac{h'(0)r}{1+(\mu+\nu)n+\mu\nu n^2} + \sum_{n=2}^{\infty}\frac{|h'(0)|\frac{1}{n!}\prod_{k=2}^{n}(k-2\alpha)}{1+(\mu+\nu)n+\mu\nu n^2}r^n. \tag{3.2}$$

The function $l_\alpha : \mathbb{D} \to \mathbb{C}$ given by

$$l_\alpha(z) = \begin{cases} \frac{1-(1-z)^{2\alpha-1}}{2\alpha-1}, & \alpha \neq \frac{1}{2}, \\ -\ln(1-z), & \alpha = \frac{1}{2} \end{cases} \tag{3.3}$$

is an extremal function for the class $\mathcal{CV}(\alpha)$. Since

$$H(z) = \frac{h(z)-h(0)}{h'(0)}$$

is a normalized convex function of order α in $\mathbb{D}$, it follows that the function l_α provides the case of equality in the following growth inequality satisfied by convex functions of order α:

$$-l_\alpha(-r) \leq |H(re^{i\theta})| \leq l_\alpha(r).$$

So that

$$d(0, \partial H(\mathbb{D})) \geq -l_\alpha(-1),$$

which yields

$$d(h(0), \partial h(\mathbb{D})) = \inf_{\zeta\in\partial\mathbb{D}} |h(\zeta)-h(0)| \geq -|h'(0)|l_\alpha(-1). \tag{3.4}$$

By (2.4) and (3.4), it follows that

$$\sum_{n=1}^{\infty}|a_n|r^n \leq -\frac{d(h(0),\partial h(\mathbb{D}))}{l_\alpha(-1)}\left(\frac{r}{1+(\mu+\nu)n+\mu\nu n^2} + \sum_{n=2}^{\infty}\frac{\frac{1}{n!}\prod_{k=2}^{n}(k-2\alpha)}{1+(\mu+\nu)n+\mu\nu n^2}r^n\right).$$

Thus, the Bohr radius r^* is the smallest positive root of the equation given by

$$\frac{r}{1+(\mu+\nu)n+\mu\nu n^2} + \sum_{n=2}^{\infty}\frac{\frac{1}{n!}\prod_{k=2}^{n}(k-2\alpha)}{1+(\mu+\nu)n+\mu\nu n^2}r^n = -l_\alpha(-1).$$

The result is sharp for the function $f(z) := q(z)$ and $h(z) := l(z) = l_\alpha(z)$, where $q(z)$ and $l_\alpha(z)$ are, respectively, as defined in (3.1) and (3.3), that is

$$f(z) = q(z) = (\phi_\nu * \phi_\mu) * h(z) = \frac{z}{1+(\mu+\nu)n+\mu\nu n^2} + \sum_{n=2}^{\infty}\frac{\frac{1}{n!}\prod_{k=2}^{n}(k-2\alpha)}{1+(\mu+\nu)n+\mu\nu n^2}z^n.$$

For $|z| = r^*$,

$$\begin{aligned}\sum_{n=1}^{\infty}|a_n z^n| &= \frac{r^*}{1+(\mu+\nu)n+\mu\nu n^2}+\sum_{n=2}^{\infty}\frac{\frac{1}{n!}\prod_{k=2}^{n}(k-2\alpha)}{1+(\mu+\nu)n+\mu\nu n^2}(r^*)^n\\ &= -l_\alpha(-1)\\ &= -|h'(0)|l_\alpha(-1)\\ &= d(h(0),\partial h(\mathbb{D})).\end{aligned}$$

□

For $\alpha = 0$, Theorem 3.1 reduces to the following result.

Corollary 3.2 ([1, Theorem 3.1, p. 129]) *Let* $f(z) = \sum_{n=0}^{\infty} a_n z^n \in \mathcal{R}(\beta,\gamma,h)$, *and* h *be convex. Then*

$$\sum_{n=1}^{\infty}|a_n z^n| \le d(h(0),\partial h(\mathbb{D}))$$

for all $|z| \le r^*$, *where* $r^* \in (0,1]$ *is the smallest positive root of the equation*

$$\sum_{n=1}^{\infty}\frac{r^n}{1+(\mu+\nu)n+\mu\nu n^2}=\frac{1}{2}.$$

Theorem 3.3 *Let* $f(z) = \sum_{n=0}^{\infty} a_n z^n \in \mathcal{R}(\beta,\gamma,h)$, *and* h *be starlike of order* α. *Then*

$$\sum_{n=1}^{\infty}|a_n z^n| \le d(h(0),\partial h(\mathbb{D}))$$

for all $|z| \le r^*$, *where* $r^* \in (0,1]$ *is the smallest positive root of the equation*

$$\frac{r}{1+(\mu+\nu)n+\mu\nu n^2}+\sum_{n=2}^{\infty}\frac{\frac{1}{(n-1)!}\prod_{k=2}^{n}(k-2\alpha)}{1+(\mu+\nu)n+\mu\nu n^2}r^n=\frac{1}{2^{2(1-\alpha)}}.$$

The result is sharp.

Proof Let $F \in \mathcal{R}(\beta,\gamma,h)$. Then

$$F(z) = f(z)+\beta z f'(z)+\gamma z^2 f''(z) = \sum_{n=0}^{\infty}[1+\beta n+\gamma n(n-1)]a_n z^n \prec h(z),\ (a_1=1)$$

and

$$\frac{1}{h'(0)}\sum_{n=1}^{\infty}[1+\beta n+\gamma n(n-1)]a_n z^n = \frac{F(z)-F(0)}{h'(0)} \prec \frac{h(z)-h(0)}{h'(0)} = H(z).$$

Since $H(z)$ is a normalized starlike function of order α in $\mathbb{D}$, by [18], we have

$$\left|\frac{1+\beta n+\gamma n(n-1)}{h'(0)}\right| |a_n| \leq \frac{1}{(n-1)!}\prod_{k=2}^{n}(k-2\alpha),\ n\geq 2,$$

which yields

$$\sum_{n=1}^{\infty}|a_n|r^n \leq \left|\frac{h'(0)}{1+(\mu+\nu)n+\mu\nu n^2}\right| r+\sum_{n=2}^{\infty}\frac{|h'(0)|\frac{1}{(n-1)!}\prod_{k=2}^{n}(k-2\alpha)}{1+(\mu+\nu)n+\mu\nu n^2}r^n. \tag{3.5}$$

It follows from [18], that

$$d(0,\partial H(\mathbb{D}))\geq \frac{1}{2^{2(1-\alpha)}},$$

which yields

$$d(h(0),\partial h(\mathbb{D}))=\inf_{\zeta\in\partial\mathbb{D}}|h(\zeta)-h(0)|\geq \frac{|h'(0)|}{2^{2(1-\alpha)}}. \tag{3.6}$$

By (3.5) and (3.6), it follows that

$$\sum_{n=1}^{\infty}|a_n|r^n \leq d(h(0),\partial h(\mathbb{D}))2^{2(1-\alpha)}\left(\frac{r}{1+(\mu+\nu)n+\mu\nu n^2}+\sum_{n=2}^{\infty}\frac{\frac{1}{(n-1)!}\prod_{k=2}^{n}(k-2\alpha)}{1+(\mu+\nu)n+\mu\nu n^2}r^n\right).$$

Thus, the Bohr radius r^* is the smallest positive root of the equation is

$$\frac{r}{1+(\mu+\nu)n+\mu\nu n^2}+\sum_{n=2}^{\infty}\frac{\frac{1}{(n-1)!}\prod_{k=2}^{n}(k-2\alpha)}{1+(\mu+\nu)n+\mu\nu n^2}r^n=\frac{1}{2^{2(1-\alpha)}}.$$

The result is sharp for the function $f(z):=q(z)$, where $q(z)$ as defined in (3.1), and $h(z):=l(z)=z/(1-z)^{2(1-\alpha)}$. Sharpness can be proved as in Theorem 3.1.

□

For $\alpha=0$, Theorem 3.3 reduces to the following result.

Corollary 3.4 ([1, Theorem 3.3, p. 131]) *Let* $f(z)=\sum_{n=0}^{\infty}a_nz^n\in\mathcal{R}(\beta,\gamma,h)$, *and* h *be starlike. Then*

$$\sum_{n=1}^{\infty}|a_nz^n|\leq d(h(0),\partial h(\mathbb{D}))$$

for all $|z|\leq r^*$, *where* $r^*\in(0,1]$ *is the smallest positive root of the equation*

$$\sum_{n=1}^{\infty}\frac{n}{1+(\mu+\nu)n+\mu\nu n^2}r^n=\frac{1}{4}.$$

The result is sharp.

References

1. Y. Abu Muhanna, R.M. Ali, Z.C. Ng, S.F.M. Hasni, Bohr radius for subordinating families of analytic functions and bounded harmonic mappings. J. Math. Anal. Appl. **420**(1), 124–136 (2014)
2. L. Aizenberg, A. Aytuna, P. Djakov, Generalization of a theorem of Bohr for bases in spaces of holomorphic functions of several complex variables. J. Math. Anal. Appl. **258**(2), 429–447 (2001)
3. R.M. Ali, S.K. Lee, K.G. Subramanian, A. Swaminathan, A third-order differential equation and starlikeness of a double integral operator. Abstr. Appl. Anal. **2011** Art. ID 901235, 10
4. R.M. Ali, Z. Abdulhadi, Z.C. Ng, The Bohr radius for starlike logharmonic mappings. Complex Var. Elliptic Equ. **61**(1), 1–14 (2016)
5. A. Aytuna, P. Djakov, Bohr property of bases in the space of entire functions and its generalizations. Bull. Lond. Math. Soc. **45**(2), 411–420 (2013)
6. B. Bhowmik, N. Das, Bohr phenomenon for subordinating families of certain univalent functions. J. Math. Anal. Appl. **462**(2), 1087–1098 (2018)
7. O. Blasco, The Bohr radius of a Banach space, in *Vector Measures, Integration And Related Topics* Operator Theory: Advances and Applications, vol. 201 (Birkhäuser, Basel, pp. 59–64)
8. H. Bohr, A theorem concerning power series. Proc. London Math. Soc. **13**(2), 1–5 (1914)
9. P.G. Dixon, Banach algebras satisfying the non-unital von Neumann inequality. Bull. London Math. Soc. **27**(4), 359–362 (1995)
10. A. Defant, C. Prengel, Harald Bohr meets Stefan Banach, in *Methods in Banach space theory*, London Mathematical Society Lecture Note Series, vol. 337 (Cambridge University Press, Cambridge) pp. 317–339
11. A. Defant, D. García, M. Maestre, Bohr's power series theorem and local Banach space theory. J. Reine Angew. Math. **557**, 173–197 (2003)
12. P.K. Kulshrestha, Distortion of spiral-like mappings. Proc. Roy. Irish Acad. Sect. A **73**, 1–5 (1973)
13. R.J. Libera, Univalent α-spiral functions. Canad. J. Math. **19**, 449–456 (1967)
14. W. C. Ma and D. Minda, A unified treatment of some special classes of univalent functions, in *Proceedings of the Conference on Complex Analysis Tianjin*, Proceedings of the Lecture Notes Analyses, I. (International Press, Cambridge, 1992), pp. 157–169
15. V.I. Paulsen, G. Popescu, D. Singh, On Bohr's inequality. Proc. London Math. Soc. (3) **85**(2), 493–512 (2002)
16. V.I. Paulsen, D. Singh, Bohr's inequality for uniform algebras. Proc. Am. Math. Soc. **132**(12), 3577–3579 (2004)
17. V.I. Paulsen, D. Singh, Extensions of Bohr's inequality. Bull. London Math. Soc. **38**(6), 991–999 (2006)
18. M.I.S. Robertson, On the theory of univalent functions. Ann. Math. (2) **37**(2), 374–408 (1936)
19. St. Ruscheweyh, T. Sheil-Small, Hadamard products of Schlicht functions and the Pólya-Schoenberg conjecture. Comment. Math. Helv. **48** (1973), 119–135
20. S. Sidon, Über einen Satz von Herrn Bohr. Math. Z. **26**(1), 731–732 (1927)
21. R. Szász, The sharp version of a criterion for starlikeness related to the operator of Alexander. Ann. Polon. Math. **94**(1), 1–14 (2008)
22. M. Tomić, Sur un théorème de H. Bohr, Math. Scand. **11**, 103–106 (1962)
23. D.-G. Yang, J.-L. Liu, On a class of analytic functions with missing coefficients. Appl. Math. Comput. **215**(9), 3473–3481 (2010)

Third Hankel Determinant for Certain Classes of Analytic Functions

Virendra Kumar, Sushil Kumar and V. Ravichandran

Abstract There exists a rich literature on the Hankel determinants in the field of geometric function theory. Particularly, it is not easy to find out the sharp bound on the third Hankel determinant as compared to calculate the sharp bound on the second Hankel determinant. The present paper is an attempt to improve certain existing bound on the third Hankel determinant for some classes of analytic functions by using the concept of subordination.

Keywords Analytic functions · Hankel determinant · Starlike functions w.r.t. symmetric points

2010 Mathematics Subject Classification 30C45 · 30C80 · 30C50

1 Introduction

The class $\mathcal{S}$ consists of univalent analytic functions f defined on the unit disk $\mathbb{D} := \{z \in \mathbb{C} : |z| < 1\}$ normalized by the conditions $f(0) = 0$ and $f'(0) = 1$. The subclasses starlike and convex functions of $\mathcal{S}$ are denoted by $\mathcal{S}^*$ and $\mathcal{K}$, respectively. Analytically, these classes are defined by $\mathcal{S}^* := \{f \in \mathcal{S} : \mathrm{Re}(zf'(z)/f(z)) > 0, z \in \mathbb{D}\}$ and $\mathcal{K} := \{f \in \mathcal{S} : \mathrm{Re}(1 + (zf''(z)/f'(z))) > 0, z \in \mathbb{D}\}$. Let $\mathcal{S}_s^*$ denote the class of starlike univalent functions with respect to symmetric points, introduced and

V. Kumar (✉)
Department of Mathematics, Ramanujan College, University of Delhi, Delhi 110019, India
e-mail: vktmaths@yahoo.in

S. Kumar
Bharati Vidyapeeth's College of Engineering, Delhi 110063, India
e-mail: sushilkumar16n@gmail.com

V. Ravichandran
Department of Mathematics, National Institute of Technology, Tiruchirappalli 620015, India
e-mail: ravic@nitt.edu; vravi68@gmail.com

Department of Mathematics, University of Delhi, Delhi 110007, India

N. Deo et al. (eds.), *Mathematical Analysis I: Approximation Theory*,
Springer Proceedings in Mathematics & Statistics 306,
https://doi.org/10.1007/978-981-15-1153-0_19

studied by Sakaguchi [18]. Analytically, $f \in \mathcal{S}_s^*$ if and only if $\mathrm{Re}\,\big(2zf'(z)/(f(z) - f(-z))\big) > 0$ for all $z \in \mathbb{D}$. In similar fashion, Das and Singh [5] introduced and investigated the class $\mathcal{K}_s$ which consists of the convex univalent functions with respect to symmetric points. Analytically, we say $f \in \mathcal{K}_s$ if it satisfies the condition $\mathrm{Re}\,\big(2(zf'(z))'/(f(z) - f(-z))'\big) > 0$ for all $z \in \mathbb{D}$. Using analytic representation similar to the classes of starlike and convex functions, for some λ $(\lambda > 1)$, Nishiwaki and Owa [13] considered and investigated two subclasses $\mathcal{M}(\lambda)$ and $\mathcal{N}(\lambda)$ of the class $\mathcal{S}$ consisting of functions f satisfying, respectively

$$\mathrm{Re}\left(\frac{zf'(z)}{f(z)}\right) < \lambda \quad \text{and} \quad \mathrm{Re}\left(1 + \frac{zf''(z)}{f'(z)}\right) < \lambda \quad (z \in \mathbb{D}).$$

The coefficient bounds yield information regarding the geometric properties of univalent functions. In 1916, Bieberbach [2] computed an estimate for the second coefficient of normalized univalent analytic function and this bound provides the growth, distortion, and covering theorems. Similarly, using the Hankel determinants (which also deals with the bound on coefficients), Cantor [3] proved that "if ratio of two bounded analytic functions in $\mathbb{D}$, then the function is rational". For given natural numbers n, q and $a_1 = 1$, the Hankel determinant $H_{q,n}(f)$ of a function $f \in \mathcal{A}$ is defined by means of the following determinant:

$$H_{q,n}(f) := \begin{vmatrix} a_n & a_{n+1} & \cdots & a_{n+q-1} \\ a_{n+1} & a_{n+2} & \cdots & a_{n+q} \\ \vdots & \vdots & \vdots & \vdots \\ a_{n+q-1} & a_{n+q} & \cdots & a_{n+2(q-1)} \end{vmatrix}.$$

For some specific values of q and n, the quantities $H_{2,1}(f) = a_3 - a_2^2$ and $H_{2,2}(f) := a_2a_4 - a_3^2$ are known as Fekete–Szegö functional and second Hankel determinant, respectively. The third Hankel determinant is defined as

$$H_{3,\,1}(f) := a_3(a_2a_4 - a_3^2) - a_4(a_4 - a_2a_3) + a_5(a_3 - a_2^2).$$

Finding the sharp estimate on the second Hankel determinant $H_{2,2}(f)$ is rather easier than finding the sharp estimate on the third Hankel determinant. The usual way of finding estimate on the third Hankel determinant is to calculate sharp bounds on the initial coefficients, second Hankel determinant and that on the Fekete–Szegö functional using the triangle inequality. The sharp bound on the second Hankel determinant for the class of starlike and convex functions were investigated by Janteng et al. [6]. Babalola [1] proved the estimates $|H_{3,1}(f)| \le 16$ and $|H_{3,1}(f)| \le (32 + 33\sqrt{3})/72\sqrt{3} \approx 0.71$ for the classes $\mathcal{S}^*$ and $\mathcal{K}$, respectively. Later, the bound on the third Hankel determinant for the class of starlike and convex functions were proved as $|H_{3,1}(f)| \le 1$ and $|H_{3,1}(f)| \le 49/540$, respectively, see [24]. In 2018, Kowalczyk et al. [8] proved that the bound $|H_{3,\,1}(f)| \le 4/135$ is sharp for the class of convex functions. However, the best known estimate for starlike functions is $|H_{3,1}(f)| \leq 4/135$ due to Kwon et al. [10]. Later, Lecko et al. [11] proved that the bound $|H_{3,1}(f)| \leq 1/9$ is sharp for starlike function of order 1/2. In the year 2018, Kowalczyk et al. [7] have found sharp bound of the third kind for the class $T(\alpha) := \{f \in \mathcal{A} : \mathrm{Re}(f(z)/z) > \alpha; z \in \mathbb{D}\}$ when $\alpha = 0$ and $\alpha = 1/2$. Vamshee et al. [21] estimated $|H_{3,1}(f)| \leq 5/2$ and $|H_{3,1}(f)| \leq 19/135$ for starlike and convex functions with respect to symmetric points, respectively. In 2017, Prajapat et al. [16] computed the estimates $|H_{3,1}(f)| \leq (81 + 16\sqrt{3})/216 \approx 0.5033$ and $|H_{3,1}(f)| \leq 139/5760 \approx 0.0241$ for the classes $\mathcal{M} := \mathcal{M}(3/2)$ and $\mathcal{N} := \mathcal{N}(3/2)$, respectively. For recent development of Hankel determinant, see [4, 12, 14, 15, 19, 20, 22–24].

Motivated by these works, in this paper, an attempt has been made to improve the existing bound on the third Hankel determinant for the classes $\mathcal{S}_s^*$, $\mathcal{K}_s$, $\mathcal{M}$ and $\mathcal{N}$.

2 Third Hankel Determinant

The following lemmas will be needed to derive our main results in this section:

Lemma 2.1 ([17, Lemma 2.3, p. 507]) *Let $p \in \mathcal{P}$. Then for all $n, m \in \mathbb{N}$,*

$$|\mu p_n p_m - p_{m+n}| \leq \begin{cases} 2, & 0 \leq \mu \leq 1; \\ 2|2\mu - 1|, & \textit{elsewhere.} \end{cases}$$

If $0 < \mu < 1$, then the inequality is sharp for the function $p(z) = (1 + z^{m+n})/(1 - z^{m+n})$. In the other cases, the inequality is sharp for the function $\hat{p}_0(z) = (1 + z)/(1 - z)$.

Lemma 2.2 ([9, Lemma 1]) *Let $p(z) = 1 + p_1 z + p_2 z^2 + p_3 z^3 + \cdots \in \mathcal{P}$. Then, for any real number μ,*

$$\left|\mu p_3 - p_1^3\right| \leq \begin{cases} 2|\mu - 4| & \left(\mu \leq \dfrac{4}{3}\right) \\[2ex] 2\mu\sqrt{\dfrac{\mu}{\mu - 1}} & \left(\dfrac{4}{3} < \mu\right). \end{cases}$$

The result is sharp. If $\mu \leq \frac{4}{3}$, then equality holds for the function

$$p_0(z) := \frac{1 + z}{1 - z}$$

and if $\mu > \frac{4}{3}$, then equality holds for the function

$$p_1(z) := \frac{1 - z^2}{z^2 - 2\sqrt{\frac{\mu}{\mu-1}}\, z + 1}.$$

The following theorem gives an improvement to the existing estimate on the third Hankel determinant related to the starlike and convex functions with respect to the symmetric points.

Theorem 2.3 *The third Hankel determinant for the functions in the classes $\mathcal{S}_s^*$ and $\mathcal{K}_s$ are $5/4$ and $91/1728$, respectively.*

Proof The proof will be accomplished in two parts.

(a) Let $f \in \mathcal{S}_s^*$. Then we can associate a function $p(z) = 1 + p_1 z + p_2 z^2 + p_3 z^3 + \cdots \in \mathcal{P}$ such that

$$\frac{2zf'(z)}{f(z) - f(-z)} = p(z).$$

On comparing coefficients on both sides of the above equation, we have

$$a_2 = \frac{p_1}{2}, \quad a_3 = \frac{p_2}{2}, \quad a_4 = \frac{1}{8}(p_1 p_2 + 2p_3) \text{ and } a_5 = \frac{1}{8}\left(p_2^2 + 2p_4\right).$$

Using the above we can write

$$\begin{aligned} H_{3,1}(f) &= a_5\left(a_3 - a_2^2\right) + a_3\left(a_2 a_4 - a_3^2\right) - a_4(a_4 - a_2 a_3) \\ &= \frac{1}{64}\left(p_1^2\left(p_2^2 - 4p_4\right) + 4p_1 p_2 p_3 - 4\left(p_2^3 - 2p_2 p_4 + p_3^2\right)\right) \end{aligned}$$

Further, by suitably arranging the terms, we have

$$\begin{aligned} |64H_{3,1}(f)| &= |8p_2 p_4 - 4p_1^2 p_4 + p_1^2 p_2^2 - 4p_2^3 + 4p_1 p_2 p_3 - 4p_3^2| \\ &\leq |4p_4(2p_2 - p_1^2)| + |p_2^2(p_1^2 - 4p_2)| + |4p_3(p_1 p_2 - 4p_3)|. \end{aligned} \tag{2.1}$$

By Lemma 2.1, we see that

$$|4p_4(2p_2 - p_1^2)| \leq 32, \quad |p_2^2(p_1^2 - 4p_2)| \leq 32 \text{ and } |4p_3(p_1 p_2 - 4p_3)| \leq 16. \tag{2.2}$$

Thus, using (2.2) and (2.1), we have

$$|H_{3,1}(f)| \leq \frac{80}{64} = \frac{5}{4} < \frac{5}{2}. \tag{2.3}$$

This is the desired estimate.

(b) Let $f \in \mathcal{K}_s$. Then, we can associate a function $p(z) = 1 + p_1 z + p_2 z^2 + p_3 z^3 + \cdots \in \mathcal{P}$ such that

$$\frac{2(zf'(z))'}{(f(z)-f(-z))'} = p(z).$$

On comparing coefficients on both sides of the above equation, we have

$$a_2 = \frac{p_1}{4}, \quad a_3 = \frac{p_2}{6}, \quad a_4 = \frac{1}{32}(p_1p_2 + 3p_3) \text{ and } a_5 = \frac{1}{40}\left(p_2^2 + 2p_4\right).$$

Now, a computation using the above gives

$$H_{3,1}(f) = \frac{9p_1^2\left(p_2^2 - 48p_4\right) + 180p_1p_2p_3 - 64p_2^3 + 1152p_2p_4 - 540p_3^2}{138240}.$$

By suitably arranging terms, we have

$$138240H_{3,1}(f) = 1152p_4\left(p_2 - \frac{432}{1152}p_1^2\right) + 64p_2^2\left(\frac{9}{64}p_1^2 - p_2\right) + 540p_3\left(\frac{180}{540}p_1p_2 - p_3\right). \tag{2.4}$$

Now using Lemma 2.1 and the fact $|p_i| \leq 2$, we have

$$1152\left|p_4\left(p_2 - \frac{432}{1152}p_1^2\right)\right| \leq 4608, \tag{2.5}$$

$$64\left|p_2^2\left(\frac{9}{64}p_1^2 - p_2\right)\right| \leq 512, \tag{2.6}$$

and

$$540\left|p_3\left(\frac{180}{540}p_1p_2 - p_3\right)\right| \leq 2160. \tag{2.7}$$

Now by using Eqs. (2.4), (2.5), (2.6), and (2.7), we get

$$|H_{3,1}(f)| \leq \frac{4608 + 512 + 2160}{138240} = \frac{91}{1728} \approx 0.052662 < \frac{19}{135} \approx 0.140741. \tag{2.8}$$

This completes the proof. □

Remark 2.4 It is important to note that Krishna et al. [21, Corollary 3.4, p. 43] prove that $\left|H_{3,1}(f)\right| \leq 5/2$ for function $f \in \mathcal{S}_s^*$. Thus, estimate (2.3) improves the exiting result derived in [21, Corollary 4.3, p. 43]. Similarly, the estimate in (2.8) provides an improvement over the estimate $\left|H_{3,1}(f)\right| \leq 19/135$ for function $f \in \mathcal{K}_s$, see [21, Corollary 3.8, p. 45].

The following theorem yields an improvement to the existing estimate on the third Hankel determinant related to the classes $\mathcal{M}$ and $\mathcal{N}$.

Theorem 2.5 *The third Hankel determinant for the functions in the classes* $\mathcal{M}$ *and* $\mathcal{N}$ *are bounded by* $(579 + 8\sqrt{3})/1728$ *and* $(144431 + 96\sqrt{141})/6497280$*, respectively.*

Proof The proof will be accomplished in two parts.

(a) Let $f \in \mathcal{M}$. Then, we can associate a function $p(z) = 1 + p_1 z + p_2 z^2 + p_3 z^3 + \cdots \in \mathcal{P}$ such that

$$\frac{zf'(z)}{f(z)} = \frac{1}{2}(3 - p(z)). \tag{2.9}$$

On comparing the coefficients on both sides on (2.9), we get

$$a_2 = -\frac{p_1}{2}, \quad a_3 = \frac{1}{8}\left(p_1^2 - 2p_2\right), \quad a_4 = \frac{1}{48}\left(-p_1^3 + 6p_1p_2 - p_3\right)$$

and

$$a_5 = \frac{1}{384}\left(p_1^4 - 12p_1^2p_2 + 4p_1p_3 + 12\left(p_2^2 - 4p_4\right)\right).$$

Further computation gives

$$H_{3,1}(f) = \frac{-p_1^6 - 6p_1^4p_2 + 4p_1^3p_3 - 36p_1^2\left(p_2^2 - 4p_4\right) - 24p_1p_2p_3 + 72p_2^3 + 288p_2p_4 - 4p_3^2}{9216}.$$

By suitably arranging the terms, we can write

$$\begin{aligned} -9216H_{3,1}(f) = 36p_1^2p_2^2 - 144p_1^2p_4 + p_1^6 - 4p_1^3p_3 + 24p_1p_2p_3 - 288p_2p_4 \\ + 6p_1^4p_2 - 72p_2^3 + 4p_3^2. \end{aligned} \tag{2.10}$$

The Inequalities (2.11), (2.13), and (2.14) are obtained by using Lemma 2.1 and the fact $|p_i| \leq 2$, whereas the Inequality (2.12) is a consequence of Lemma 2.2

$$\left|36p_1^2p_2^2 - 144p_1^2p_4\right| = 144|p_1^2|\left|\frac{1}{4}p_2^2 - p_4\right| \leq 1152, \tag{2.11}$$

$$\left|p_1^6 - 4p_1^3p_3\right| = |p_1^3|\left|p_1^3 - 4p_3\right| \leq \frac{128}{\sqrt{3}}, \tag{2.12}$$

$$|24p_1p_2p_3 - 288p_2p_4| = 288|p_2|\left|\frac{1}{12}p_1p_3 - p_4\right| \leq 1152, \tag{2.13}$$

and

$$\left|6p_1^4p_2 - 72p_2^3 + 4p_3^2\right| \leq \left|6p_1^4p_2\right| + \left|72p_2^3\right| + 4\left|p_3^2\right| \leq 784. \tag{2.14}$$

By using triangle inequality and (2.10), (2.11), (2.12), (2.13), and (2.14), we get

$$|H_{3,1}(f)| \leq \frac{579 + 8\sqrt{3}}{1728} \approx 0.343088 < \frac{81 + 16\sqrt{3}}{216} \approx 0.5033. \tag{2.15}$$

(b) Let $f \in \mathcal{N}$. Then, we can associate a function $p(z) = 1 + p_1 z + p_2 z^2 + p_3 z^3 + \cdots \in \mathcal{P}$ such that

$$1 + \frac{zf''(z)}{f'(z)} = \frac{1}{2}\left(3 - p(z)\right). \tag{2.16}$$

On comparing the coefficients on both sides on (2.16), we get

$$a_2 = -\frac{p_1}{4}, \quad a_3 = \frac{1}{12}\left(\frac{p_1^2}{2} - p_2\right), \quad a_4 = \frac{1}{192}\left(-p_1^3 + 6p_1p_2 - 8p_3\right)$$

and

$$a_5 = \frac{p_1^4 - 12p_1^2p_2 + 32p_1p_3 + 12\left(p_2^2 - 4p_4\right)}{1920}.$$

Using the above further computation, we get

$$H_{3,1}(f) = \frac{-p_1^6 - 12p_1^4p_2 + 48p_1^3p_3 + p_1^2\left(288p_4 - 84p_2^2\right) - 288p_1p_2p_3 + 32\left(p_2^3 + 36p_2p_4 - 30p_3^2\right)}{552960}.$$

After suitable arrangement of terms, we have

$$-552960H_{3,1}(f) = 84p_1^2p_2^2 - 288p_1^2p_4 + p_1^6 - 48p_1^3p_3 + 288p_1p_2p_3 - 1152p_2p_4 + 12p_1^4p_2 - 32p_2^3 + 960p_3^2. \tag{2.17}$$

Now using Lemma 2.1 and the fact $|p_i| \le 2$, we have the Inequalities (2.18), (2.20), and (2.21). Further, an application of Lemma 2.2 gives (2.19).

$$|84p_1^2p_2^2 - 288p_1^2p_4| = 288|p_1^2|\left|\frac{84}{288}p_2^2 - p_4\right| \le 2304, \tag{2.18}$$

$$|p_1^6 - 48p_1^3p_3| = |p_1^3|\left|p_1^3 - 48p_3\right| \le 384\sqrt{\frac{3}{47}}, \tag{2.19}$$

$$|288p_1p_2p_3 - 1152p_2p_4| = 1152|p_2|\left|\frac{288}{1152}p_1p_3 - p_4\right| \le 4608 \tag{2.20}$$

and

$$|12p_1^4p_2 - 32p_2^3 + 960p_3^2| \le 12|p_1^4p_2| + 32|p_2^3| + 960|p_3^2| \le 4480. \tag{2.21}$$

From (2.17), (2.18), (2.19), (2.20), and (2.21), we have

$$H_{3,1}(f)| \le \frac{144431 + 96\sqrt{141}}{6497280} \approx 0.0224049 < \frac{139}{5760} \approx 0.0241319 \tag{2.22}$$

This completes the proof. □

Remark 2.6 Prajapat et al. [16, Theorem 2.4, p. 190] proved that the third Hankel determinant $H_{3,1}(f)$ for function $f \in \mathcal{M}$ is bounded by 185/512. Clearly, the bound derived in (2.15) for the function $f \in \mathcal{M}$ is better than that of derived by Prajapat et al. [16, Theorem 3.4, p. 190]. Furthermore, the estimate (2.22) improves over the result [16, Theorem 2.8, p. 193] for the function $f \in \mathcal{N}$.

Conjecture 2.7 The sharp bound on the third Hankel determinant for the classes of starlike and convex functions with respect to symmetric point are 1/4 and 4/135, respectively. Equality, for the class $\mathcal{S}_s^*$, holds in case of the function f_0 defined by

$$\frac{2zf_0'(z)}{f_0(z) - f_0(z)} = \frac{1+z^3}{1-z^3}.$$

Further, for the class $\mathcal{K}_s$, equality holds in case of the function f_1 defined by

$$\frac{2(zf_1'(z))'}{(f_1(z) - f_1(-z))'} = \frac{1+z^2}{1-z^2}.$$

Conjecture 2.8 The sharp bound on the third Hankel determinant for the classes $\mathcal{M}$ and $\mathcal{N}$ are 119/576 and 19/2160, respectively. Equality, for the class $\mathcal{M}$, holds in case of the function f_2 defined by (2.9) with the choice $p(z) = (1+z)/(1-z)$ whereas, for the class $\mathcal{N}$, equality holds in case of the function f_4 defined by (2.16) with the choice of function $p(z) = (1+z^2)/(1-z^2)$.

Acknowledgements The authors would like to express their gratitude to the referees for many valuable suggestions regarding a previous version of this manuscript.

References

1. K.O. Babalola, On third order Hankel determinant for some classes of univalent functions. Inequal. Theory Appl. **6**, 1–7 (2010)
2. L. Bieberbach, Uber die Koeffizienten derjenigen Potenzreihen, welche eine schlichte Abbildung des Einheitskreises vermitteln. S.-B. Preuss. Akad. Wiss. **38**, 940–955 (1916)
3. D.G. Cantor, Power series with integral coefficients. Bull. Am. Math. Soc. **69**, 362–366 (1963)
4. N.E. Cho, B. Kowalczyk, O.S. Kwon, A. Lecko, Y.J. Sim, The bounds of some determinants for starlike functions of order alpha. Bull. Malays. Math. Sci. Soc. **41**(1), 523–535 (2018)
5. R.N. Das, P. Singh, On subclasses of schlicht mapping. Indian J. Pure Appl. Math. **8**(8), 864–872 (1977)
6. A. Janteng, S.A. Halim, M. Darus, Hankel determinant for starlike and convex functions. Int. J. Math. Anal. (Ruse) **1**(13–16), 619–625 (2007)
7. B. Kowalczyk, A. Lecko, M. Lecko, Y.J. Sim, The sharp bound of the third Hankel determinant for some classes of analytic functions. Bull. Korean Math. Soc. **55**(6), 1859–1868 (2018)
8. B. Kowalczyk, A. Lecko, Y.J. Sim, The sharp bound for the Hankel determinant of the third kind for convex functions. Bull. Aust. Math. Soc. **97**(3), 435–445 (2018)

9. V. Kumar, N.E. Cho, V. Ravichandran, H.M. Srivastava, Sharp coefficient bounds for starlike functions associated with the bell numbers. Math. Slovaca, **69**(5), 1–12 (2019)
10. O.S. Kwon, A. Lecko, Y.J. Sim, The bound of the Hankel determinant of the third kind for starlike functions. Bull. Malays. Math. Sci. Soc. **42**(2), 767–780 (2019)
11. A. Lecko, Y.J. Sim, B. Śmiarowska, The sharp bound of the Hankel determinant of the third kind for starlike functions of order 1/2. Complex Anal. Oper. Theory (2018). https://doi.org/10.1007/s11785-018-0819-0
12. S.K. Lee, V. Ravichandran, S. Supramaniam, Bounds for the second Hankel determinant of certain univalent functions. J. Inequal. Appl. **2013**(281), 17 Art. 281 (2013)
13. J. Nishiwaki, S. Owa, Coefficient inequalities for certain analytic functions. Int. J. Math. Math. Sci. **29**(5), 285–290 (2002)
14. H. Orhan, P. Zaprawa, Third Hankel determinants for starlike and convex functions of order alpha. Bull. Korean Math. Soc. **55**(1), 165–173 (2018)
15. C. Pommerenke, On the coefficients and Hankel determinants of univalent functions. J. Lond. Math. Soc. **41**, 111–122 (1966)
16. J.K. Prajapat, D. Bansal, S. Maharana, Bounds on third Hankel determinant for certain classes of analytic functions. Stud. Univ. Babeş-Bolyai Math. **62**(2), 183–195 (2017)
17. V. Ravichandran, S. Verma, Bound for the fifth coefficient of certain starlike functions. C. R. Math. Acad. Sci. Paris **353**(6), 505–510 (2015)
18. K. Sakaguchi, On a certain univalent mapping. J. Math. Soc. Jpn **11**, 72–75 (1959)
19. J. Sokól, D.K. Thomas, The second Hankel determinant for alpha-convex functions. Lith Math J **58** 212 (2018). https://doi.org/10.1007/s10986-018-9397-0
20. D. Vamshee Krishna, B. Venkateswarlu, T. RamReddy, Third Hankel determinant for certain subclass of p-valent functions. Complex Var. Elliptic Equ. **60**(9), 1301–1307 (2015)
21. D. Vamshee Krishna, B. Venkateswarlu, T. RamReddy, Third Hankel determinant for starlike and convex functions with respect to symmetric points. Ann. Univ. Mariae Curie-Skłodowska Sect. A **70**(1), 37–45 (2016)
22. B. Venkateswarlu, N. Rani, Third Hankel determinant for the inverse of reciprocal of bounded turning functions has a positive real part of order alpha. Ufa Math. J. **9**(2), 109–118 (2017)
23. T. Yavuz, Upper bound on the third Hankel determinant for functions defined by Ruscheweyh derivative operator. Commun. Korean Math. Soc. **33**(2), 437–444 (2018)
24. P. Zaprawa, Third Hankel determinants for subclasses of univalent functions. Mediterr. J. Math. **14**(1) Art. 19 10 (2017)

μ-Statistical Convergence of Sequences in Probabilistic n-Normed Spaces

Rupam Haloi and Mausumi Sen

Abstract In this article, using the notion of a two-valued measure μ, we propose the ideas of μ-statistical convergence and μ-density convergence in probabilistic n-normed spaces and study some of their properties in probabilistic n-normed spaces. Further, a condition for equality of the sets of μ-statistical convergent and μ-density convergent sequences in the space have been established. The definition of μ-statistical Cauchy sequence in the space has also been introduced and some results have been established. Finally, we propose the notion of μ-statistical limit points in these new settings and studied some properties.

Keywords Probabilistic n-normed linear space · μ-Statistical convergence · μ-Density convergence · μ-Statistical Cauchy sequence

MSC 2010 40A15 · 40G15 · 60B99

1 Introduction

As an important generalization of the concept of distance as proposed by Fréchet [1] in 1906, Menger [2] developed the idea of a statistical metric space, now called probabilistic metric space. Employing the idea of probabilistic metric and simplifying the concept of ordinary normed linear space, Sherstnev [3] proposed the concept of probabilistic normed space (in short PN-space) in 1962, in which the norm of a vector was described by a distribution function rather than by a positive number. Tripathy and Goswami [4–7], Tripathy et al. [8] and others have introduced different classes of sequences using the notion of probabilistic norm and have investigated

---

R. Haloi (✉) · M. Sen
Department of Mathematics, National Institute of Technology Silchar, Silchar 788010, Assam, India
e-mail: rupam.haloi15@gmail.com

M. Sen
e-mail: senmausumi@gmail.com

N. Deo et al. (eds.), *Mathematical Analysis I: Approximation Theory*, Springer Proceedings in Mathematics & Statistics 306,
https://doi.org/10.1007/978-981-15-1153-0_20

their different algebraic and topological properties. The situation where crisp norm fails to measure the length of a vector precisely, the notion of probabilistic norm happens to be very much useful. The theory of PN-space is decisive as a conclusion of deterministic results of normed linear spaces and furnish us some decisive tools relevant to the study of convergence of random variables, continuity properties, linear operators, geometry of nuclear physics, topological spaces, etc. This space was further generalized into the theory of probabilistic n-normed spaces (abbreviated as PnN-spaces) by Rahmat and Noorani [9] and many authors. As an important generalization to the theory of convergence, Fast [10] initially proposed the idea of statistical convergence and then studied by many researchers. Karakus [11] has extended idea of statistical convergence into probabilistic normed space 2007. As an interesting generalization of statistical convergence, Connor [12, 13] introduced the idea of statistical convergence with the help of a complete $\{0,1\}$ valued measure μ defined on an algebra of subsets of $\mathbb{N}$. Some works in this field can be found in [14–17]. The notion of statistical limit points was first introduced by Fridy [18]. The aim of this article is to introduce and study the concepts of μ-statistical convergence and μ-density convergence in PnN-spaces.

A brief sketch of the article is as follows: IP Sect. 2 contains some basic definitions that are relevant for subsequent sections. We have introduced the definitions of μ-statistical convergence and μ-density convergence in PnN-spaces and discussed some of their properties in Sect. 3. Section 4 deals with the concept of μ-statistical limit points in PnN-space and their properties. Finally, a brief conclusion to the article follows in Sect. 5.

2 Preliminaries

Throughout the paper, $\mathbb{R}$, $\mathbb{N}$, and $\mathbb{R}^+$ denote the sets of real, natural, and nonnegative real numbers, respectively.

Definition 1 ([19]) A function $f : \mathbb{R}^+ \to [0, 1]$ is called a distribution function if it is nondecreasing, left-continuous with $\inf_{t\in\mathbb{R}^+} f(t) = 0$ and $\sup_{t\in\mathbb{R}^+} f(t) = 1$.

Throughout D denotes the set of all distribution functions.

Definition 2 ([19]) A binary operation $* : [0, 1] \times [0, 1] \to [0, 1]$ is said to be a continuous t-norm if it satisfies the following conditions, for all $a, b, c, d \in [0, 1]$:

1. $a * 1 = a$,
2. $a * b = b * a$,
3. $a * b \le c * d$, whenever $a \le c$ and $b \le d$,
4. $(a * b) * c = a * (b * c)$.

Definition 3 ([9]) A triplet $(Y, M, *)$ is called a probabilistic n-normed space (in short a PnN-space) if Y is a real vector space of dimension $d \ge n$, M a mapping from Y^n into D and $*$ is a t-norm satisfying the following conditions for every $y_1, y_2, \ldots, y_n \in Y$ and $s, t > 0$:

1. $M((y_1, y_2, \ldots, y_n), t) = 1$ if and only if $y_1, y_2, \ldots, y_n$ are linearly dependent,
2. $M((y_1, y_2, \ldots, y_n), t)$ is invariant under any permutations of $y_1, y_2, \ldots, y_n$,
3. $M((y_1, y_2, \ldots, \alpha y_n), t) = M\left((y_1, y_2, \ldots, y_n), \frac{t}{|\alpha|}\right)$ for all $\alpha \in \mathbb{R} \setminus \{0\}$,
4. $M((y_1, y_2, \ldots, y_n + y_n'), s + t) \geq M((y_1, y_2, \ldots, y_n), s) * M((y_1, y_2, \ldots, y_n'), t)$.

Example 4 [9] Let $(Y, ||\cdot, \ldots, \cdot||)$ be a n-normed linear space. Let $a * b = \min\{a, b\}$, for all $a, b \in [0, 1]$ and $M((y_1, y_2, \ldots, y_n), t) = \dfrac{t}{t + ||(y_1, y_2, \ldots, y_n)||}$, $t \geq 0$. Then $(Y, M, *)$ is a PnN-space.

Definition 5 ([9]) A sequence $y = (y_k)$ in a PnN-space $(Y, M, *)$ is said to be convergent to $y_0 \in Y$ in terms of the probabilistic n-norm M^n, if for every $\varepsilon > 0$, $\lambda \in (0, 1)$ and $z_1, z_2, \ldots, z_{n-1} \in Y$, there exists a positive integer k_0 such that

$$M((z_1, z_2, \ldots, z_{n-1}, y_k - y_0), \varepsilon) > 1 - \lambda,$$

whenever $k \geq k_0$. In this case, we write $M^n - \lim y = y_0$.

Definition 6 ([9]) A sequence $y = (y_k)$ in a PnN-space $(Y, M, *)$ is said to be Cauchy sequence, if for every $\varepsilon > 0$, $\lambda \in (0, 1)$ and $z_1, z_2, \ldots, z_{n-1} \in Y$, there exists a positive integer k_0 such that

$$M((z_1, z_2, \ldots, z_{n-1}, y_k - y_m), \varepsilon) > 1 - \lambda,$$

for all $k, m \geq k_0$.

Definition 7 ([9]) A sequence $y = (y_k)$ in a PnN-space $(Y, M, *)$ is said to be bounded in terms of the probabilistic n-norm M^n, if for every $z_1, z_2, \ldots, z_{n-1} \in Y$, there exists an $\varepsilon > 0$ such that

$$M((z_1, z_2, \ldots, z_{n-1}, y_k), \varepsilon) > 1 - \lambda,$$

for every $\lambda \in (0, 1)$ and for all $k \in \mathbb{N}$.

3 μ-Statistical Convergence and μ-Density Convergence in PnN-Spaces

Right through the article, by μ we represent a complete $\{0, 1\}$-valued finitely additive measure defined on a field Γ of all finite subsets of $\mathbb{N}$ and suppose that $\mu(P) = 0$, if $|P| < \infty$; if $P \subset Q$ and $\mu(Q) = 0$, then $\mu(P) = 0$; and $\mu(\mathbb{N}) = 1$.

Definition 8 A sequence $y = (y_k)$ is said to be μ-statistically convergent to y_o in terms of the probabilistic n-norm M^n, if for every $\lambda \in (0, 1)$, $\varepsilon > 0$ and $z_1, z_2, \ldots, z_{n-1} \in Y$,

$$\mu(\{k \in \mathbb{N} : M((z_1, z_2, \ldots, z_{n-1}, y_k - y_0), \varepsilon) \leq 1 - \lambda\}) = 0.$$

It is written as $\mu - stat_{M(n)} - \lim y = y_0$.

In view of the Definition 3.1 and other properties of measure, we state the following result without proof.

Theorem 9 *Let $(Y, M, *)$ be a PnN-space. Then for every $\lambda \in (0, 1)$, $\varepsilon > 0$ and $z_1, z_2, \ldots, z_{n-1} \in Y$, the following statements are equivalent:*

1. $\mu - stat_{M(n)} - \lim y = y_0$,
2. $\mu(\{k \in \mathbb{N} : M((z_1, z_2, \ldots, z_{n-1}, y_k - y_0), \varepsilon) \leq 1 - \lambda\}) = 0$,
3. $\mu(\{k \in \mathbb{N} : M((z_1, z_2, \ldots, z_{n-1}, y_k - y_0), \varepsilon) > 1 - \lambda\}) = 1$,
4. $\mu - stat - \lim M((z_1, z_2, \ldots, z_{n-1}, y_k - y_0), \varepsilon) = 1$.

The following results are consequences of Theorem 9.

Corollary 10 *Let $(Y, M, *)$ be a PnN-space. If a sequence (y_k) is μ-statistically convergent in terms of the probabilistic n-norm M^n, then $\mu - stat_{M(n)} - \lim y$ is unique.*

Corollary 11 *Let $(Y, M, *)$ be a PnN-space. If $M^n - \lim y = y_0$, then $\mu - stat_{M(n)} - \lim y = y_0$, but not necessarily conversely.*

The converse of the Corollary 11 does not hold always, which can be shown from the following example.

Example 12 Let us consider $Y = \mathbb{R}^n$ with usual norm. Let $p * q = pq$ for $p, q \in [0, 1]$ and $M((z_1, z_2, \ldots, z_{n-1}, y), t) = \dfrac{t}{t + ||(z_1, z_2, \ldots, z_{n-1}, y)||}$, where $(z_1, z_2, \ldots, z_{n-1}, y) \in \mathbb{R}^n$ and $t \geq 0$. Then $(\mathbb{R}^n, M, *)$ is a PnN-space. Let $A \subset \mathbb{N}$ be such that $\mu(A) = 0$. We define a sequence $y = (y_k)$ as follows:

$$y_k = \begin{cases} (k, 0, \ldots, 0) \in \mathbb{R}^n, \text{ if } k = j^2, \ j \in \mathbb{N} \\ (0, 0, \ldots, 0) \in \mathbb{R}^n, \text{ otherwise.} \end{cases}$$

Then we can easily verify that the sequence (y_k) is μ-statistically convergent in terms of the probabilistic n-norm M^n, but the sequence (y_k) is not convergent in terms of the probabilistic n-norm M^n, as it is not convergent in the space $(\mathbb{R}, \| \cdot \|)$.

We now introduce the concept of μ-statistical Cauchy sequence on probabilistic n-normed space and provide a characterization.

Definition 13 Let $(Y, M, *)$ be a PnN-space. We say that a sequence $y = (y_k)$ is μ-statistically Cauchy in terms of the probabilistic n-norm M^n, provided that for every $\lambda \in (0, 1)$, $\varepsilon > 0$ and $z_1, z_2, \ldots, z_{n-1} \in Y$, there exists a positive integer $m \in \mathbb{N}$ satisfying

$$\mu(\{k \in \mathbb{N} : M((z_1, z_2, \ldots, z_{n-1}, y_k - y_m), \varepsilon) \leq 1 - \lambda\}) = 0.$$

Theorem 14 *Let $(Y, M, *)$ be a PnN-space. If a sequence $y = (y_k)$ is μ-statistically convergent in terms of the probabilistic n-norm M^n, then it is μ-statistically Cauchy in terms of the probabilistic n-norm M^n.*

Definition 15 A sequence (y_k) is said to be μ-density convergent to $y_0 \in Y$ in terms of the probabilistic n-norm M^n, if there exists an $A \in \Gamma$ with $\mu(A) = 1$ such that $(y_k - y_0)_{k \in A}$ is convergent to 0 in terms of the probabilistic n-norm M^n.

By $\omega(Y, M, *)$, we denote the space of all sequences with elements from the PnN-space $(Y, M, *)$ and by $\ell_\infty(Y, M, *)$, the space of all bounded sequences with elements from the probabilistic n-normed space $(Y, M, *)$.

Theorem 16 *Let $y \in \omega(Y, M, *)$. If y is μ-density convergent to r in terms of the probabilistic n-norm M^n, then y is μ-statistically convergent to r in terms of the probabilistic n-norm M^n.*

Proof Let $y = (y_k) \in \omega(Y, M, *)$. Let $A \subset \mathbb{N}$ such that $(y_k - r)_{k \in A}$ is convergent to 0 in terms of the probabilistic n-norm M^n and $\mu(A) = 1$. Let $\varepsilon > 0$ be given and $z_1, z_2, \ldots, z_{n-1} \in Y$. Then it is observed that

$$\{k \in \mathbb{N} : M((z_1, z_2, \ldots, z_{n-1}, y_k - r), \varepsilon) \leq 1 - \lambda\}$$

contains at most finitely many terms of $A \subset \mathbb{N}$. Thus, we have

$$\mu(\{k \in A : M((z_1, z_2, \ldots, z_{n-1}, y_k - r), \varepsilon) \leq 1 - \lambda\}) = 0.$$

Now,

$$\begin{aligned} C &= \{k \in \mathbb{N} : M((z_1, z_2, \ldots, z_{n-1}, y_k - r), \varepsilon) \leq 1 - \lambda\} \\ &\subseteq \{k \in A : M((z_1, z_2, \ldots, z_{n-1}, y_k - r), \varepsilon) \leq 1 - \lambda\} \cup A^c. \end{aligned}$$

Thus, we have $\mu(C) = 0$, and consequently

$$\mu(\{k \in \mathbb{N} : M((z_1, z_2, \ldots, z_{n-1}, y_k - r), \varepsilon) \leq 1 - \lambda\}) = 0,$$

which shows that $y = (y_k)$ is μ-statistically convergent in terms of the probabilistic n-norm M^n. □

Definition 17 *(APO condition*[12]*)* A measure μ is said to have the additive property of null sets or the APO condition, if given a collection $\{A_i\}_{i \in \mathbb{N}} \subseteq \Gamma$ of mutually disjoint μ-null sets (i.e., $\mu(A_i) = 0$, for all $i \in \mathbb{N}$) such that $A_i \cap A_j = \phi$, for $i \neq j$, then there exists a collection $\{B_i\}_{i\in\mathbb{N}} \subseteq \Gamma$ with $|A_i \bigtriangleup B_i| < \infty$, for each $i \in \mathbb{N}$ and $B = \cup_i B_i \in \Gamma$ with $\mu(B) = 0$.

Let Y be any set, M be the probabilistic n-norm and $y = (y_k)$ be any sequence in Y. Let us define two sets as follows:

1. $D_\mu(Y, M, *) = \{y \in \ell_\infty(Y, M, *) : y$ is μ-density convergent to 0 in terms of the probabilistic n-norm $M^n\}$,
2. $S_\mu(Y, M, *) = \{y \in \ell_\infty(Y, M, *) : y$ is μ-statistically convergent to 0 in terms of the probabilistic n-norm $M^n\}$.

Definition 18 Let $(Y, M, *)$ be a PnN-space. For $\varepsilon > 0$, the open ball $B(y, s, \varepsilon)$ with center y and radius $s \in (0, 1)$ is defined by

$$B(y, s, \varepsilon) = \{x \in Y : M((z_1, z_2, \dots, z_{n-1}, x - y), \varepsilon) > 1 - s, \\ \forall\, z_1, z_2, \dots, z_{n-1} \in Y\}.$$

Theorem 19 *$S_\mu(Y, M, *)$ is closed in $\ell_\infty(Y, M, *)$ and $\overline{D}_\mu(Y, M, *) = S_\mu(Y, M, *)$.*

Proof Clearly, $S_\mu(Y, M, *) \subset \overline{S}_\mu(Y, M, *)$. Now, we will show that $\overline{S}_\mu(Y, M, *) \subset S_\mu(Y, M, *)$. Let $x = (x_k) \in \overline{S}_\mu(Y, M, *)$. Let $\varepsilon > 0$ be given and $\lambda \in (0, 1)$. Since $B(x, r, \varepsilon/2) \cap S_\mu(Y, M, *) \neq \phi$, there is an $y \in B(x, r, \varepsilon/2) \cap S_\mu(Y, M, *)$. Choose $r \in (0, 1)$ such that $(1 - r) * (1 - r) > 1 - \lambda$. Since $y \in B(x, r, \varepsilon/2) \cap S_\mu(Y, M, *)$, so $\mu - stat_{M(n)} - \lim y = 0$. We define

$$A = \{k \in \mathbb{N} : M((z_1, z_2, \dots, z_{n-1}, y_k), \varepsilon/2) > 1 - r\},$$

for $z_1, z_2, \dots, z_{n-1} \in Y$. Then, we have $\mu(A) = 1$. Now for each $k \in A$ and $z_1, z_2, \dots, z_{n-1} \in Y$,

$$\begin{aligned} &M((z_1, z_2, \dots, z_{n-1}, x_k), \varepsilon) \\ &= M((z_1, z_2, \dots, z_{n-1}, (x_k - y_k) + y_k), \varepsilon/2 + \varepsilon/2) \\ &\geq M((z_1, z_2, \dots, z_{n-1}, x_k - y_k), \varepsilon/2) \\ &\qquad\qquad * M((z_1, z_2, \dots, z_{n-1}, y_k), \varepsilon/2) \\ &> (1 - r) * (1 - r) \\ &> (1 - \lambda). \end{aligned}$$

Therefore, $x = (x_k) \in S_\mu(Y, M, *)$ and so $\overline{S}_\mu(Y, M, *) \subset S_\mu(Y, M, *)$. Thus, $S_\mu(Y, M, *)$ is closed in $\ell_\infty(Y, M, *)$.

Now for the second part, it is clearly seen that $D_\mu(Y, M, *) \subseteq S_\mu(Y, M, *)$ which implies that $\overline{D}_\mu(Y, M, *) \subseteq S_\mu(Y, M, *)$. Thus, it is adequate to prove that $S_\mu(Y, M, *) \subseteq \overline{D}_\mu(Y, M, *)$.

Let $x = (x_k) \in S_\mu(Y, M, *)$. Then, for $\lambda \in (0, 1)$, $\varepsilon > 0$ and $z_1, z_2, \ldots, z_{n-1} \in Y$, we have

$$\mu(A) = \mu(\{k \in \mathbb{N} : M((z_1, z_2, \ldots, z_{n-1}, x_k), \varepsilon) \leq 1 - \lambda\}) = 0.$$

We define $y = (y_k)$ by

$$y_k = \begin{cases} x_k, & \text{if } k \in A \\ 0, & \text{otherwise.} \end{cases}$$

Then, $y \in D_\mu(Y, M, *)$ since $\mu(A^c) = 1$ and $y \in B(x, \lambda, \varepsilon)$. Thus, $S_\mu(Y, M, *) \subseteq \overline{D}_\mu(Y, M, *)$ and hence the proof. □

Theorem 20 *Let μ be a measure. Then $S_\mu(Y, M, *) = D_\mu(Y, M, *)$ if and only if μ has the APO condition.*

Proof Let μ be a measure with the APO condition. From Theorem 16, it is clearly seen that for any measure μ, $D_\mu(Y, M, *) \subseteq S_\mu(Y, M, *)$. Then it is adequate to prove that $S_\mu(Y, M, *) \subseteq D_\mu(Y, M, *)$. Let $y = (y_k) \in S_\mu(Y, M, *)$, then $\mu - stat_{M(n)} - \lim y = 0$. So, for every $\lambda \in (0, 1)$, $\varepsilon > 0$ and $z_1, z_2, \ldots, z_{n-1} \in Y$, we have

$$\mu(\{k \in \mathbb{N} : M((z_1, z_2, \ldots, z_{n-1}, y_k), \varepsilon) \leq 1 - \lambda\}) = 0.$$

Now, for $\varepsilon > 0$, $j \in \mathbb{N}$ and $z_1, z_2, \ldots, z_{n-1} \in Y$, we define

$$A_j = \left\{k \in \mathbb{N} : 1 - \frac{1}{j} \leq M((z_1, z_2, \ldots, z_{n-1}, y_k), \varepsilon) < 1 - \frac{1}{j+1}\right\}.$$

Then $\{A_j\}_{j\in\mathbb{N}}$ is a countable family of disjoint μ-null sets. Thus by APO condition, there exists a family $\{B_j\}_{j\in\mathbb{N}}$ such that $|A_j \triangle B_j| < \infty$, for all $j \in \mathbb{N}$ and $B = \bigcup_{j\in\mathbb{N}} B_j \in \Gamma$ with $\mu(B) = 0$. Let $A = \mathbb{N} \setminus B$, then $\mu(A) = 1$. We claim that $(y_k)_{k\in A}$ is convergent to 0 in terms of probabilistic n-norm M^n.

Let $\eta \in (0, 1)$ and $\varepsilon > 0$ be given and $z_1, z_2, \ldots, z_{n-1} \in Y$. We choose a positive integer N such that $\frac{1}{N} < \eta$. Then, we observe that

$$\begin{aligned} &\{k \in \mathbb{N} : M((z_1, z_2, \ldots, z_{n-1}, y_k), \varepsilon) \leq 1 - \eta\} \\ &\subset \left\{k \in \mathbb{N} : M((z_1, z_2, \ldots, z_{n-1}, y_k), \varepsilon) \leq 1 - \frac{1}{N}\right\} \\ &\subset \bigcup_{j=1}^{N-1} A_j. \end{aligned}$$

Since $A_j \triangle B_j$ is a finite set for each $j = 1, 2, \ldots, N-1$, so there is an $k_0 \in \mathbb{N}$ such that

$$\left(\bigcup_{j=1}^{N-1} B_j\right) \cap \{k \in \mathbb{N} : k \geq k_0\}$$
$$= \left(\bigcup_{j=1}^{N-1} A_j\right) \cap \{k \in \mathbb{N} : k \geq k_0\}.$$

If $k \in A$ and $k \geq k_0$, then $k \notin B$, which implies $k \notin \bigcup_{j=1}^{N-1} B_j$ and so $k \notin \bigcup_{j=1}^{N-1} A_j$. Hence, for every $k \geq k_0$, $k \in A$ and $z_1, z_2, \ldots, z_{n-1} \in Y$, we have

$$M((z_1, z_2, \ldots, z_{n-1}, y_k), \varepsilon) > 1 - \eta.$$

So $y = (y_k) \in D_\mu(Y, M, *)$. Thus, $S_\mu(Y, M, *) \subseteq D_\mu(Y, M, *)$.

Conversely, suppose $S_\mu(Y, M, *) = D_\mu(Y, M, *)$, for a measure μ. We need to show that μ has the APO. We choose a monotone sequence $x = (x_k)$ of distinct nonzero elements of Y such that $M^n - \lim y = 0$. Then for every $\varepsilon > 0$ and $z_1, z_2, \ldots, z_{n-1} \in Y$, $\{M((z_1, z_2, \ldots, z_{n-1}, x_k), \varepsilon)\}$ is an increasing sequence converging to 1. Let $\{A_i\}_{i \in \mathbb{N}}$ be a family such that $A_i \cap A_j = \phi$ for $i \neq j$ with $\mu(A_i) = 0$, for all $i \in \mathbb{N}$. We define a sequence (y_k) as follows:

$$y_k = \begin{cases} x_i, & \text{if } k \in A_i \\ 0, & \text{otherwise.} \end{cases}$$

Let $\lambda \in (0, 1)$ be given. We choose $k \in \mathbb{N}$ such that $M((z_1, z_2, \ldots, z_{n-1}, x_k), \varepsilon) > 1 - \lambda$ for each nonzero $z_1, z_2, \ldots, z_{n-1} \in Y$. Then

$$\begin{aligned} K(\varepsilon, \lambda) &= \{k \in \mathbb{N} : M((z_1, z_2, \ldots, z_{n-1}, y_k), \varepsilon) \leq 1 - \lambda\} \\ &\subseteq A_1 \cup A_2 \cup \ldots \cup A_k. \end{aligned}$$

So $\mu(\{K(\varepsilon, \lambda)\}) = 0$ and hence $\mu - stat_{M(n)} - \lim y = 0$. So, $(y_k) \in S_\mu(Y, M, *)$ which implies that $(y_k) \in D_\mu(Y, M, *)$. Therefore, there exists $P \subseteq \mathbb{N}$ with $\mu(P) = 1$ such that $\{y_k\}_{k \in P}$ is μ-density convergent to 0 in terms of the probabilistic n-norm M^n. Let $C = \mathbb{N} \setminus P$. Then $\mu(C) = 0$. Define $B_i = A_i \cap C$. Then $\bigcup_{i=1}^{\infty} B_i \subseteq C$ and so, $\mu\left(\bigcup_{i=1}^{\infty} B_i\right) = 0$, *i.e.*, $\mu(B) = 0$, where $B = \bigcup_{i=1}^{\infty} B_i$.

Finally, we show that $A_i \triangle B_i$ is finite. Now,

$$A_i \triangle B_i = A_i \cap P,$$

which is finite, otherwise if $A_i \cap P$ is infinite, then $y_k = x_i$, for infinite number of $k \in P$, which is a contradiction to the fact that (y_k) is $\mu-$statistically convergent to 0 with respect to probabilistic n-norm M^n. Hence $A_i \triangle B_i$ is finite, and hence the proof. □

Definition 21 A sequence $y = (y_k)_{k\in\mathbb{N}}$ in a PnN-space $(Y, M, *)$ is said to be Cauchy sequence in μ-density if there is a set $C \subseteq \mathbb{N}$ with $\mu(C) = 1$ such that $(y_k)_{k\in C}$ is a usual Cauchy sequence in PnN-space.

Theorem 22 *In a PnN-space $(Y, M, *)$, if a sequence is a Cauchy sequence in μ-density, then it is always a μ-statistically Cauchy sequence.*

Proof Let $y = (y_k)_{k\in\mathbb{N}}$ be a Cauchy sequence in μ-density. Then there exists $A \subseteq \mathbb{N}$ with $\mu(A) = 1$, such that $(y_k)_{k\in A}$ is a usual Cauchy sequence in the PnN-space $(Y, M, *)$. Then for every $\lambda \in (0, 1)$, $\varepsilon > 0$ and $z_1, z_2, \ldots, z_{n-1} \in Y$ there is a $k_1 \in \mathbb{N}$ such that

$$M((z_1, z_2, \ldots, z_{n-1}, y_k - y_m), \varepsilon) > 1 - \lambda,$$

for all $k, m \geq k_1$ and $k, m \in A$. Choose $m_0 \in A$ with $m_0 \geq k_1$. Then clearly

$$M((z_1, z_2, \ldots, z_{n-1}, y_k - y_{m_0}), \varepsilon) > 1 - \lambda,$$

for all $k, m_0 \geq k_1$ and $z_1, z_2, \ldots, z_{n-1} \in Y$. Hence,

$$\{k \in \mathbb{N} : M((z_1, z_2, \ldots, z_{n-1}, y_k - y_{m_0}), \varepsilon) \leq 1 - \lambda\} \subseteq A^c.$$

Therefore,

$$\mu(\{k \in \mathbb{N} : M((z_1, z_2, \ldots, z_{n-1}, y_k - y_{m_0}), \varepsilon) \leq 1 - \lambda\}) = 0$$

Hence, y is μ-statistically Cauchy. □

4 μ-Statistical Limit Points in PnN-Spaces

Definition 23 Let $(Y, M, *)$ be a PnN-space. A number $L \in Y$ is called a limit point of the sequence $y = (y_k)$ in terms of the probabilistic n-norm M^n, if there exists a subsequence of y that converges to L, in terms of the probabilistic n-norm M^n.

Let $L_{M(n)}(y)$ denotes the set of all limit points of the sequence y in terms of the probabilistic n-norm M^n.

Definition 24 Let $(Y, M, *)$ be a PnN-space. Then $\gamma \in Y$ is called a μ-statistical limit point of sequence $y = (y_k)$ in terms of the probabilistic n-norm M^n, if there exists a set $M = \{m_1 < m_2 < \cdots\} \subset \mathbb{N}$ such that $\mu(M) \neq 0$ and $M^n - \lim y_{m_k} = \gamma$.

Let $\Lambda^{\mu}_{M(n)}(y)$ denotes the set of all μ-$stat_{M(n)}$-limit points of the sequence y in terms of the probabilistic n-norm M^n.

Theorem 25 *Let $(Y, M, *)$ be a PnN-space. For a sequence $y = (y_k)$, if $\mu - stat_{M(n)} - \lim y = y_0$, then $\Lambda^{\mu}_{M(n)}(y) = y_0$.*

Proof Let $y = (y_k)$ be a sequence such that $\mu - stat_{M(n)} - \lim y = y_0$. Suppose that $\Lambda^{\mu}_{M(n)}(y) = \{y_0, z_0\}$ such that $y_0 \neq z_0$. Then there exists two sets

$$M = \{m_1 < m_2 < \cdots\} \subset \mathbb{N} \text{ and } L = \{l_1 < l_2 < \cdots\} \subset \mathbb{N}$$

such that

$$\mu(M) \neq 0, \ \mu(L) \neq 0$$

and

$$M^n - \lim y_{m_j} = y_0, \ M^n - \lim y_{l_i} = z_0.$$

Therefore, for every $\lambda \in (0, 1)$, $\varepsilon > 0$ and $z_1, z_2, \ldots, z_{n-1} \in Y$, we have

$$\mu(\{l_i \in L : M((z_1, z_2, \ldots, z_{n-1}, y_{l_i} - z_0), \varepsilon) \leq 1 - \lambda\}) = 0.$$

Then, we observe that

$$\begin{aligned}
&\{l_i \in L : i \in \mathbb{N}\} \\
&= \{l_i \in L : M((z_1, z_2, \ldots, z_{n-1}, y_{l_i} - z_0), \varepsilon) > 1 - \lambda\} \\
&\cup \{l_i \in L : M((z_1, z_2, \ldots, z_{n-1}, y_{l_i} - z_0), \varepsilon) \leq 1 - \lambda\},
\end{aligned}$$

which implies

$$\mu(\{l_i \in L : M((z_1, z_2, \ldots, z_{n-1}, y_{l_i} - z_0), \varepsilon) > 1 - \lambda\}) \neq 0. \quad (1)$$

Since $\mu - stat_{M(n)} - \lim y = y_0$, so, we have

$$\mu(\{k \in \mathbb{N} : M((z_1, z_2, \ldots, z_{n-1}, y_k - y_0), \varepsilon) \leq 1 - \lambda\}) = 0, \quad (2)$$

for every $\varepsilon > 0$ and $z_1, z_2, \ldots, z_{n-1} \in Y$. Therefore, we can write

$$\mu(\{k \in \mathbb{N} : M((z_1, z_2, \ldots, z_{n-1}, y_k - y_0), \varepsilon) > 1 - \lambda\}) \neq 0.$$

Now, for every $y_0 \neq z_0$, we have

$$\begin{aligned}
&\{l_i \in L : M((z_1, z_2, \ldots, z_{n-1}, y_{l_i} - z_0), \varepsilon) > 1 - \lambda\} \\
&\quad \cap \{k \in \mathbb{N} : M((z_1, z_2, \ldots, z_{n-1}, y_k - y_0), \varepsilon) > 1 - \lambda\} = \phi.
\end{aligned}$$

Thus,

$$\begin{aligned}
\{l_i \in &L : M((z_1, z_2, \ldots, z_{n-1}, y_{l_i} - z_0), \varepsilon) > 1 - \lambda\} \\
&\subseteq \{k \in \mathbb{N} : M((z_1, z_2, \ldots, z_{n-1}, y_k - y_0), \varepsilon) \leq 1 - \lambda\},
\end{aligned}$$

which implies

$$\mu(\{l_i \in L : M((z_1, z_2, \ldots, z_{n-1}, y_{l_i} - z_0), \varepsilon) > 1 - \lambda\}) = 0.$$

This contradicts the Eq. (1) and hence $\Lambda^{\mu}_{M(n)}(y) = \{y_0\}$. □

5 Conclusion

In the article, we have introduced the concepts of μ-statistical convergence and μ-density convergence of a sequence in a probabilistic n-normed space and investigated their various characterizations. We have also introduced the notion of Cauchy sequence in μ-density and μ-statistical limit point of a sequence in a probabilistic n-normed space and established some results regarding these concepts. Since every classical norm induces a probabilistic n-norm, so the results established here are the straightforward generalization of the corresponding results of the ordinary normed space.

Acknowledgements This work has been supported by the Research Project SB/S4/MS:887/14 of SERB-Department of Science and Technology, Govt. of India.

References

1. M.M. Fréchet, Rend. Del Circ. Mat. Di Palermo (1884–1940) **22**, 1 (1906)
2. K. Menger, Proc. Natl. Acad. Sci. **28**, 535 (1942)
3. A.N. Sherstnev, Uchenye zapiski kazanskogo universiteta. Seriya Fiz.-Mat. Nauk. **122**, 3 (1962)
4. B.C. Tripathy, R. Goswami, Proyecciones (Antofagasta) **33**, 157 (2014)
5. B.C. Tripathy, R. Goswami, Afr. Mat. **26**, 753 (2015)
6. B.C. Tripathy, R. Goswami, Afr. Mat. **26**, 1281 (2015)
7. B.C. Tripathy, R. Goswami, Politehn. Univ. Bucharest Sci. Bull. Ser. A Appl. Math. Phys. **78**, 83 (2016)
8. B.C. Tripathy, M. Sen, S. Nath, Soft Comput. **16**, 1021 (2012)
9. M.R. Rahmat, M.S.M. Noorani, Int. J. Stat. Econ. **1**, 20 (2008)
10. H. Fast, in *Colloquium Mathematicae*, vol. 2 (1951), pp. 241–244
11. S. Karakus, Math. Commun. **12**, 11 (2007)
12. J. Connor, Analysis **10**, 373 (1990)
13. J. Connor, Proc. Am. Math. Soc. **115**, 319 (1992)
14. P. Das, S. Bhunia, Czechoslov. Math. J. **59**, 1141 (2009)
15. P. Das, E. Savas, S. Bhunia, Czechoslov. Math. J. **61**, 809 (2011)
16. E. Savas, Iran. J. Sci. Technol. (Sci.) **36**, 341 (2014)
17. E. Savaş, J. Inequal. Appl. **2013**, 347 (2013)
18. J. Fridy, Proc. Am. Math. Soc. **118**, 1187 (1993)
19. B. Schweizer, A. Sklar et al., Pacific J. Math. **10**, 313 (1960)

Lacunary Statistical Convergence of Order α for Generalized Difference Sequences and Summability Through Modulus Function

A. K. Verma and Sudhanshu Kumar

Abstract In this paper, we define the space $S_\theta^\alpha(\Delta_v^m)$ of all Δ_v^m-lacunary statistical convergent sequences of order α and the space $N_\theta^\alpha(\Delta_v^m, p)$ of all strongly $N_\theta(\Delta_v^m, p)$-summable sequences of order α, where p is a positive real number. Some inclusion relations between these spaces have been obtained. We have studied the space $\omega_\theta^\alpha(\Delta_v^m, f, p)$ of all strongly $\omega_\theta(\Delta_v^m, f, p)$-summable sequences of order α by using modulus function f and bounded sequence (p_k) of positive real numbers with $\inf_k p_k > 0$. The inclusion relations between spaces $\omega_\theta^\alpha(\Delta_v^m, f, p)$ and $S_\theta^\alpha(\Delta_v^m)$ are also obtained.

Keywords Statistical convergence · Lacunary sequence · Difference sequence space · Modulus function

1 Introduction

In 1951, Steinhaus [25] and Fast [8] introduced the concept of statistical convergence. Later on, Schoenberg [20] studied this concept independently in 1959. Further, it was investigated from the point of view of sequence spaces and related with summability theory by Connor [3], Fridy [10], Salat [19], Maddox [15], Rath and Tripathy [17], and many others.

Let w denotes the space of all sequences of complex numbers.

Kizmaz [12] introduced difference operator Δ for l_∞, c and c_0, Further, Colak [4] generalized the notion of difference operator Δ, by

$$X(\Delta^m) = \left\{x = (x_k) \in w : \Delta^m x \in X\right\} \text{ for } X = l_\infty, c \text{ and } c_0,$$

A. K. Verma (✉) · S. Kumar
Department of Mathematics and Statistics, H.S. Gour Central University, Sagar 470003, Madhya Pradesh, India
e-mail: arvindamathematica@gmail.com

S. Kumar
e-mail: sudhanshu_tomar@yahoo.com

N. Deo et al. (eds.), *Mathematical Analysis I: Approximation Theory*, Springer Proceedings in Mathematics & Statistics 306,
https://doi.org/10.1007/978-981-15-1153-0_21

where m is fixed positive integer, $\Delta^m x = (\Delta^m x_k) = (\Delta^{m-1}x_k - \Delta^{m-1}x_{k+1})$, for $m \geqslant 1$ and $\Delta^0 x = (x_k)$.

Et and Esi [5] generalized the space $X(\Delta^m)$ by taking the sequence $v = (v_k)$ of nonzero complex numbers. They defined the sequence space $X(\Delta_v^m)$ as follows:

$$X(\Delta_v^m) = \{x = (x_k) \in w : \Delta_v^m x \in X\} \text{ for } X = l_\infty, c \text{ and } c_0,$$

where $\Delta_v^0 x_k = (v_k x_k)$ and

$$\Delta_v^m x_k = \sum_{i=0}^{m} (-1)^i \binom{m}{i} v_{k+i} x_{k+i}, \text{ for } m \geqslant 1.$$

A sequence $x = (x_k) \in w$ is said to be statistical convergent to the number l if for every $\varepsilon > 0$,

$$\lim_n \frac{1}{n} |\{k \leqslant n : |x_k - l| \geq \varepsilon\}| = 0,$$

where $|A|$ denotes the cardinality of set A [10].

Colak[2] introduced the concept of statistical convergence of order α by as follows:
The sequence $x = (x_k)$ is statistically convergent of order α to a number l if for every $\varepsilon > 0$,

$$\lim_n \frac{1}{n^\alpha} |\{k \leqslant n : |x_k - l| \geq \varepsilon\}| = 0,$$

where $0 < \alpha \leqslant 1$.

Lacunary sequence means an increasing integer sequence $\theta = (k_r)$ such that $k_0 = 0$ and $h_r = (k_r - k_{r-1}) \to \infty$ as $r \to \infty$. In this paper, we denote I_r and q_r by an interval $(k_{r-1}, k_r]$ and the ratio $\frac{k_r}{k_{r-1}}$, respectively.

In 1978, Freedman et al.[9] introduced space N_θ using lacunary sequence θ as

$$N_\theta = \left\{x = (x_k) \in w : \text{there exists } l \text{ such that } h_r^{-1} \sum_{k \in I_r} |x_k - l| \to 0\right\}.$$

In 1993, Fridy and Orhan[11] introduced the concept of lacunary statistical convergence as follows:

A sequence x is said to be lacunary statistical convergent to l if for every $\varepsilon > 0$,

$$\lim_r \frac{1}{h_r} |\{k \in I_r : |x_k - l| \geqslant \varepsilon\}| = 0.$$

In this case we write S_θ-$\lim x = l$ or $x_k \to l(S_\theta)$. The space of all lacunary statistical convergent sequence is given by S_θ.

In 2005, Tripathy and Et [26] introduced lacunay statistical convergence using mth-order difference operator. The sequence $x = (x_k)$ is Δ^m-lacunary statistical convergent to l if for every $\varepsilon > 0$,

$$\lim_r \frac{1}{h_r} |\{k \in I_r : |\Delta^m x_k - l| \geqslant \varepsilon\}| = 0.$$

In 2014, Şengül and Et [23] introduced lacunary statistical convergent sequence of order α. The sequence $x = (x_k)$ is said to be S_θ^α-statistically convergent to l if for every $\varepsilon > 0$,

$$\lim_r \frac{1}{h_r^\alpha} |\{k \in I_r : |x_k - l| \geqslant \varepsilon\}| = 0,$$

where $h^\alpha = (h_r^\alpha) = (h_1^\alpha, h_2^\alpha, \ldots, h_r^\alpha, \ldots)$.

Also, a sequence x is said to be strongly $N_\theta^\alpha(p)$-summable if there exists l such that

$$\lim_r \frac{1}{h_r^\alpha} \sum_{k \in I_r} |x_k - l|^p = 0, \quad \text{where p is a positive real number.}$$

The idea of modulus function was introduced by Nakano [16]. Later, Ruckle [18] generalized the idea by constructing a class of FK-spaces. Following Ruckle [18] and Maddox [14], we recall that a modulus f is a function from $[0, \infty)$ to $[0, \infty)$ such that

(i) $f(x) = 0$ if and only if $x = 0$,
(ii) $f(x + y) \leqslant f(x) + f(y)$,
(iii) f is increasing,
(iv) f is continuous from the right at 0.

Modulus function is frequently used by many authors to construct scalar and vector valued sequence spaces.

Et and Şengül [6] studied various inclusion relations between space of S_θ^α-convergent sequences, $N_\theta^\alpha(p)$-summable space and space $\omega_\theta^\alpha(f, p)$, which is defined by

$$\omega_\theta^\alpha(f, p) = \left\{ x = (x_k) : \lim_r \frac{1}{h_r^\alpha} \sum_{k \in I_r} [f(|x_k - l|)]^{p_k} = 0, \text{ for some } l \right\},$$

where (p_k) is bounded sequence of positive real numbers such that $\inf_k p_k > 0$.

In 2015, Altin et al. [1] defined Δ_v^m−statistical convergence of order α and introduced space $w_p^\alpha(\Delta_v^m, f)$ defined as follows:

$$w_p^\alpha(\Delta_v^m, f) = \left\{ x = (x_k) : \lim_n \frac{1}{n^\alpha} \sum_{k=1}^n \left[f(|\Delta_v^m x_k - l|) \right]^{p_k} = 0, \text{ for some } l \right\}.$$

Recently, lacunary statistical convergence of order α has been studied in ([7, 21, 22, 24]) in relation to I-lacunary statistical convergence, lacunary statistical convergence of order (α, β), Wijsman I-lacunary statistical convergence and f-lacunary statistical convergence of order α.

In this paper, we have generalized Δ_v^m−statistical convergence of order α by taking lacunary sequence θ. We also obtained some inclusion relations between the sequence spaces $S_\theta^\alpha(\Delta_v^m)$, $N_\theta^\alpha(\Delta_v^m, p)$, and $\omega_\theta^\alpha(\Delta_v^m, f, p)$. Throughout this paper, let (v_k) be fixed sequence of nonzero complex number, m be a fixed positive integer, $\theta = (k_r)$ be a lacunary sequence, and α be a real number lies in the interval $(0, 1]$.

2 Definitions Related to Our Work

Definition 2.1 Let $\theta = (k_r)$ be a lacunary sequence and $\alpha \in (0, 1]$. A sequence $x = (x_k) \in w$ is called $S_\theta^\alpha(\Delta_v^m)$-statistically convergent to l if for every $\varepsilon > 0$,

$$\lim_r \frac{1}{h_r^\alpha} |\left\{k \in I_r : |\Delta_v^m x_k - l| \geqslant \varepsilon\right\}| = 0.$$

In this case, we write $S_\theta^\alpha(\Delta_v^m)$-$\lim x_k = l$. The set of all Δ_v^m-lacunary statistical convergent sequence of order α is denoted by $S_\theta^\alpha(\Delta_v^m)$.

If we take $\theta = (2^r)$, then space $S_\theta^\alpha(\Delta_v^m)$ becomes $S^\alpha(\Delta_v^m)$, which was studied by Altin et al. [1]. If $\alpha = 1, m = 0$, and $(v_k) = (1, 1, 1, \ldots)$, then $S_\theta^\alpha(\Delta_v^m)$-statistically convergence of order α coincides with classical lacunary statistical convergence, which is discussed by Fridy and Orhan [11] in 1993. We write $S(\Delta_v^m)$ if $\theta = (2^r)$ and $\alpha = 1$.

Definition 2.2 Let p be a positive real number. A sequence $x = (x_k)$ is said to be strongly $N_\theta(\Delta_v^m, p)$-summable of order α if there is number l such that

$$\lim_r \frac{1}{h_r^\alpha} \left(\sum_{k \in I_r} |\Delta_v^m x_k - l|^p \right) = 0.$$

In this case, we write $N_\theta^\alpha(\Delta_v^m, p)$-$\lim x_k = l$. The set of all strongly $N_\theta(\Delta_v^m, p)$-summable sequences of order α is denoted by $N_\theta^\alpha(\Delta_v^m, p)$.

Definition 2.3 Let f be a modulus function and $p = (p_k)$ be bounded sequence of positive real numbers with $\inf_k p_k > 0$. A sequence $x = (x_k)$ is said to be strongly $\omega_\theta(\Delta_v^m, p)$-summable of order α with respect to modulus function f if there is number l such that

$$\lim_r \frac{1}{h_r^\alpha} \sum_{k \in I_r} [f(|\Delta_v^m x_k - l|)]^{p_k} = 0.$$

In this case, we write $\omega_\theta^\alpha(\Delta_v^m, f, p)$-$\lim x_k = l$. The set of all strongly $\omega_\theta(\Delta_v^m, f, p)$-summable sequences of order α is denoted by $\omega_\theta^\alpha(\Delta_v^m, f, p)$. If $f(x) = x$ and $p_k = p$, for all $k \in \mathbb{N}$, then we have obtained space $N_\theta^\alpha(\Delta_v^m, p)$.

3 Main Results on $S_\theta^\alpha(\Delta_v^m)$ & $N_\theta^\alpha(\Delta_v^m, p)$

In view of Theorem 2.2 of Sengul and Et [23], we formulate the following result without proof.

Theorem 3.1 *Let $x = (x_k)$ and $y = (y_k)$ be any two sequences. Then*

(i) If $S_\theta^\alpha(\Delta_v^m)$-$\lim x_k = x_0$ and $c \in \mathbb{C}$, then $S_\theta^\alpha(\Delta_v^m)$-$\lim cx_k = cx_0$.
(ii) If $S_\theta^\alpha(\Delta_v^m)$-$\lim x_k = x_0$ and $S_\theta^\alpha(\Delta_v^m)$-$\lim y_k = y_0$, then $S_\theta^\alpha(\Delta_v^m)$-$\lim(x_k + y_k) = x_0 + y_0$.

Theorem 3.2 *Let α and β be real numbers such that $0 < \alpha \leqslant \beta \leqslant 1$. Then $S_\theta^\alpha(\Delta_v^m) \subseteq S_\theta^\beta(\Delta_v^m)$ and the inclusion is strict for some α and β.*

Proof Let $0 < \alpha \leqslant \beta \leqslant 1$. Then for every $\varepsilon > 0$,

$$\frac{1}{h_r^\beta}\left|\left\{k \in I_r : |\Delta_v^m x_k - l| \geqslant \varepsilon\right\}\right| \leqslant \frac{1}{h_r^\alpha}\left|\left\{k \in I_r : |\Delta_v^m x_k - l| \geqslant \varepsilon\right\}\right|.$$

Above inequality gives $S_\theta^\alpha(\Delta_v^m) \subseteq S_\theta^\beta(\Delta_v^m)$.

For strictness of inclusion, let $m = 0$, $(v_k) = (1, 1, 1, \ldots)$ and sequence $x = (x_k)$ be defined by

$$x_k = \begin{cases} [\sqrt{h_r}], & k = 1, 2, 3, \ldots, [\sqrt{h_r}], \\ 0, & \text{otherwise.} \end{cases}$$

Then, for $\frac{1}{2} < \beta \leqslant 1$, $\frac{1}{h_r^\beta}\left|\left\{k \in I_r : |\Delta_v^m x_k - 0| \geqslant \varepsilon\right\}\right| = \frac{[\sqrt{h_r}]}{h_r^\beta} \to 0$ as $r \to \infty$.

This means that $x \in S_\theta^\beta(\Delta_v^m)$.

But, for $0 < \alpha \leqslant \frac{1}{2}$, $\frac{[\sqrt{h_r}]}{h_r^\alpha} \nrightarrow 0$ as $r \to \infty$ and hence $x \notin S_\theta^\alpha(\Delta_v^m)$. Thus the inclusion is strict.

Corollary 1 *If a sequence is $S_\theta^\alpha(\Delta_v^m)$-statistically convergent to l, then it is $S_\theta(\Delta_v^m)$-statistically convergent to l.*

Theorem 3.3 *Let α and β be fixed real numbers such that $0 < \alpha \leqslant \beta \leqslant 1$. Then $N_\theta^\alpha(\Delta_v^m, p) \subseteq S_\theta^\beta(\Delta_v^m)$ and the inclusion is strict for some α and β.*

Proof Let $x = (x_k) \in N_\theta^\alpha(\Delta_v^m, p)$. Then for a given $\varepsilon > 0$, we have

$$\begin{aligned}\sum_{k\in I_r} |\Delta_v^m x_k - l|^p &= \sum_{\substack{k\in I_r \\ |\Delta_v^m x_k - l| \geqslant \varepsilon}} |\Delta_v^m x_k - l|^p + \sum_{\substack{k\in I_r \\ |\Delta_v^m x_k - l| < \varepsilon}} |\Delta_v^m x_k - l|^p \\ &\geqslant \sum_{\substack{k\in I_r \\ |\Delta_v^m x_k - l| \geqslant \varepsilon}} |\Delta_v^m x_k - l|^p \\ &\geqslant \left|\left\{k \in I_r : |\Delta_v^m x_k - l| \geqslant \varepsilon\right\}\right| \varepsilon^p.\end{aligned}$$

Since h_r is an increasing sequence and $\alpha \leqslant \beta$, so above inequality reduces to

$$\frac{1}{h_r^\alpha}\sum_{k\in I_r}|\Delta_v^m x_k - l|^p \geqslant \frac{1}{h_r^\beta}|\left\{k \in I_r : |\Delta_v^m x_k - l| \geqslant \varepsilon\right\}|\varepsilon^p.$$

Hence we get the inclusion $N_\theta^\alpha(\Delta_v^m, p) \subseteq S_\theta^\beta(\Delta_v^m)$. For strictness of inclusion, let $m = 0$, $(v_k) = (1, 1, \ldots)$, $p = 1$ and construct sequence (x_k) same as Theorem 3.2.

Then for $\frac{1}{2} < \beta \leqslant 1$, $x \in S_\theta^\beta(\Delta_v^m)$.

But, for $0 < \alpha < 1$, $\frac{1}{h_r^\alpha}\sum_{k\in I_r}|x_k| = \frac{[\sqrt{h_r}][\sqrt{h_r}]}{h_r^\alpha} \to \infty$, which means that $x \notin N_\theta^\alpha(\Delta_v^m, p)$.

Corollary 2 *If a sequence is strongly $N_\theta^\alpha(\Delta_v^m, p)$-summable to l, then it is $S_\theta^\alpha(\Delta_v^m)$-statistically convergent to l.*

Theorem 3.4 *Let $\theta = (k_r)$ be lacunary sequence and $0 < \alpha \leqslant 1$.*

(i) If $\liminf_r q_r > 1$, then $S^\alpha(\Delta_v^m) \subseteq S_\theta^\alpha(\Delta_v^m)$.
(ii) If $\limsup_r q_r < \infty$, then $S_\theta^\alpha(\Delta_v^m) \subseteq S^\alpha(\Delta_v^m)$.
(iii) If $\liminf_{r\to\infty} \frac{h_r^\alpha}{k_r} > 0$, then $S(\Delta_v^m) \subseteq S_\theta^\alpha(\Delta_v^m)$.

Proof The proof of (i) and (ii) are similar to that of Theorems 2.9 and 2.10 of Sengul and Et [23]. So we omit it.
(iii) For any $\varepsilon > 0$, we have

$$\{k \leqslant k_r : |\Delta_v^m x_k - l| \geqslant \varepsilon\} \supseteq \{k \in I_r : |\Delta_v^m x_k - l| \geqslant \varepsilon\}.$$

Therefore,

$$\begin{aligned}\frac{1}{k_r}|\{k \leqslant k_r : |\Delta_v^m x_k - l| \geqslant \varepsilon\}| &\geqslant \frac{1}{k_r}|\{k \in I_r : |\Delta_v^m x_k - l| \geqslant \varepsilon\}|\\ &= \frac{h_r^\alpha}{k_r}\frac{1}{h_r^\alpha}|\{k \in I_r : |\Delta_v^m x_k - l| \geqslant \varepsilon\}|.\end{aligned}$$

On taking $r \to \infty$ in both sides of above inequality and using $\liminf_r \frac{h_r^\alpha}{k_r} > 0$, we can obtain the required inclusion.

Theorem 3.5 *Let α and β be fixed real numbers such that $0 < \alpha \leqslant \beta \leqslant 1$. Then*

(i) $N_\theta^\alpha(\Delta_v^m, p) \subseteq N_\theta^\beta(\Delta_v^m, p)$ and the inclusion is strict for some α and β.
(ii) $N_\theta^\beta(\Delta_v^m, p) \subseteq N_\theta(\Delta_v^m, p)$.

Proof (i) We have $\frac{1}{h_r^\beta}\sum_{k\in I_r}|\Delta_v^m x_k - l|^p \leqslant \frac{1}{h_r^\alpha}\sum_{k\in I_r}|\Delta_v^m x_k - l|^p$.
From above inequality, we get the required inclusion. For strictness of inclusion, let $m = 0$, $p = 1$, $(v_k) = (1, 1, 1, \ldots)$, and define $x = (x_k)$ by

$$x_k = \begin{cases} 1, & \text{if } k \text{ is a perfect square number}, \\ 0, & \text{otherwise}. \end{cases}$$

Now, for $\frac{1}{2} < \beta \leqslant 1$, $x \in N_\theta^\beta(\Delta_v^m, p)$, but $x \notin N_\theta^\alpha(\Delta_v^m, p)$ for $0 < \alpha \leqslant \frac{1}{2}$.
(ii) Inclusion follows directly if we put $\alpha = \beta$ and $\beta = 1$ in (i).

4 Results on Space $\omega_\theta^\alpha(\Delta_v^m, f, p)$

In this section, we obtain representation of $S_\theta^\alpha(\Delta_v^m)$-statistical convergence in terms of strongly $\omega_\theta(\Delta_v^m, f, p)$-summability with respect to modulus function f by establishing some inclusion relations between space $S_\theta^\alpha(\Delta_v^m)$ and $\omega_\theta^\alpha(\Delta_v^m, f, p)$.

Theorem 4.1 *If $x = (x_k)$ is strongly $\omega_\theta(\Delta_v^m, f, p)$-summable of order α, then the limit l of x is unique.*

Proof Suppose $\omega_\theta^\alpha(\Delta_v^m, f, p)$ lim-$x_k = l_1$ and $\omega_\theta^\alpha(\Delta_v^m, f, p) - \lim x_k = l_2$. Then by the subadditive property of modulus function,

$$\frac{1}{h_r^\alpha}\sum_{k\in I_r}[f(|l_1 - l_2|)]^{p_k} \leqslant \frac{1}{h_r^\alpha}\sum_{k\in I_r}[f(|\Delta_v^m x_k - l_1|) + f(|\Delta_v^m x_k - l_2|)]^{p_k}.$$

Using the inequality $|a_k + b_k|^{p_k} \leqslant T\{|a_k|^{p_k} + |b_k|^{p_k}\}$ (one may refer to Maddox [13]), we get

$$\frac{1}{h_r^\alpha}\sum_{k\in I_r}[f(|l_1 - l_2|)]^{p_k} \leqslant \frac{T}{h_r^\alpha}\sum_{k\in I_r}[f(|\Delta_v^m x_k - l_1|)]^{p_k} + \frac{T}{h_r^\alpha}\sum_{k\in I_r}[f(|\Delta_v^m x_k - l_2|)]^{p_k},$$

where $T = \max(1, 2^{H-1})$ and $H = \sup_k p_k$.

Taking limit as $r \to \infty$ on both sides of above inequality, we have

$$\lim_r \frac{1}{h_r^\alpha}\sum_{k\in I_r}[f(|l_1 - l_2|)]^{p_k} = 0.$$

Since (p_k) is bounded sequence, so above equality is possible only when $l_1 - l_2 = 0$. Thus, the limit of x is unique.

Theorem 4.2 *Let α and β be fixed real numbers such that $0 < \alpha \leqslant \beta \leqslant 1$. Then $\omega_\theta^\alpha(\Delta_v^m, f, p) \subseteq S_\theta^\beta(\Delta_v^m)$.*

Proof Let $x \in \omega_\theta^\alpha(\Delta_v^m, f, p)$. For any $\varepsilon > 0$, we have

$$\begin{aligned}
\frac{1}{h_r^\alpha}\sum_{k\in I_r}[f(|\Delta_v^m x_k - l|)]^{p_k} &\geqslant \frac{1}{h_r^\beta}\sum_{k\in I_r}[f(|\Delta_v^m x_k - l|)]^{p_k}\\
&= \frac{1}{h_r^\beta}\Bigg[\sum_{\substack{k\in I_r\\ |\Delta_v^m x_k - l|\geqslant\varepsilon}}[f(|\Delta_v^m x_k - l|)]^{p_k} + \sum_{\substack{k\in I_r\\ |\Delta_v^m x_k - l|<\varepsilon}}[f(|\Delta_v^m x_k - l|)]^{p_k}\Bigg]\\
&\geqslant \frac{1}{h_r^\beta}\Bigg[\sum_{\substack{k\in I_r\\ |\Delta_v^m x_k - l|\geqslant\varepsilon}}[f(\varepsilon)]^{p_k}\Bigg]\\
&\geqslant \frac{1}{h_r^\beta}\Bigg[\sum_{\substack{k\in I_r\\ |\Delta_v^m x_k - l|\geqslant\varepsilon}}\min\Big([f(\varepsilon)]^h, [f(\varepsilon)]^H\Big)\Bigg],\ h = \inf_k p_k \ \&\ H = \sup_k p_k\\
&= \frac{1}{h_r^\beta}|\{k \in I_r : |\Delta_v^m x_k - l| \geqslant \varepsilon\}|\min\Big([f(\varepsilon)]^h, [f(\varepsilon)]^H\Big).
\end{aligned}$$

Above inequality gives, $x \in S_\theta^\beta(\Delta_v^m)$ and hence establishes the required inclusion.

Theorem 4.3 *If the modulus function f is bounded and $\lim_r \frac{h_r}{h_r^\alpha} = 1$, then $S_\theta^\alpha(\Delta_v^m) \subseteq \omega_\theta^\alpha(\Delta_v^m, f, p)$.*

Proof Since f is bounded, so there exists a positive integer M such that $f(x) \leqslant M$, for all $x \geqslant 0$. Let $x \in S_\theta^\alpha(\Delta_v^m)$. For any $\varepsilon > 0$, we have

$$\begin{aligned}
\frac{1}{h_r^\alpha}\sum_{k\in I_r}[f(|\Delta_v^m x_k - l|)]^{p_k} &= \frac{1}{h_r^\alpha}\Bigg[\sum_{\substack{k\in I_r\\ |\Delta_v^m x_k - l|\geqslant\varepsilon}}[f(|\Delta_v^m x_k - l|)]^{p_k} + \sum_{\substack{k\in I_r\\ |\Delta_v^m x_k - l|<\varepsilon}}[f(|\Delta_v^m x_k - l|)]^{p_k}\Bigg]\\
&\leqslant \frac{1}{h_r^\alpha}\sum_{\substack{k\in I_r\\ |\Delta_v^m x_k - l|\geqslant\varepsilon}} M^{p_k} + \frac{1}{h_r^\alpha}\sum_{\substack{k\in I_r\\ |\Delta_v^m x_k - l|<\varepsilon}}[f(\varepsilon)]^{p_k}\\
&\leqslant \frac{1}{h_r^\alpha}\sum_{\substack{k\in I_r\\ |\Delta_v^m x_k - l|\geqslant\varepsilon}}\max\Big(M^h, M^H\Big) + \frac{1}{h_r^\alpha}\sum_{\substack{k\in I_r\\ |\Delta_v^m x_k - l|<\varepsilon}}[f(\varepsilon)]^{p_k}\\
&\leqslant \max\big(M^h, M^H\big)\frac{1}{h_r^\alpha}|\{k \in I_r : |\Delta_v^m x_k - l| \geqslant \varepsilon\}| + \frac{h_r}{h_r^\alpha}\max\big(f(\varepsilon)^h, f(\varepsilon)^H\big).
\end{aligned}$$

Since f is a continuous function, so taking $r \to \infty$ and using $\lim_r \frac{h_r}{h_r^\alpha} = 1$ in the above inequality, we can obtained $x \in \omega_\theta^\alpha(\Delta_v^m, f, p)$.

Theorem 4.4 *Let $\theta_1 = (k_r)$ and $\theta_2 = (s_r)$ be two lacunary sequence such that $I_r \subseteq J_r$ ($I_r = (k_{r-1}, k_r]$ and $l_r = (s_{r-1}, s_r]$) for all $r \in \mathbb{N}$. Suppose α and β be fixed real numbers such that $0 < \alpha \leqslant \beta \leqslant 1$.*

(i) If $\liminf_r \frac{h_r^\alpha}{l_r^\beta} > 0$, then $\omega_{\theta_2}^\beta(\Delta_v^m, f, p) \subseteq \omega_{\theta_1}^\alpha(\Delta_v^m, f, p)$.

(ii) If the modulus function f is bounded and $\lim_r \frac{l_r}{h_r^\beta} = 1$, then $\omega_{\theta_1}^\alpha(\Delta_v^m, f, p) \subseteq \omega_{\theta_2}^\beta(\Delta_v^m, f, p)$.

Proof (i) Let $x \in \omega_{\theta_2}^\beta(\Delta_v^m, f, p)$. We can write

$$\begin{aligned}
\frac{1}{l_r^\beta}\sum_{k\in J_r}[f(|\Delta_v^m x_k - l|)]^{p_k} &= \frac{1}{l_r^\beta}\sum_{k\in J_r - I_r}[f(|\Delta_v^m x_k - l|)]^{p_k} + \frac{1}{l_r^\beta}\sum_{k\in I_r}[f(|\Delta_v^m x_k - l|)]^{p_k}\\
&\geqslant \frac{1}{l_r^\beta}\sum_{k\in I_r}[f(|\Delta_v^m x_k - l|)]^{p_k}\\
&\geqslant \frac{h_r^\alpha}{l_r^\beta}\Big(\frac{1}{h_r^\alpha}\sum_{k\in I_r}[f(|\Delta_v^m x_k - l|)]^{p_k}\Big).
\end{aligned}$$

Since $\liminf_r \frac{h_r^\alpha}{l_r^\beta} > 0$, so the above inequality gives $x \in \omega_{\theta_1}^\alpha(\Delta_v^m, f, p)$.
(ii) Let $x \in \omega_{\theta_1}^\alpha(\Delta_v^m, f, p)$. Since f is bounded, so there exists a positive integer K such that $f(x) \leqslant K$, for all $x \geqslant 0$. Also, $I_r \subseteq J_r$ implies that $h_r \leqslant l_r$ for all $r \in \mathbb{N}$. Now, for any $r \in \mathbb{N}$, we have

$$\begin{aligned}
\frac{1}{l_r^\beta}\sum_{k\in J_r}[f(|\Delta_v^m x_k - l|)]^{p_k} &= \frac{1}{l_r^\beta}\sum_{k\in J_r - I_r}[f(|\Delta_v^m x_k - l|)]^{p_k} + \frac{1}{l_r^\beta}\sum_{k\in I_r}[f(|\Delta_v^m x_k - l|)]^{p_k}\\
&\leqslant \frac{1}{l_r^\beta}\sum_{k\in J_r - I_r}[K]^{p_k} + \frac{1}{l_r^\beta}\sum_{k\in I_r}[f(|\Delta_v^m x_k - l|)]^{p_k}\\
&\leqslant \Big(\frac{l_r - h_r}{l_r^\beta}\Big)K^H + \frac{1}{l_r^\beta}\sum_{k\in I_r}[f(|\Delta_v^m x_k - l|)]^{p_k}\\
&\leqslant \Big(\frac{l_r - h_r^\beta}{h_r^\beta}\Big)K^H + \frac{1}{h_r^\beta}\sum_{k\in I_r}[f(|\Delta_v^m x_k - l|)]^{p_k}\\
&\leqslant \Big(\frac{l_r}{h_r^\beta} - 1\Big)K^H + \frac{1}{h_r^\alpha}\sum_{k\in I_r}[f(|\Delta_v^m x_k - l|)]^{p_k}.
\end{aligned}$$

Finally, taking limit $r \to \infty$ and using condition $\lim_r \frac{l_r}{h_r^\beta} = 1$, we can obtain the required inclusion.

Acknowledgements Author would like to thank funding agency University Grant Commission (UGC) of the Government of India for providing financial support during the research work, in the form of CSIR-UGC NET-JRF.

References

1. Y. Altin, H. Altinok, R. Colak, Statistical convergence of order α for difference sequences. Quaest. Math. **38**(4), 505–514 (2015)
2. R. Colak, Statistical Convergence of Order α. *Modern Methods in Analysis and Its Applications* (Anamaya Publication, New Delhi, 2010), pp. 121–129
3. J.S. Connor, The statistical and strong p-Cesaro convergence of sequences. Analysis **8**, 47–63 (1988)
4. M. Et, R. Colak, On some generalized difference sequence spaces. Soochow J. Math. **21**(4), 377–386 (1995)
5. M. Et, A. Esi, On Köthe-Toeplitz duals of generalized difference sequence spaces. Bull. Malays. Math. Sci. Soc. **23**(1), 25–32 (2000)
6. M. Et, H. Şengül, Some Cesaro-type summability spaces of order α and lacunary statistical convergence of order α. Filomat **28**(8), 1593–1602 (2014)
7. M. Et, H. Şengül, On (Δ^m, I)-lacunary statistical convergence of order α. J. Math. Anal. **7**(5), 78–84 (2016)
8. H. Fast, Sur la convergence statistique. Colloq. Math. **2**, 241–244 (1951)
9. A.R. Freedman, J.J. Sember, M. Raphael, Some Cesaro type summability spaces, in *Proceedings of the London Mathematical Society*, vol. 37 (1978), pp. 508–520
10. J.A. Fridy, On statistical convergence. Analysis **5**, 301–313 (1985)
11. J.A. Fridy, C. Orhan, Lacunary statistical convergence. Pac. J. Math. **160**(1), 43–51 (1993)
12. H. Kizmaz, On certain sequence spaces. Can. Math. Bull. **24**(2), 169–176 (1981)
13. I.J. Maddox, *Elements of Functional Analysis* (Cambridge University Press, Cambridge, 1970)
14. I.J. Maddox, Sequence spaces defined by a modulus. Math. Proc. Camb. Philos. Soc. **100**, 161–166 (1986)
15. I.J. Maddox, Statistical convergence in a locally convex space. Math. Proc. Camb. Philos. Soc. **104**, 141–145 (1988)
16. H. Nakano, Concave modulars. J. Math. Soc. Jpn. **5**, 29–49 (1953)
17. D. Rath, B.C. Tripathy, On statically convergent and statically Cauchy sequences. Indian J. Pure Appl. Math. **25**(4), 381–386 (1994)
18. W.H. Ruckle, FK-spaces in which the sequence of coordinate vectors is bounded. Can. J. Math. **25**(5), 973–978 (1973)
19. T. Salat, On statistically convergent sequences of real numbers. Math. Slovaca **30**, 139–150 (1980)
20. I.J. Schoenberg, The integrability of certain functions and related summability methods. Am. Math. Mon. **66**, 361–375 (1959)
21. H. Şengül, On $S_\alpha^\beta(\theta)$-convergence and strong $N_\alpha^\beta(\theta, p)$-summability. J. Nonlinear Sci. Appl. **10**, 5108–5115 (2017)
22. H. Şengül, On Wijsman I-lacunary statistical equivalence of order (η, μ). J. Inequal. Spec. Funct. **9**(2), 92–101 (2018)
23. H. Şengül, M. Et, On lacunary statistical convergence of order α. Acta Math. Sci. Ser. B Engl. Ed. **34**(2), 473–482 (2014)
24. H. Şengül, M. Et, f-lacunary statistical convergence and strong f-lacunary summability of order α. Filomat **32**(13), 4513–4521 (2018)
25. H. Steinhaus, Sur la convergence ordinaire et la convergence asymptotique. Colloq. Math. **2**, 73–74 (1951)
26. B.C. Tripathy, M. Et, On generalized difference lacunary statistical convergence. Stud. Univ. Babes-Bolyai Math. **50**(1), 119–130 (2005)

Convergence of Three-Step Iterative Process for Generalized Asymptotically Quasi-nonexpansive Mappings in CAT(0) Spaces

Ritika

Abstract We consider a class of generalized asymptotically quasi-nonexpansive mappings introduced by Imnang and Suantai (Abstr Appl Anal 728510: 1–14, 2009, [7]) and seen as a generalization of asymptotically quasi-nonexpansive mappings introduced by Liu (J Math Anal Appl 259:1–7, 2001, [9]). We prove some strong convergence theorems for approximating fixed points of such mappings under suitable conditions in CAT(0) spaces. Our results generalize those of Thakur et al. (Filomat 30(10):2711–2720, 2016, [17]) to the case of this kind of mappings. Our results generalize the corresponding results of many authors.

Keywords CAT(0) spaces · Generalized asymptotically quasi-nonexpansive mappings · Strong convergence

AMS (MOS) Subject Classification 54E40 · 54H25 · 47H10

1 Introduction

Let C be a nonempty subset of a CAT(0) space X and $T : C \to C$ be a mapping. A point $x \in C$ is called a fixed point of T if $Tx = x$. Denote by $F(T)$ the set of fixed points of T, i.e., $F(T) = \{x \in C : Tx = x\}$.

Let (X, d) be a metric space. A geodesic path joining $x \in X$ to $y \in X$ (or, more briefly, a geodesic from x to y) is a map c from a closed interval $[0, l] \subset R$ to X such that $c(0) = x$, $c(l) = y$ and $d(c(t), c(t')) = |t - t'|$ for all $t, t' \in [0, l]$. In particular, c is an isometry and $d(x, y) = l$. The image α of c is called a geodesic (or metric) segment joining x and y. When it is unique this geodesic segment is denoted by $[x, y]$. The space (X, d) is said to be a geodesic space if every two points of X are joined by a geodesic and X is said to be uniquely geodesic if there

Ritika (✉)
Department of Mathematics, Pt. N. R. S. Government College, M. D. University, Rohtak 124001, Haryana, India
e-mail: math.riti@gmail.com

N. Deo et al. (eds.), *Mathematical Analysis I: Approximation Theory*, Springer Proceedings in Mathematics & Statistics 306,
https://doi.org/10.1007/978-981-15-1153-0_22

is exactly one geodesic joining x and y for each $x, y \in X$. A subset $Y \subseteq X$ is said to be convex if Y includes every geodesic segment joining any two of its points. A geodesic triangle $\Delta(x_1, x_2, x_2)$ in a geodesic metric space (X, d) consists of three points x_1, x_2, x_3 in X (the vertices of Δ) and a geodesic segment between each pair of vertices (the edges of Δ). A comparison triangle for the geodesic triangle $\Delta(x_1, x_2, x_3)$ in (X, d) is a triangle $\bar{\Delta}(x_1, x_2, x_3) = \Delta(\bar{x}_1, \bar{x}_2, \bar{x}_3)$ in the Euclidean plane $\mathbb{E}^2$ such that $d_{\mathbb{E}^2}(\bar{x}_i, \bar{x}_j) = (x_i, x_j)$ for $i, j \in \{1, 2, 3\}$.

A geodesic space is said to be a CAT(0) space if all geodesic triangles satisfy the following comparison axiom:

Let Δ be a geodesic triangle in X and let $\bar{\Delta}$ be a comparison triangle for Δ. Then Δ is said to satisfy the CAT(0) inequality if for all $x, y \in \Delta$ and all comparison points $\bar{x}, \bar{y} \in \bar{\Delta}$, $d(x, y) \leq d_{\mathbb{E}^2}(\bar{x}, \bar{y})$.

If x, y_1, y_2 are points in a CAT(0) space and if y_0 is the midpoint of the segment $[y_1, y_2]$, then the CAT(0) inequality implies

$$d(x, y_0)^2 \leq \frac{1}{2}d(x, y_1)^2 + \frac{1}{2}d(x, y_2)^2 - \frac{1}{4}d(y_1, y_2)^2 \qquad \text{(CN)}$$

This is the (CN) inequality of Bruhat and Tits [2]. In fact, a geodesic space is a CAT(0) space only if it satisfy (CN) inequality.

Definition 1.1 Let C be nonempty subset of a CAT(0) space X and $T : C \to C$ be a mapping. Then T is said to be

1. nonexpansive if $d(Tx, Ty) \leq d(x, y)$ for all $x, y \in C$
2. quasi-nonexpansive [4] if $d(Tx, p) \leq d(x, p)$ for all $x \in C$ and $p \in F(T)$.
3. asymptotically nonexpansive [6] if there exists a sequence $r_n \in [0, \infty)$ with the property $\lim_{n\to\infty} r_n = 0$ and such that $d(T^n x, T^n y) \leq (1 + r_n)d(x, y)$ for all $x, y \in C$ and $n = 1, 2, 3, \ldots$.
4. asymptotically quasi-nonexpansive [5, 8] if there exists a sequence $r_n \in [0, \infty)$ with the property $\lim_{n\to\infty} r_n = 0$ and such that $d(T^n x, p) \leq (1 + r_n)d(x, p)$ for all $x \in C$, $p \in F(T)$ and $n = 1, 2, 3, \ldots$.
5. generalized asymptotically quasi-nonexpansive [7] if $F(T) \neq \phi$ and there exist two sequences r_n, s_n with the property $\lim_{n\to\infty} r_n = 0 = \lim_{n\to\infty} s_n$ such that $d(T^n x, p) \leq (1 + r_n)d(x, p) + s_n$ for all $x \in C$, $p \in F(T)$ and $n = 1, 2, 3, \ldots$.

Definition 1.2 ([12]) Let $\{x_n\}$ be a sequence in X and C be a subset of X. We say that $\{x_n\}$ is

1. of monotone type (A) with respect to C if for each $p \in C$, there exist two sequences $\{a_n\}$ and $\{b_n\}$ of nonnegative real numbers such that $\sum_{n=1}^{\infty} a_n < \infty$, $\sum_{n=1}^{\infty} b_n < \infty$ and $d(x_{n+1}, p) \leq (1 + a_n)d(x_n, p) + b_n$.
2. of monotone type (B) with respect to C if there exist two sequences $\{a_n\}$ and $\{b_n\}$ of nonnegative real numbers such that $\sum_{n=1}^{\infty} a_n < \infty$, $\sum_{n=1}^{\infty} b_n < \infty$ and $d(x_{n+1}, C) \leq (1 + a_n)d(x_n, C) + b_n$ (See also [18]).

In 2016, Thakur et al. [17] established a new three-step iterative process in Banach spaces. Now we modify this iterative process into a CAT(0) space as follows: Let C be a nonempty closed convex subset of a complete CAT(0) space X and $T : C \to C$ be a mapping. Then we define the sequence $\{x_n\}$ in C iteratively as

$$\begin{aligned} x_{n+1} &= (1-\alpha_n)Tz_n \oplus \alpha_n T y_n, \\ y_n &= (1-\beta_n)z_n \oplus \beta_n T z_n, \\ z_n &= (1-\gamma_n)x_n \oplus \gamma_n T x_n, \quad \text{for all } n \geq 1, \end{aligned} \tag{1.1}$$

where $\{\alpha_n\}_{n=1}^{\infty}, \{\beta_n\}_{n=1}^{\infty}, \{\gamma_n\}_{n=1}^{\infty}$ are sequences of positive numbers in $(0, 1)$.

Our aim of this paper is to provide convergence results for the above iterative process (1.1) for the generalized asymptotically quasi-nonexpansive mappings given by definition 1.1 (v) in CAT(0) space setting. Our results extend and improve many results in the existing literature due to [1, 9–17] and many others.

2 Main Results

Before going to our main result, we need the following useful lemma.

Lemma 2.1 ([3]) *Let X be a CAT(0) space. Then*

$$d((1-t)x \oplus ty, z) \leq (1-t)d(x, z) + td(y, z)$$

for all $x, y, z \in X$ and $t \in [0, 1]$.

Lemma 2.2 ([18]) *Suppose that $\{p_n\}$, $\{q_n\}$ and $\{r_n\}$ are three sequences of non-negative real numbers satisfying the following conditions: $p_{n+1} \leq (1+q_n)p_n + r_n$, $n = 1, 2, 3, \ldots$ and $\sum_{n=1}^{\infty} q_n < \infty$ and $\sum_{n=1}^{\infty} r_n < \infty$. Then*

1. *$\lim_{n\to\infty} p_n$ exists.*
2. *In addition, if $\liminf_{n\to\infty} p_n = 0$, then $\lim_{n\to\infty} p_n = 0$.*

Theorem 2.3 *Let C be a nonempty closed bounded and convex subset of a complete $CAT(0)$ space X and $T : C \to C$ be a generalized asymptotically quasi-nonexpansive mapping with $\{r_n\}, \{s_n\} \subset [0, \infty)$ such that $\sum_{n=1}^{\infty} r_n < \infty$ and $\sum_{n=1}^{\infty} s_n < \infty$. For a given $x_1 \in C$ and $n = 1, 2, 3, \ldots$, define the sequence $\{x_n\}$ by*

$$\begin{aligned} x_{n+1} &= (1-\alpha_n)T^n z_n \oplus \alpha_n T^n y_n, \\ y_n &= (1-\beta_n)z_n \oplus \beta_n T^n z_n, \\ z_n &= (1-\gamma_n)x_n \oplus \gamma_n T^n x_n. \end{aligned} \tag{2.1}$$

Then, the sequence $\{x_n\}$ is of monotone type (A) and monotone type (B) with respect to $F(T)$. Moreover, the sequence $\{x_n\}$ converges strongly to a fixed point p of *the mapping T if and only if* $\liminf_{n\to\infty} d(x_n, F(T)) = 0$, *where* $d(x, F(T)) = inf_{p\in F(T)} d(x, p)$.

Proof The necessary condition is quite obvious and so here we prove only the sufficient condition. Let $p \in F(T)$. Consider

$$\begin{aligned} d(z_n, p) &= d((1-\gamma_n)x_n \oplus \gamma_n T^n x_n, p) \\ &\leq (1-\gamma_n)d(x_n, p) + \gamma_n d(T^n x_n, p) \\ &\leq (1-\gamma_n)d(x_n, p) + \gamma_n[(1+r_n)d(x_n, p) + s_n] \\ &= (1+\gamma_n r_n)d(x_n, p) + \gamma_n s_n \\ &\leq (1+r_n)d(x_n, p) + s_n \end{aligned} \tag{2.2}$$

Now,

$$\begin{aligned} d(y_n, p) &= d((1-\beta_n)z_n \oplus \beta_n T^n z_n, p) \\ &\leq (1-\beta_n)d(z_n, p) + \beta_n d(T^n z_n, p) \\ &\leq (1-\beta_n)d(z_n, p) + \beta_n[(1+r_n)d(z_n, p) + s_n] \\ &= (1+\beta_n r_n)d(z_n, p) + \beta_n s_n \\ &\leq (1+\beta_n r_n)[(1+r_n)d(x_n, p) + s_n] + \beta_n s_n \\ &\leq (1+r_n)(1+r_n)d(x_n, p) + [1+\beta_n r_n]s_n + \beta_n s_n \\ &\leq (1+r_n)^2 d(x_n, p) + [1+\beta_n r_n + \beta_n]s_n \\ &\leq (1+r_n)^2 d(x_n, p) + [1+(1+r_n)\beta_n]s_n \\ &\leq (1+r_n)^2 d(x_n, p) + (2+r_n)s_n \end{aligned} \tag{2.3}$$

Now,

$$\begin{aligned} d(x_{n+1}, p) &= d((1-\alpha_n)T^n z_n \oplus \alpha_n T^n y_n, p) \\ &\leq (1-\alpha_n)d(T^n z_n, p) + \alpha_n d(T^n y_n, p) \\ &\leq (1-\alpha_n)[(1+r_n)d(z_n, p) + s_n] + \alpha_n[(1+r_n)d(y_n, p) + s_n] \\ &\leq (1-\alpha_n)[(1+r_n)[(1+r_n)d(x_n, p) + s_n] + s_n] \\ &+ \alpha_n[(1+r_n)[(1+r_n)^2 d(x_n, p) + (2+r_n)s_n] + s_n] \\ &= (1-\alpha_n)(1+r_n)^2 d(x_n, p) + (1-\alpha_n)s_n(2+r_n) \\ &+ \alpha_n(1+r_n)^3 d(x_n, p) + [(1+r_n)(2+r_n) + 1]\alpha_n s_n \\ &= (1+r_n)^2[(1-\alpha_n) + \alpha_n(1+r_n)]d(x_n, p) \\ &+ [(1-\alpha_n)(2+r_n) + \alpha_n[(1+r_n)(2+r_n) + 1]]s_n \\ &\leq (1+r_n)^3 d(x_n, p) \\ &+ [2+\alpha_n + r_n + 2\alpha_n r_n + \alpha_n {r_n}^2]s_n \\ &\leq (1+3r_n+3{r_n}^2+{r_n}^3)d(x_n, p) + \end{aligned}$$

$$+ [3 + r_n + r_n(2 + r_n)]s_n$$
$$= (1 + R_n)d(x_n, p) + S_n, \tag{2.4}$$

where $R_n = 3r_n + 3{r_n}^2 + {r_n}^3$ and $S_n = [3 + r_n + r_n(2 + r_n)]s_n$.

Since by hypothesis, $\sum_{n=1}^{\infty} r_n < \infty$ and $\sum_{n=1}^{\infty} s_n < \infty$, it follows that $\sum_{n=1}^{\infty} R_n < \infty$ and $\sum_{n=1}^{\infty} S_n < \infty$.

Now, from (2.4), we get

$$d(x_{n+1}, p) \le (1 + R_n)d(x_n, p) + S_n \tag{2.5}$$

and

$$d(x_{n+1}, F(T)) \le (1 + R_n)d(x_n, F(T)) + S_n \tag{2.6}$$

These inequalities, respectively, prove that $\{x_n\}$ is a sequence of monotone type (A) and monotone type (B) with respect to $F(T)$.

Now, we prove that $\{x_n\}$ converges strongly to a fixed point of the mapping T if and only if $\liminf_{n\to\infty} d(x_n, F(T)) = 0$.

If $\{x_n\} \to p \in F(T)$, then $\lim_{n\to\infty} d(x_n, p) = 0$. Since $0 \le d(x_n, F(T)) \le d(x_n, p)$, we have that $\liminf_{n\to\infty} d(x_n, F(T)) = 0$.

Conversely, suppose that $\liminf_{n\to\infty} d(x_n, F(T)) = 0$. Applying Lemma 2.2 to (2.6), we have that $\lim_{n\to\infty} d(x_n, F(T))$ exists. Also from hypothesis $\liminf_{n\to\infty} d(x_n, F(T)) = 0$, so we conclude that

$$\lim_{n\to\infty} d(x_n, F(T)) = 0. \tag{2.7}$$

Now, we prove that $\{x_n\}$ is a Cauchy sequence.

Since $(1 + x) \le e^x$ for some $x \ge 0$.

Thus

$$\begin{aligned}
d(x_{n+m}, p) &\le (1 + R_{n+m-1})d(x_{n+m-1}, p) + S_{n+m-1} \\
&\le e^{R_{n+m-1}} d(x_{n+m-1}, p) + S_{n+m-1} \\
&\le e^{R_{n+m-1}}[e^{R_{n+m-2}} d(x_{n+m-2}, p) + S_{n+m-2}] + S_{n+m-1} \\
&\le e^{R_{n+m-1}+R_{n+m-2}} d(x_{n+m-2}, p) + e^{R_{n+m-1}}[S_{n+m-2} + S_{n+m-1}] \\
&\le \cdots \\
&\le e^{\sum_{k=n}^{n+m-1} R_k} d(x_n, p) + e^{\sum_{k=n+1}^{n+m-1} R_k} \sum_{k=n}^{n+m-1} S_k \\
&\le e^{\sum_{k=n}^{n+m-1} R_k} d(x_n, p) + e^{\sum_{k=n}^{n+m-1} R_k} \sum_{k=n}^{n+m-1} S_k
\end{aligned}$$

Let $e^{\sum_{k=n}^{n+m-1} R_k} = M$. Thus, there exists a constant $M > 0$ such that

$$d(x_{n+m}, p) \leq Md(x_n, p) + M\left\{\sum_{k=n}^{n+m-1} S_k\right\}$$

for all $n, m \in N$ and $p \in F(T)$.

Since $\lim_{n\to\infty} d(x_n, F(T)) = 0$, therefore, for each $\epsilon > 0$, there exists $n_1 \in N$ such that $d(x_n, F(T)) < \frac{\epsilon}{8M}$ and $\sum_{k=n}^{n+m-1} S_k < \frac{\epsilon}{4M}$ for all $n > n_1$. Thus there exists $p_1 \in F(T)$ such that $d(x_n, p_1) < \frac{\epsilon}{4M}$ for all $n > n_1$ and $m \geq 1$, we have

$$\begin{aligned}
d(x_{n+m}, x_n) &\leq d(x_{n+m}, p_1) + d(p_1, x_n) \\
&\leq Md(x_{n_1}, p_1) + M(\sum_{k=n_1}^{\infty} S_k) + Md(p_1, x_{n_1}) + M(\sum_{k=n_1}^{\infty} S_k) \\
&= 2M(d(x_{n_1}, p_1) + \sum_{k=n_1}^{\infty} S_k) \\
&\leq 2M(\frac{\epsilon}{4M} + \frac{\epsilon}{4M}) = \epsilon
\end{aligned}$$

for all $m, n > n_1$. This proves that $\{x_n\}$ is a Cauchy sequence in C. Since the set C is complete, the sequence $\{x_n\}$ must converge to a fixed point in C.

Let $\lim_{n\to\infty} x_n = y$. Since C is closed, therefore $y \in C$.

Next, we show that $y \in F(T)$.

Now, these two inequalities

$$d(y, p) \leq d(y, x_n) + d(x_n, p)$$

for all $p \in F(T), n = 1, 2, 3, \ldots$ and

$$d(y, x_n) \leq d(y, p) + d(x_n, p)$$

for all $p \in F(T), n = 1, 2, 3, \ldots$ give

$$-d(y, x_n) \leq d(y, F(T)) - d(x_n, F(T)) \leq d(y, x_n),$$

for all $n \geq 1$. That is, $|d(y, F(T)) - d(x_n, F(T))| \leq d(y, x_n)$ for $n \geq 1$.

As $\lim_{n\to\infty} x_n = y$ and $\lim_{n\to\infty} d(x_n, F(T)) = 0$, we conclude that $y \in F(T)$. This completes the proof. □

Corollary 2.4 *Let C be a nonempty closed bounded and convex subset of a complete CAT(0) space X and $T : C \to C$ be a generalized asymptotically quasi-nonexpansive mapping with $\{r_n\}, \{s_n\} \subset [0, \infty)$ such that $\sum_{n=1}^{\infty} r_n < \infty$ and $\sum_{n=1}^{\infty} s_n < \infty$. Suppose that $F(T)$ is closed. Let $\{x_n\}$ be the iteration sequence defined by (2.1). Then, the sequence $\{x_n\}$ converges strongly to a fixed point p of the* *mapping T if and only if there exists a subsequence $\{x_{n_j}\}$ of $\{x_n\}$ which converges to $p \in F(T)$.*

Corollary 2.5 *Let C be a nonempty closed bounded and convex subset of a complete CAT(0) space X and $T : C \to C$ be an asymptotically quasi-nonexpansive mapping with $\{r_n\} \subset [0, \infty)$ such that $\sum_{n=1}^{\infty} r_n < \infty$. Suppose that $F(T)$ is closed. Let $\{x_n\}$ be the iteration sequence defined by (2.1). Then, the sequence $\{x_n\}$ converges strongly to a fixed point p of the mapping T if and only if $\liminf_{n\to\infty} d(x_n, F(T)) = 0$.*

References

1. I. Beg, M. Abbas, An iterative process for a family of asymptotically quasi-nonexpansive mappings in CAT(0) spaces. Novi Sad J. Math. **41**(2), 149–157 (2011)
2. F. Bruhat, J. Tits, Groupes reductifs sur un corps local. I. Donnees Radicielles values. Inst. Hautes Etudes Sci. Publ. Math. **41**, 5–251 (1972)
3. S. Dhompongsa, B. Panyanak, On Δ-convergence theorems in CAT(0) spaces. Comput. Math. Appl. **56**, 2572–2579 (2008)
4. J.B. Diaz, F.T. Metcalf, On the structure of the set of subsequential limit points of successive approximations. Bull. Am. Math. Soc. **73**, 516–519 (1967)
5. H. Fukhar-ud-din, S.H. Khan, Convergence of iterates with errors of asymptotically quasi-nonexpansive mappings and applications. J. Math. Anal. Appl. **328**, 821–829 (2007)
6. K. Goebel, W.A. Kirk, Iteration processes for non-expansive mappings, in *Topological Methods In Nonlinear Functional Analysis (Toronto, Ontario, 1982)*, vol. 21 (Contemporary Mathematics, American Mathematical Society, Rhode Island, 1983), pp. 115–123
7. S. Imnang, S. Suantai, Common fixed points of multi-step noor iterations with errors for a finite family of generalized asymptotically quasi-nonexpansive mappings. Abstr. Appl. Anal. **728510**, 1–14 (2009)
8. A.R. Khan, M.A. Khamsi, H. Fukhar-ud-din, Strong convergence of a general iteration scheme in CAT(0) spaces. Nonlinear Anal. **74**, 783–791 (2011)
9. Q. Liu, Iterative sequences for asymptotically quasi-nonexpansive mappings. J. Math. Anal. Appl. **259**, 1–7 (2001)
10. Y. Niwongsa, B. Panyanak, Noor iterations for asymptotically non-expansive mappings in CAT(0) spaces. Int. J. Math. Anal. **4**(13), 645–656 (2010)
11. S. Rathee, Ritika, Convergence theorems for SP-iteration procedure in CAT(0) spaces. Int. J. Math. Arch. **6**(3), 39–45 (2015)
12. Ritika, S.H. Khan, Strong convergence of a new three step iterative process for generalized asymptotically quasi-nonexpansive mappings in CAT(0) spaces. Submitt. Thai J. Math. (2017)
13. Ritika, S. Rathee, Convergence of an iterative process in CAT(0) spaces. Int. J. Math. Arch. **6**(2), 31–34 (2015)
14. A. Sahin, M. Basarir, On the strong convergence of modified S-iteration process for asymptotically quasi-nonexpansive mappings in CAT(0) space. Fixed Point Theory Appl. **2013**, 12 (2013). https://doi.org/10.1186/1687-1812-2013-12
15. S. Suantai, Weak and strong convergence criteria of noor iterations for asymptotically nonexpansive mappings. J. Math. Anal. Appl. **311**, 506–517 (2005)
16. K.K. Tan, H.K. Xu, Approximating fixed points of nonexpansive mapping by ihe ishikawa iteration process. J. Math. Anal. Appl. **178**, 301–308 (1993)
17. B.S. Thakur, D. Thakur, M. Postolache, A new iteration scheme for approximating fixed points of nonexpansive mappings. Filomat **30**(10), 2711–2720 (2016). https://doi.org/10.2298/FIL1610711T
18. H.Y. Zhou, G.L. Gao, G.T. Guo, Y.J. Cho, Some general convergence principles with applications. Bull. Korean Math. Soc. **40**(3), 351–363 (2003)

MIX
Papier aus verantwortungsvollen Quellen
Paper from responsible sources
FSC® C105338

If you have any concerns about our products,
you can contact us on
ProductSafety@springernature.com

In case Publisher is established outside the EU,
the EU authorized representative is:
Springer Nature Customer Service Center GmbH
Europaplatz 3, 69115 Heidelberg, Germany

Printed by Libri Plureos GmbH
in Hamburg, Germany